Use R!

Series Editors:

Robert Gentleman
Kurt Hornik
Giovanni Parmigiani

T0181015

For other titles published in this series, go to
http://www.springer.com/series/6991

Eric A. Suess • Bruce E. Trumbo

Introduction to Probability Simulation and Gibbs Sampling with R

 Springer

Eric A. Suess
Department of Statistics and Biostatistics
California State University, East Bay
Hayward, CA 94542-3087
USA
eric.suess@csueastbay.edu

Bruce E. Trumbo
Department of Statistics and Biostatistics
California State University, East Bay
Hayward, CA 94542-3087
USA
bruce.trumbo@csueastbay.edu

ISBN 978-0-387-40273-4   e-ISBN 978-0-387-68765-0
DOI 10.1007/978-0-387-68765-0
Springer New York Dordrecht Heidelberg London

Library of Congress Control Number: 2010928331

Printed on acid-free paper

Springer is part of Springer Science+Business Media (www.springer.com)

We dedicate this book to our students.

Preface

Our primary motivation in writing this book is to provide a basic introduction to Gibbs sampling. However, the early chapters are written so that the book can also serve as an introduction to simulating probability models or to computational aspects of Markov chains.

Prerequisites are a basic course in statistical inference (including data description, and confidence intervals using t and chi-squared distributions) and a post-calculus course in probability theory (including binomial, Poisson, normal, gamma, and beta families of distributions, conditional and joint distributions, and the Law of Large Numbers). Accordingly, the target audience is upper-division BS or first-year MS students and practitioners of statistics with a knowledge of these prerequisites.

Specific Topics. Many students at the target level lack the full spectrum of experience necessary to understand Gibbs sampling, and thus much of the book is devoted to laying an appropriate foundation. Here are some specifics.

The first four chapters introduce the ideas of *random number generation* and *probability simulation*. Fruitful use of simulation requires the right mixture of confidence that it can often work quite well, consideration of margins of error, and skepticism about whether necessary assumptions are met. Most of our early examples of simulation contain a component for which exact analytical results are available as a reality check. Some theoretical justifications based on the Law of Large Numbers and the Central Limit Theorem are provided. Later examples rely increasingly on various diagnostic methods.

Because Gibbs sampling is a computational tool for estimation, Chapter 5 introduces *screening tests* and poses the problem of estimating prevalence of a trait from test data. This topic provides realistic examples where traditional methods of estimation are problematic but for which a Bayesian framework with Gibbs sampling is later shown to provide useful results. An important pedagogical by product of this chapter is to start the reader thinking about practical applications of conditional probability models.

Most meaningful uses of Gibbs sampling involve limiting distributions of *ergodic Markov chains* with continuous state spaces. Chapter 6 introduces the theory of Markov dependence for chains with two states, material that may be a review for some readers. Then Chapter 7 considers computations and simulations for chains with finite, countable, and continuous state spaces. Examples of absorbing and "almost absorbing" chains illustrate the importance of diagnostic methods. We have found that students can benefit from these computational treatments of Markov chains either before or after a traditional theory-based undergraduate course in stochastic processes that includes Markov chains.

Because Gibbs sampling is mainly used to compute Bayesian posterior distributions, Chapter 8 provides an elementary introduction to some aspects of *Bayesian estimation*. Then Chapter 9 shows a variety of applications of *Gibbs samplers* in which the appropriate partial conditional distributions are programmed in R to allow simulation of the desired posterior distributions. Examples involving screening tests and variance components show how Gibbs sampling in a Bayesian framework can provide useful interval estimates in cases where traditional methods do not. Also, both flat and informative priors are investigated. Chapter 10 briefly illustrates how the use of *BUGS software* can sometimes minimize the programming necessary for Gibbs sampling.

Our treatment of Gibbs sampling is introductory. Readers who will use this important computational method beyond the scope of our examples should refer to any one of the many other more advanced books on the topic of Bayesian modeling.

Computation. R software is used throughout. We provide and explain R programs for all examples and many of the problems. Text files for all but the briefest bits of code are also available online at

`www.sci.csueastbay.edu/~esuess/psgs.htm`

The goal has been to provide code that is simple to understand, taking advantage of the object-oriented structure of R whenever feasible, but avoiding user-defined functions, data files, and libraries beyond the "base." Readers who have not used R before should begin with Chapter 11, which focuses on just the terminology and procedures of R needed in our first few chapters.

Figures and Problems. Two important features of the book are the extraordinarily large number of illustrations made using R and the extensive and varied problems.

Throughout the book, *figures* are widely used to illustrate ideas and as diagnostic tools. Although most of the figures have been embellished to show well in print, the basic R code is provided for making images similar to those in almost all figures.

Problems take up about a third of the pages in the book. They range from simple drill problems to more difficult or advanced ones. Some of the problems introduce optional topics and some can be easily expanded into more extensive student projects. A few of the problems lead the reader through formal proofs of probability results that are stated without proof in the text.

Frequentist and Bayesian Courses. We have used parts of Chapters 1–7 of this book in teaching two essentially non-Bayesian courses, one a beginning course in probability simulation and the other a heavily computational course on Markov chains and other applied probability models. In these courses that do not cover the Bayesian and Gibbs-sampling topics, we have made heavier use of the problems.

We have also used parts of Chapters 8–10 of this book in teaching Bayesian Statistics and Mathematical Statistics at the undergraduate and beginning graduate levels. In these courses, we present the ideas of Bayesian modeling and the use of R and WinBUGS for the estimation of these models. In the graduate mathematical statistics classes, Bayesian modeling is presented along with maximum likelihood estimation and bootstrapping to show the connection between the frequentist and Bayesian interpretations of statistics.

Acknowledgments. In writing this book, we have been indebted to authors of many books on related topics. In each chapter, we have tried to cite the references that may be most useful for our readers. Also, we thank the many students who have made corrections and helpful comments as they used drafts of our manuscript.

Eric A. Suess
Bruce E. Trumbo
Hayward, CA, February 2010

Contents

1

Introductory Examples: Simulation, Estimation, and Graphics

1.1 Simulating Random Samples from Finite Populations

Because simulation is a major topic of this book, it seems appropriate to start with a few simple simulations. For now, we limit ourselves to simulations based on the R-language function `sample`.

In our first example, we use combinatorial methods to find an exact solution to an easy problem—and then, as a demonstration, we use simulation to find an approximate solution. Throughout the book we have many opportunities to see that simulation is *useful*. In Example 1.1 we are content to show that simulation can produce a result we know is reasonably accurate.

Example 1.1. Sampling Computer Chips. Suppose there are 100 memory chips in a box, of which 90 are "good" and 10 are "bad." We withdraw five of the 100 chips at random to upgrade a computer. What is the probability that all five chips are good?

Let the random variable X be the number of good chips drawn. We seek $P\{X = 5\}$. Altogether there are $\binom{100}{5} = 75\,287\,520$ equally likely ways in which to select five chips from the box. Of these possible samples, $\binom{90}{5} = 43\,949\,268$ consist entirely of good chips. Thus the answer is

$$P(\text{All Good}) = P\{X = 5\} = \binom{90}{5}/\binom{100}{5} = 0.5838.$$

This can be evaluated in R as `choose(90,5)/choose(100,5)`.

Now we use the `sample` function to approximate $P\{X = 5\}$ by simulation. This function takes a random sample from a finite population. Of course, in any particular sampling procedure, we must specify the population and the sample size. We also have to say whether sampling is to be done with or without replacement (that is, with or without putting each selected item back into the population before the next item is chosen). The following arguments of the `sample` function are used to make these specifications:

E.A. Suess and B.E. Trumbo, *Introduction to Probability Simulation and Gibbs Sampling with R,* Use R!, DOI 10.1007/978-0-387-68765-0_1, © Springer Science+Business Media, LLC 2010

- The argument given first specifies the population. For our simulation, we represent the population as the vector 1:100, which has the numbers 1 through 100 as its elements. (Of these, we regard the ten chips numbered 91 through 100 to be the bad ones.)
- The second argument is the number of items to be chosen, in our case, 5.
- To implement sampling with replacement, one would have to include the argument repl=T. But in taking chips from the box, we want to sample without replacement. This is the default sampling mode in R, so the effect is the same whether we include or omit the argument repl=F.

Accordingly, we use the statement sample(1:100, 5) to sample five different chips at random from the box of 100. Because this statement is a random one, it is likely to give a different result each time it is used. Listed below are the results we obtained from using it three times:

```
> sample(1:100, 5)
[1] 46 85 68 59 81
> sample(1:100, 5)
[1] 17 43 36 99 84
> sample(1:100, 5)
[1] 58 51 57 81 43
```

Our second simulated sample contains one bad chip (numbered 99). Only good chips are found in the first and third samples. The three simulated values of our random variable are $X = 5$, 4, and 5, respectively.

For a useful simulation, we must generate many samples of five chips, and we need an automated way to count the good chips in each sample. We can do this by determining how many chips have numbers 90 or smaller.

Here is one method, suppose pick = c(17, 43, 36, 99, 84). Then the vector (pick <= 90) has elements TRUE, TRUE, TRUE, FALSE, and TRUE, respectively. If we take the sum of these elements, then R "coerces" (interprets) FALSE as 0 and TRUE as 1. This amounts to taking the sum $1 + 1 + 1 + 0 + 1 = 4$. In general, sum(pick <= 90) returns the number of good chips in pick. (Problem 1.2 asks you to explore an alternative method of counting the good chips in a sample.)

With this background, we are ready to use a large-scale simulation to approximate $P\{X = 5\}$. We simulate $m = 100\,000$ samples of five chips from the box (see Figure 1.1). In the following R program, we loop through these m samples, counting the number of good items found in each. The proportion of the m samples consisting entirely of good chips approximates $P\{X = 5\}$. In Chapter 3, we discuss the theoretical justification for this method of approximation in terms of the Law of Large Numbers. For now, it is enough to look at some cases in which the method works.

In the program below, the statement good = numeric(m) initializes the object good as a vector of m elements, all of which are 0s. Then, on the ith passage through the loop, the ith element of good is replaced by the actual

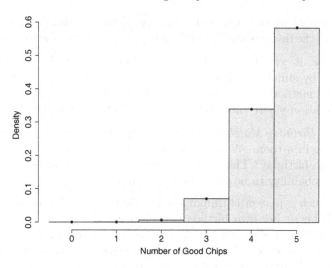

Figure 1.1. Based on 100 000 iterations, the histogram shows the simulated distribution of the number of good computer chips in a random sample of 5 from a box of 100 of which 90 are good (see Example 1.1). The heavy dots are centered on exact values from the corresponding hypergeometric distribution (see Problem 1.3).

number of good items seen in the ith simulated sample. (For a similar program, see Example 11.7.) Beneath the program, we show the result of one run.

```
# set.seed(1237)          # this seed for exact result shown
m = 100000                # number of samples to simulate
good = numeric(m)         # initialize for use in loop
for (i in 1:m)
{
  pick = sample(1:100, 5)    # vector of 5 items from ith box
  good[i] = sum(pick <= 90)  # number Good in ith box
}
mean(good == 5)           # approximates P{All Good}

> mean(good == 5)
[1] 0.58293
```

The vector **good** has $m = 100\,000$ elements, each the number of good chips in one of the simulated samples. Averaging the elements of the logical vector (**good == 5**) amounts to finding the proportion 0.58293 of the m samples in which only good chips were chosen. This particular simulation run approximates the probability $P\{X = 5\} = 0.5838$ correct to two decimal places.

If you omit the statement **set.seed** at the beginning of the program, you will get a slightly different answer each time you run it. For example, in four additional runs we got 0.58385, 0.58390, 0.58345, and 0.58195. Approximations that are wrong by more than 0.004 are very rare. If you use the same seed and software we did, you will repeat precisely the simulation displayed above and get the result 0.58293. (In Chapter 2, we explain how seeds work.) ◇

In practice, it would not make sense to approximate the probability in Example 1.1 by simulation because it is so easy to find an exact value by combinatorial methods. (See Problem 1.3 for details.) The next two examples discuss a situation where simulation turns out to be useful.

Example 1.2. Birthday Matches—Combinatorial Approach. Suppose there are $n = 25$ people in a room. What is the probability that two or more of them have the same birthday? This is an intriguing problem because many people expect the probability to be a lot smaller than it is.

- Perhaps such a person is thinking it is unlikely for someone in the room to match his or her own birthday, overlooking that there are $\binom{25}{2} = 300$ possible pairs of people who might have matching birthdays.
- Perhaps such a person observes that there would have to be 367 people in the room to be absolutely sure of a match, and so imagines a match must be unlikely in a room with only 25 people. But the relationship between room size and probability is far from linear (see Figure 1.2).

In order to get a useful solution to this problem by elementary combinatorial methods, we make some assumptions:

1. Ignore leap years and pretend there are only 365 days in a year.
2. Assume that births are uniformly distributed throughout the year.
3. Assume that the people in the room are randomly chosen. Clearly, the answer to our problem would be much different if the people in the room were attending a convention of twins or of people born in December.

In most applied probability models, it is necessary to make some simplifying assumptions. Such assumptions must be made with care and their effects verified when possible. In our case, we know the first two assumptions are false but hope they make little difference in our answer. We hope the third assumption is true, or nearly so.

Under these assumptions, any one of 365^{25} equally likely sequences of birthdays might occur in the room. Also, $_{365}P_{25} = 365!/(365 - 25)!$ of these possible outcomes avoid birthday matches. Therefore,

$$P(\text{No Match}) = \frac{_{365}P_{25}}{365^{25}} = \prod_{i=0}^{24}\left(1 - \frac{i}{365}\right) = 0.4313\,. \qquad (1.1)$$

So $p = P(\text{At Least One Match}) = 1 - 0.4313 = 0.5687$. In R, $P(\text{No Match})$ can be evaluated as `prod((365:(365-24))/365)` or as `prod(1 - (0:24)/365)`.

Intuitively, it seems the probability p of getting a match must increase as the number n of people in the room increases. The following program explores the relationship between n and p. It loops through rooms of sizes $n = 1$ to 60, finding p for each room size. Then it plots p against n. The result is shown in Figure 1.2.

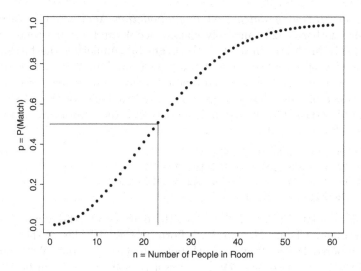

Figure 1.2. As the number of people in a room increases, so does the probability of matching birthdays among them. In a room with 23 people, the probability of such a match slightly exceeds 1/2; with 60 or more, matches are almost certain.

```
n = 1:60                      # vector of room sizes
p = numeric(60)               # initialize vector, all 0s

for (i in n)                  # index values for loop
{
   q = prod(1 - (0:(i-1))/365)  # P(No match) if i people in room
   p[i] = 1 - q                 # changes ith element of p
}

plot(n, p)                    # plot of p against n
```

Of course, the probability of getting at least one match is 0 if $n = 1$. From Figure 1.2 or, more precisely, from the additional code p[c(22, 23, 60)], we see that p first exceeds 50% at $n = 23$ and that, in a room with $n = 60$ people, the probability of at least one birthday match is 0.994. (Making labels and reference lines requires additional code; for simplicity, we do not usually show the code used to embellish graphs for publication.) ◇

In Example 1.3, we solve the birthday matching problem by simulation. One disadvantage of simulation is that it yields no formula, such as (1.1), that can be generalized to any number n of people. Furthermore, simulation does not even provide an exact answer for a particular number of people. However, this example shows an elementary problem in which simulation turns out to have important advantages.

Example 1.3. Birthday Matches—Using Simulation. Again, we focus attention on a room with $n = 25$ randomly chosen people and assume there are 365 equally likely birthdays in a year. We begin by simulating the birthdays in one room. Numbering the days of the year from 1 to 365, we can use the **sample** function to get a list of 25 random birthdays in a room. Of course, we sample *with* replacement here because we need to allow for the possibility of matching birthdays. One use of this function with the appropriate parameters gives the following result:

```
> b = sample(1:365, 25, repl=T);  b
 [1]  74 251 335 104  39 256 193 295 350  41
[11] 100 180 117 205  96  74 142 325 203 308
[21] 325 264  78  83  52
```

In this simulated room of 25 people, there are two birthday matches: The people numbered 1 and 16 were both born on the 74th day of the year, and those numbered 18 and 21 were both born on the 325th day. We would also have said there are two birthday matches if person 18 had been born on the 74th day.

The function **unique** can be used to automate the counting of birthday matches. For example, the vector **unique(b)** is the same as **b**, except that it has only 23 elements. The second occurrences of birthdays 74 and 325 are removed. Thus **x = 25 - length(unique(b))** computes $25 - 23 = 2$. This is the number X of birthday matches (redundant birthdays) among the 25 people in the "room" simulated above.

In the program below, we simulate a large number m of rooms. Almost as if taking a poll, we "ask" each room, "How many birthday matches do you have?" In a large poll, we anticipate that the fraction of rooms with no matches will be very nearly $P(\text{No Matches}) = P\{X = 0\} = 0.4313$, as obtained with combinatorics in Example 1.2. We also anticipate that the mean number of birthday matches in the m rooms will approximate the expected number $E(X)$ of birthday matches. This expected value is difficult to evaluate by combinatorial methods.

The last three lines of the program estimate $P\{X = 0\}$ (by **mean(x == 0)**) and $E(X)$ (by **mean(x)**) and plot a histogram that gives a good idea of the shape of the distribution of X.

```
set.seed(1237)
m = 100000;  n = 25              # iterations; people in room
x = numeric(m)                   # vector for numbers of matches
for (i in 1:m)
{
  b = sample(1:365, n, repl=T)   # n random birthdays in ith room
  x[i] = n - length(unique(b))   # no. of matches in ith room
}
mean(x == 0);  mean(x)           # approximates P{X=0}; E(X)
cutp = (0:(max(x)+1)) - .5       # break points for histogram
hist(x, breaks=cutp, prob=T)     # relative freq. histogram
```

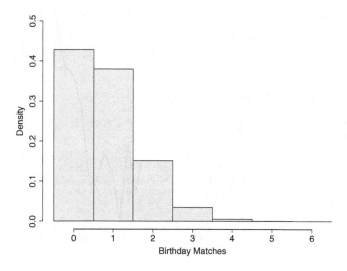

Figure 1.3. A histogram based on 100 000 simulated rooms, each with 25 people. It approximates the distribution of the number of birthday matches, which would be difficult to obtain by combinatorial methods. The probability 0.43 of no matches (height of the leftmost bar) agrees with a combinatorial result in Example 1.2.

To two places, the resulting approximations are $P\{X = 0\} \approx 0.43$ and $E(X) \approx 0.81$. Because the former agrees with the known value 0.4313, we believe that the simulation is performing as we intended—and so also that the latter value is a useful approximation of $E(X)$.

See Figure 1.3 for the histogram. The parameter `prob=T` puts a density scale on the vertical axis. Thus we can see that the bar of the histogram at $x = 0$ is approximately 0.43 units high. It also seems reasonable that the balance point of the histogram is consistent with $E(X) = 0.81$. (We used the parameter `breaks=cutp` to specify breakpoints for the histogram because we did not like the histogram we obtained with the default breakpoints.)

To end this example, we discuss the assumption that birthdays are uniformly distributed throughout the year. For people born in the United States, this assumption is not exactly correct. For example, there are more births in summer months than in winter (see Figure 1.4).

An important advantage of simulation is the ease with which one may account for such variations and thus test the effect of this departure from the uniformity assumption. Nonuniformity tends to make birthday matches more likely. However, in Problem 1.8 you can verify that the actual pattern of U.S. birthrates gives very nearly the same probability of birthday matches as does a uniform pattern. For practical purposes, the assumption that birthdays are uniformly distributed does no harm. (These simulation results for nonuniform birthrates agree with analytic approximations derived by methods considerably beyond the mathematical level of this book. See [Nun92] and [PC00].) ◇

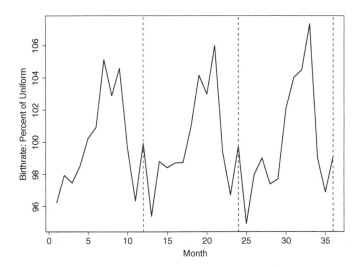

Figure 1.4. Empirical U.S. birthrates for 36 consecutive months. The vertical scale emphasizes seasonal variations, but simulation shows they do not seriously affect the probability of birthday matches (see Problem 1.8). Data are for January 1997 through December 1999, From the U.S. National Center for Health Statistics.

Many of the methods we discuss in this book use simulation. Simulation requires a reliable and plentiful supply of numbers that can be treated as random. Therefore, in Chapter 2 we give a brief introduction to methods for generating such numbers on a computer—and discuss some of the reasons it is wise to be cautious about the software that provides them.

1.2 Coverage Probabilities of Binomial Confidence Intervals

Now that we have shown some introductory examples of simulation, we turn in this section to estimation—another major topic of this book. Later in this book, we show that modern methods of estimation often require simulation, but simulation plays no essential role in this section. The modern aspect here is the enormous amount of computation required.

Suppose that n binomial trials with $\pi = P(\text{Success})$ result in X Successes. (In this book, we usually denote the probability of Success by π, making special note of the few instances where we need to use $\pi = 3.1416$.) The traditional procedure, shown in many statistics texts, for computing an approximate 95% confidence interval for estimating π is to use the formula

$$p \pm 1.96\sqrt{\frac{p(1-p)}{n}}. \tag{1.2}$$

This confidence interval is centered at the **point estimate** $p = X/n$ of the parameter π. The quantity we add and subtract to form the confidence interval is called the **margin of error**.

In discussing this formula for computing confidence intervals, it is useful to define

$$Z = \frac{p - \pi}{\sqrt{\frac{\pi(1-\pi)}{n}}} = \frac{X - n\pi}{\sqrt{n\pi(1 - \pi)}} \quad \text{and} \quad Z' = \frac{p - \pi}{\sqrt{\frac{p(1-p)}{n}}} = \frac{X - n\pi}{\sqrt{\frac{X(n-X)}{n}}}. \quad (1.3)$$

In terms of these quantities, the validity of (1.2) as a 95% confidence interval for π rests upon two assumptions:

1. Z is approximately standard normal, so that $P\{|Z| < 1.96\} \approx 95\%$.
2. π is estimated accurately enough by p that also $P\{|Z'| < 1.96\} \approx 95\%$.

The approximation symbols ($\approx$) can be replaced by equality only in the limit as $n \to \infty$. Especially for small values of n or for values of π near 0 or 1, these assumptions are worrisome.

First, if $p = 0$ or 1, then formula (1.2) gives a nonsensical "interval" of zero length. Also, as illustrated in Example 1.4, extreme values of p can give confidence intervals that extend to impossible values beyond the interval (0, 1). Because troublesome values of p are most likely to occur for extreme values of π, it seems clear that intervals of the form (1.2) should not be used in certain kinds of practical applications where Successes are either very rare or very common. Such situations occur regularly; for example, in epidemiology and quality management.

Second, binomial distributions are discrete and (except when $\pi = 1/2$) also skewed. Normal distributions are continuous and symmetrical. Especially when n is small and π near 0 or 1, it may not be realistic to expect a useful normal approximation to a binomial distribution.

Finally, the length of the confidence interval is based on p rather than π. If the interval is longer or shorter than it should be, that would affect the chance that it covers the value of π. How large must n be before p can reasonably be used as a substitute for π?

Example 1.4. Estimating the Probability that a Die Shows a Six. As an elementary illustration of confidence intervals made with formula (1.2), suppose 20 students in a class were each asked to roll a die 30 times. We note the number X of 6s observed and find the corresponding confidence interval.

Figure 1.5 shows the results. Two students (numbers 2 and 12 in the figure) were surprised to get only one 6 in 30 rolls, so each of them obtained the confidence interval $(-0.031, 0.098)$. This confidence interval does not cover the probability $\pi = 1/6$ of getting a 6 with a fair die. Also, it includes impossible negative values, and so there is some question how it should be interpreted. In contrast, student 10 rolled ten 6s; the lower end of her confidence interval $(0.165, 0.502)$ barely covered $\pi = 1/6$.

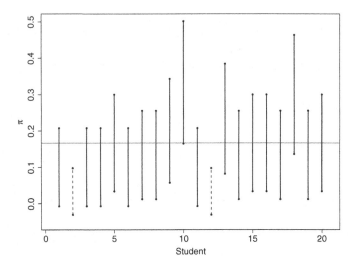

Figure 1.5. Each of 20 students rolled a die 30 times, observed the number of 6s obtained, and used formula (1.2) to make a confidence interval to estimate the probability of getting a 6. Of these 20 intervals, 18 covered the probability $\pi = 1/6$.

Thus 18 of the 20 students obtained confidence intervals covering $1/6$ and two did not. If we were to do this experiment with a very large number of students, if the dice were fair, and if the true coverage probability of confidence intervals made with formula (1.2) were 95%, then we would expect about 95% of the students to get confidence intervals covering $1/6$. ◇

The next two examples are computationally intensive investigations of the usefulness of (1.2). Each step is elementary; the complexity arises only because there are so many computations at each step.

Example 1.5. Two Coverage Probabilities. Suppose a new process for making a prescription drug is in development. Of $n = 30$ trial batches made with the current version of the process, $X = 24$ batches give satisfactory results. Then $p = 24/30 = 0.8$ estimates the population proportion $\pi = P(\text{Success})$ of satisfactory batches with the current version of the process. Wondering how near $p = 0.8$ might be to π, the investigators use (1.2) to obtain the approximate 95% confidence interval 0.8 ± 0.143 or $(0.657, 0.943)$.

The question is whether a 95% level of confidence in the resulting interval is warranted. If (1.2) is used repeatedly, in what proportion of instances does it yield an interval that covers the true value π? If (1.2) is valid here, then the simple answer ought to be 95%. Unfortunately, there is no simple answer to this question. It turns out that the coverage probability depends on the value of π.

In our situation, there are 31 possible values $0, 1, \ldots, 30$ of X, and thus of p. From (1.2) we can compute the confidence interval corresponding to each of these 31 possible outcomes, just as we computed the confidence interval $(0.657, 0.943)$ corresponding to the outcome $X = 24$ above. (This particular interval appears again in the printout of results from the program below—on the 7th row from the bottom.)

By letting the number of successes run from 0 through 30, the first five lines of R code below make a complete list of the possible confidence intervals for $n = 30$. Their lower confidence limits are in the vector lcl and their upper confidence limits in ucl.

Now choose a particular value of π, say $\pi = 0.8$, so that X has a binomial distribution with $n = 30$ trials and success probability $\pi = 0.8$. We write this as $X \sim \text{BINOM}(30, 0.8)$. The vector prob of the 31 probabilities $P\{X = x\}$ in this distribution is found with the R function dbinom(0:30, 30, .8). In the R code, we use pp, population proportion, for π because R reserves pi for the usual constant. We also use sp, sample proportion, for $p = x/n$.

Next, we determine which of the 31 confidence intervals cover the value $\pi = 0.8$. Finally, the coverage probability is computed: It is the sum of the probabilities corresponding to values of x that yield intervals covering π.

```
n = 30                               # number of trials
x = 0:n;  sp = x/n                   # n+1 possible outcomes
m.err = 1.96*sqrt(sp*(1-sp)/n)       # n+1 Margins of error
lcl = sp - m.err                     # n+1 Lower conf. limits
ucl = sp + m.err                     # n+1 Upper conf. limits

pp = .80                             # pp = P(Success)
prob = dbinom(x, n, pp)              # distribution vector
cover = (pp >= lcl) & (pp <= ucl)    # vector of 0s and 1s
round(cbind(x, sp, lcl, ucl, prob, cover), 4)  # 4-place printout
sum(dbinom(x[cover], n, pp))         # total cov. prob. at pp

> round(cbind(x, sp, lcl, ucl, prob, cov), 4)

          x      sp      lcl      ucl     prob cover
   ...
  [18,]  17  0.5667   0.3893  0.7440   0.0022     0
  [19,]  18  0.6000   0.4247  0.7753   0.0064     0
  [20,]  19  0.6333   0.4609  0.8058   0.0161     1
  [21,]  20  0.6667   0.4980  0.8354   0.0355     1
  [22,]  21  0.7000   0.5360  0.8640   0.0676     1
  [23,]  22  0.7333   0.5751  0.8916   0.1106     1
  [24,]  23  0.7667   0.6153  0.9180   0.1538     1
  [25,]  24  0.8000   0.6569  0.9431   0.1795     1
  [26,]  25  0.8333   0.7000  0.9667   0.1723     1
  [27,]  26  0.8667   0.7450  0.9883   0.1325     1
  [28,]  27  0.9000   0.7926  1.0074   0.0785     1
  [29,]  28  0.9333   0.8441  1.0226   0.0337     0
  [30,]  29  0.9667   0.9024  1.0309   0.0093     0
  [31,]  30  1.0000   1.0000  1.0000   0.0012     0
```

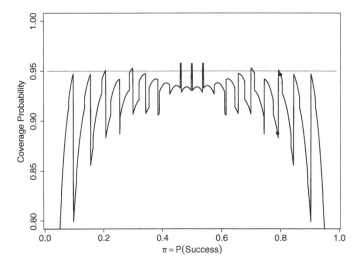

Figure 1.6. True coverage probabilities for traditional "95%" binomial confidence intervals when $n = 30$, computed according to formula (1.2). The coverage probability varies widely as π changes, seldom achieving the nominal 95%. Two heavy dots show values discussed in Example 1.5. (Compare this with Figure 1.8, p21.)

```
> sum(dbinom(x[cover], n, pp))
[1] 0.9463279
```

To save space, the first 17 lines of output are omitted; none of them involve intervals that cover π. We see that the values $x = 20, 21, \ldots, 27$ yield intervals that cover 0.30 (the ones for which cov takes the value 1). Thus the total coverage probability for $\pi = 0.30$ is

$$P(\text{Cover}) = P\{X = 20\} + P\{X = 21\} + \cdots + P\{X = 27\}$$
$$= 0.0355 + 0.0676 + \cdots + 0.0785 = 0.9463.$$

Notice that, for given n and π, the value of Z' in (1.3) depends only on the observed value of X. We have just shown by direct computation that if $n = 30$ and $\pi = 0.8$, then $P\{|Z'| < 1.96\} = P\{20 \leq X \leq 27\} = 0.9463$. This is only a little smaller than the claimed value of 95%.

In contrast, a similar computation with $n = 30$ and $\pi = 0.79$ gives a coverage probability of only 0.8876 (see Problem 1.14). This is very far below the claimed coverage probability of 95%. The individual binomial probabilities do not change much when π changes from 0.80 to 0.79. The main reason for the large change in the coverage probability is that, for $\pi = 0.79$, the confidence interval corresponding to $x = 27$ no longer covers π. ◇

For confidence intervals computed from (1.2), we have shown that a very small change in π can result in a large change in the coverage probability. The discreteness of the binomial distribution results in "lucky" and "unlucky" values of π. Perhaps of more concern, both of the coverage probabilities of traditional confidence intervals we have seen so far are smaller than 95%— one of them much smaller. Next we explore whether such "undercoverage" is exceptional or typical.

Example 1.6. Two Thousand Coverage Probabilities. To get a more comprehensive view of the performance of confidence intervals based on formula (1.2), we step through two thousand values of π from near 0 to near 1. For each value of π, we go through a procedure like that shown in Example 1.5. Finally, we plot the coverage probabilities against π.

```
n = 30                                    # number of trials
alpha = .05;   k = qnorm(1-alpha/2)       # conf level = 1-alpha
adj = 0                                   # (2 for Agresti-Coull)

x = 0:n;   sp = (x + adj)/(n + 2*adj)     # vectors of
m.err = k*sqrt(sp*(1 - sp)/(n + 2*adj))   #    length
lcl = sp - m.err                          #    n + 1
ucl = sp + m.err                          #

m = 2000                                  # no. of values of pp
pp = seq(1/n, 1 - 1/n, length=m)          # vectors
p.cov = numeric(m)                        #    of length m

for (i in 1:m)                            # loop (values of pp)
{                                         # for each pp:
   cover = (pp[i] >= lcl) & (pp[i] <= ucl)   #  1 if cover, else 0
   p.rel = dbinom(x[cover], n, pp[i])        #  relevant probs.
   p.cov[i] = sum(p.rel)                     #  total coverage prob.
}
plot(pp, p.cov, type="l", ylim=c(1-4*alpha,1))
lines(c(.01,.99), c(1-alpha,1-alpha))
```

It is clear from the resulting plot (Figure 1.6) that it is not unusual for the coverage probabilities of intervals based on (1.2) to vary rapidly as π varies in $(0, 1)$. More regrettably, the true coverage probabilities are often much lower than the claimed 95%. Furthermore, this tendency for coverage probabilities to be too low persists even for moderately large n (see Figure 1.7 and Problem 1.15). ◊

In addition to the graphs in Example 1.6, we made similar ones based on formula (1.2) for various levels of confidence $1 - \alpha$ between 90% and 99% and for various sample sizes n up to 200. These plots show that, for values of α and n frequently encountered in practice, the traditional confidence intervals cannot be relied upon to provide the promised level of confidence unless π is close to 1/2. Many ways have been proposed to improve confidence intervals for binomial proportions.

14   1 Introductory Examples

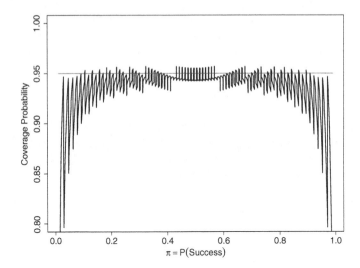

Figure 1.7. True coverage probabilities for traditional binomial confidence intervals continue to lie mainly below their target value even for moderately large n. This plot is based on a nominal confidence level of 95% and $n = 100$.

Fortunately, one of the proposed types of confidence intervals is simple to compute and has satisfactory coverage probabilities in many situations. **Agresti-Coull confidence intervals** are made by "adding two successes and two failures" to the data and then applying formula (1.2) to the adjusted data. Very briefly put, the effect of the adjustment is to shrink the distance between the point estimate and $1/2$, thus increasing the length of the confidence interval for better coverage when the point estimate is far from $1/2$. Throughout this book, we often use this type of interval when $n < 1000$. For very large n, the adjustment is not needed and has a negligible effect. See Problems 1.16, 1.17, and 1.19 for more about Agresti-Coull confidence intervals.

Perhaps you noticed the constant `adj = 0`, which played no essential role in our run of the R program in Example 1.6. If we substitute `adj = 2`, the program makes Figure 1.8 (p21), which shows coverage probabilities for 95% Agresti-Coull confidence intervals when $n = 30$.

No one style of confidence interval for binomial π seems to be satisfactory for all purposes. There is rich literature on binomial confidence intervals. The discussion and graphics in this section are largely based on papers by Agresti and Coull [AC98] and by Brown, Cai, and DasGupta [BCD01], both of which include extensive lists of references.

Because of the assumptions involved, statisticians have wondered and debated for some time about the accuracy of (1.2) in making binomial confidence intervals. However, the impact of plots such as our Figures 1.6 and 1.7 is difficult to ignore. From them, the systematic undercoverage of traditional confidence intervals is strikingly clear. The results we have reviewed in this section awaited a confluence of convenient software, modern hardware, and the ingenuity of researchers to use computer graphics to such powerful effect.

A major theme throughout this book is the use of graphical methods to better understand familiar concepts, verify the validity of statistical analyses and probability models, and explore new ideas.

1.3 Problems

Problems for Section 1.1 (Simulating Samples)

1.1   Based on Example 1.1, this problem provides some practice using R.

a) Start with `set.seed(1)`. Then execute the function `sample(1:100, 5)` five times and report the number of good chips in each.

b) Start with `set.seed(1)`, and execute `sum(sample(1:100, 5) <= 90)` five times. Report and explain the results.

c) Which two of the following four samples could not have been produced using the function `sample(1:90, 5)`? Why not?

```
[1]  2 62 84 68 60
[1] 46 39 84 16 39
[1] 43 20 79 32 84
[1] 68  2 98 20 50
```

1.2   This problem relates to the program in Example 1.1.

a) Execute the statements shown below in the order given. Explain what each statement does. Which ones produce output? What is the length of each vector? (A number is considered a vector of length 1.) Which vectors are logical, with possible elements TRUE or FALSE, and which are numeric?

```
pick = c(4, 47, 82, 21, 92);  pick <= 90;  sum(pick <= 90)
pick[1:90];  pick[pick <= 90];  length(pick[pick <= 90])
as.numeric(pick <= 90);  y = numeric(5);  y;  y[1] = 10;  y
w = c(1:5, 1:5, 1:10);  mean(w);  mean(w >= 5)
```

b) In the program, propose a substitute for the second line of code within the loop so that `good[i]` is evaluated in terms of `length` instead of `sum`. Run the resulting program, and report your result.

1.3   The random variable X of Example 1.1 has a hypergeometric distribution. In R, the hypergeometric probabilities $P\{X = x\}$ can be computed using the function `dhyper`. Its parameters are, in order, the number x of good items seen in the sample, the number of good items in the population, the number of bad items in the population, and the number of items selected without replacement. Thus each of the statements `dhyper(5, 90, 10, 5)` and `dhyper(0, 10, 90, 5)` returns 0.5837524.

a) What is the relationship between `sample(1:100,5)` and `choose(100,5)`?
b) Compute $P\{X = 2\}$ using `dhyper` and then again using `choose`.
c) Run the program of Example 1.1 followed by the statements `mean(good)` and `var(good)`. What are the numerical results of these two statements? In terms of the random variable X, what do they approximate and how good is the agreement?
d) Execute `sum((0:5)*dhyper(0:5,90,10,5))`? How many terms are being summed? What numerical result is returned? What is its connection with part (c)?

Notes: If n items are drawn at random without replacement from a box with b Bad items, g Good items, and $T = g+b$, then $E(X) = ng/T$, $V(X) = n(\frac{g}{T})(1-\frac{g}{T})(\frac{T-n}{T-1})$.

1.4   Based on concepts in the program of Example 1.3, this problem provides practice using functions in R.

a) Execute the statements shown below in the order given. Explain what each statement does. State the length of each vector. Which vectors are numeric and which are logical?

```
a = c(5, 6, 7, 6, 8, 7);  length(a);  unique(a)
length(unique(a));  length(a) - length(unique(a))
duplicated(a);  length(duplicated(a));  sum(duplicated(a))
```

b) Based on your findings in part (a), propose a way to count redundant birthdays that does not use `unique`. Modify the program to implement this method and run it. Report your results.

1.5   *Item matching.* There are ten letters and ten envelopes, a proper one for each letter. A very tired administrative assistant puts the letters into the envelopes at random. We seek the probability that no letter is put into its proper envelope and the expected number of letters put into their proper envelopes. Explain, statement by statement, how the program below approximates these quantities by simulation. Run the program with $n = 10$, then with $n = 5$, and again with $n = 20$. Report and compare the results.

```
m = 100000;  n = 10;  x = numeric(m)
for (i in 1:m) {perm = sample(1:n, n);  x[i] = sum(1:n==perm)}
cutp = (-1:n) + .5;  hist(x, breaks=cutp, prob=T)
mean(x == 0);  mean(x);  sd(x)
```

Notes: Let X be the number correct. For n envelopes, a combinatorial argument gives $P\{X = 0\} = 1/2!-1/3!+-\cdots+(-1)^n/n!$. (See [Fel57] or [Ros97].) In R, `i = 0:10`; `sum((-1)^i/factorial(i))`. For any $n > 1$, $P\{X = n - 1\} = 0$, $P\{X > n\} = 0$, $E(X) = 1$, and $V(X) = 1$. For large n, X is approximately POIS(1). Even for n as small as 10, this approximation is good to two places; to verify this, run the program above, followed by `points(0:10, dpois(0:10, 1))`.

1.6   A poker hand consists of five cards selected at random from a deck of 52 cards. (There are four Aces in the deck.)

a) Use combinatorial methods to express the probability that a poker hand has no Aces. Use R to find the numerical answer correct to five places.

b) Modify the program of Example 1.1 to approximate the probability in part (a) by simulation.

1.7   *Martian birthdays.* In his science fiction trilogy on the human colonization of Mars, Kim Stanley Robinson arranges the 669 Martian days of the Martian year into 24 months with distinct names [Rob96]. Imagine a time when the Martian population consists entirely of people born on Mars and that birthdays in the Martian-born population are uniformly distributed across the year. Make a plot for Martians similar to Figure 1.2. (You do not need to change n.) Use your plot to guess how many Martians there must be in a room in order for the probability of a birthday match just barely to exceed 1/2. Then find the exact number with min(n[p > 1/2]).

1.8   *Nonuniform birthrates.* In this problem, we explore the effect on birthday matches of the nonuniform seasonal pattern of birthrates in the United States, displayed in Figure 1.4. In this case, simulation requires an additional parameter prob of the sample function. A vector of probabilities is used to indicate the relative frequencies with which the 366 days of the year are to be sampled. We can closely reflect the annual distribution of U.S. birthrates with a vector of 366 elements:

```
p = c(rep( 96,61), rep( 98,89), rep( 99,62),
        rep(100,61), rep(104,62), rep(106,30), 25)
```

The days of the year are reordered for convenience. For example, February 29 appears last in our list, with a rate that reflects its occurrence only one year in four. Before using it, R scales this vector so that its entries add to 1.

To simulate the distribution of birthday matches based on these birthrates, we need to make only two changes in the program of Example 1.3. First, insert the line above before the loop. Second, replace the first line within the loop by b = sample(1:366, 25, repl=T, prob=p). Then run the modified program, and compare your results with those obtained in the example.

1.9   *Nonuniform birthrates (continued).* Of course, if the birthrates vary too much from uniform, the increase in the probability of birthday matches will surely be noticeable. Suppose the birthrate for 65 days we call "midsummer" is three times the birthrate for the remaining days of the year, so that the vector p in Problem 1.8 becomes p = c(rep(3, 65), rep(1, 300), 1/4).

a) What is the probability of being born in "midsummer"?

b) Letting X be the number of birthday matches in a room of 25 randomly chosen people, simulate $P\{X \geq 1\}$ and $E(X)$.

Answers: (a) sum(p[1:65])/sum(p). Why? (b) Roughly 0.67 and 1.0, respectively.

1.10   Three problems are posed about a die that is rolled repeatedly. In each case, let X be the number of different faces seen in the specified number of rolls. Using at least $m = 100\,000$ iterations, approximate $P\{X = 1\}$, $P\{X = 6\}$, and $E(X)$ by simulation. To do this **write a program** using the one in Example 1.3 as a rough guide. In what way might some of your simulated results be considered unsatisfactory? To verify that your program is working correctly, you should be able to find exact values for some, but not all, of the quantities by combinatorial methods.

a) The die is fair and it is rolled 6 times.
b) The die is fair and it is rolled 8 times.
c) The die is biased and it is rolled 6 times. The bias of the die is such that 2, 3, 4, and 5 are equally likely but 1 and 6 are each twice as likely as 2.

Answers: $P\{X = 1\} = 1/6^5$ in (a); the approximation has small absolute error but perhaps large percentage error. $P\{X = 6\} = 6!/6^6 = 5/324$ in (a), $45/4096$ in (c).

Problems for Section 1.2 (Binomial Confidence Intervals)

1.11   Suppose 40% of the employees in a very large corporation are women. If a random sample of 30 employees is chosen from the corporation, let X be the number of women in the sample.

a) For a specific x, the R function pbinom(x, 25, 0.3) computes $P\{X \leq x\}$. Use it to evaluate $P\{X \leq 17\}$, $P\{X \leq 6\}$, and hence $P\{7 \leq X \leq 17\}$.
b) Find $\mu = E(X)$ and $\sigma = SD(X)$. Use the normal approximation to evaluate $P\{7 \leq X \leq 17\}$. That is, take $Z = (X - \mu)/\sigma$ to be approximately standard normal. It is best to start with $P\{6.5 < X < 17.5\}$. Why?
c) Now suppose the proportion π of women in the corporation is unknown. A random sample of 30 employees has 20 women. Do you believe π is as small as 0.4? Explain.
d) In the circumstances of part (c), use formula (1.2) to find an approximate 95% confidence interval for π.

Hints and comments: For (a) and (b), about 0.96; you should give 4-place accuracy. The margin of error in (d) is about 0.17. Example 1.5 shows that the actual coverage probability of the confidence interval in (d) may differ substantially from 95%; a better confidence interval in this case is based on the Agresti-Coull adjustment of Problem 1.16: (0.486, 0.808).

1.12   Refer to Example 1.4 and Figure 1.5 on the experiment with a die.

a) Use formula (1.2) to verify the numerical values of the confidence intervals explicitly mentioned in the example (for students 2, 10, and 12).
b) In the figure, how many of the 20 students obtained confidence intervals extending below 0?
c) The most likely number of 6s in 30 rolls of a fair die is five. To verify this, first use i = 0:30; b = dbinom(i, 30, 1/6), and then i[b==max(b)] or round(cbind(i, b), 6). How many of the 20 students got five 6s?

1.13   Beneath the program in Example 1.1 on sampling computer chips, we claimed that the error in simulating $P\{X = 5\}$ rarely exceeds 0.004. Consider that the sample proportion 0.58298 is based on a sample of size $m = 100\,000$.

a) Use formula (1.2) to find the margin of error of a 95% confidence interval for $\pi = P\{X = 5\}$. With such a large sample size, this formula is reliable.
b) Alternatively, after running the program, you could evaluate the margin of error as `1.96*sqrt(var(good==5)/m)`. Why is this method essentially the same as in part (a)? (Ignore the difference between dividing by m and $m - 1$. Also, for a logical vector g, notice that `sum(g)` equals `sum(g^2)`.)

1.14   Modify the R program of Example 1.5 to verify that the coverage probability corresponding to $n = 30$ and $\pi = 0.79$ is 0.8876. Also, for $n = 30$, find the coverage probabilities for $\pi = 1/6 = 0.167, 0.700$, and 0.699. Then find coverage probabilities for five additional values of π of your choice. From this limited evidence, which appears to be more common—coverage probabilities below 95% or above 95%? In Example 1.4, the probability of getting a 6 is $\pi = 1/6$, and 18 of 20 confidence intervals covered π. Is this better, worse, or about the same as should be expected?

1.15   Modify the program of Example 1.6 to display coverage probabilities of traditional "95% confidence" intervals for $n = 50$ observations. Also, modify the program to show results for nominal 90% and 99% confidence intervals with $n = 30$ and $n = 50$. Comment on the coverage probabilities in each of these five cases. Finally, compare these results with Figure 1.7.

1.16   In the R program of Example 1.6, set `adj = 2` and leave $n = 30$. This adjustment implements the Agresti-Coull type of 95% confidence interval. The formula is similar to (1.2), except that one begins by "adding two successes and two failures" to the data. [Example: If we see 20 Successes in 30 trials, the 95% Agresti-Coull interval is centered at $22/34 = 0.6471$ with margin of error $1.96\sqrt{(22)(12)/34^3} = 0.1606$, and the interval is $(0.4864, 0.8077)$.]

Run the modified program, and compare your plot with Figures 1.6 (p12) and 1.8 (p21). For what values of π are such intervals too "conservative"—too long and with coverage probabilities far above 95%? Also make plots for 90% and 99% and comment. (See Problem 1.17 for more on this type of interval.)

1.17   *Algebraic derivation of alternate types of confidence intervals.* For convenience, denote the standard error of p as $\mathrm{SE}(p) = \sqrt{\pi(1 - \pi)/n}$ and its estimated standard error as $\widehat{\mathrm{SE}}(p) = \sqrt{p(1 - p)/n}$.

a) Show that $P\{|p-\pi|/\widehat{\mathrm{SE}}(p) < \kappa\} = P\{p-\kappa\,\widehat{\mathrm{SE}}(p) < \pi < p+\kappa\,\widehat{\mathrm{SE}}(p)\}$. The extreme terms in the second inequality are the endpoints of the confidence interval based on (1.2). The intended confidence level is $1 - \alpha$, and κ is defined by $P\{|Z| < \kappa\} = 1 - \alpha$ for standard normal Z.
b) We can also "isolate" π between two terms computable from observed data by using $\mathrm{SE}(p)$ instead of its estimate $\widehat{\mathrm{SE}}(p)$. Show that

$$P\{|p - \pi|/\mathrm{SE}(p) < \kappa\} = P\{\tilde{p} - E < \pi < \tilde{p} + E\},$$

where $\tilde{p} = \frac{X + \kappa^2/2}{n + \kappa^2}$, $E = \frac{\kappa}{n + \kappa^2}\sqrt{np(1 - p) + \kappa^2/4}$, and κ is as in part (a). Overall, the coverage probabilities of the **Wilson confidence interval** $\tilde{p} \pm E$ tend to be closer to $1 - \alpha$ than those of an interval based on (1.2). The Wilson interval uses the normal approximation, but it avoids estimating the standard error $\mathrm{SE}(p)$.

c) If we define $\tilde{X} = X + \kappa^2/2$ and $\tilde{n} = n + \kappa^2$, then verify that $\tilde{p} = \tilde{X}/\tilde{n}$ agrees with the $\tilde{p}$ of part (b). Also, show that $E^* = \kappa\sqrt{\tilde{p}(1 - \tilde{p})/\tilde{n}}$ is larger than E of part (b), but not by much. With these definitions of $\tilde{n}$ and $\tilde{p}$, the Agresti-Coull confidence interval $\tilde{p} \pm E^*$ is similar in form to the interval of (1.2). But, for most values of π, its coverage probabilities are closer to $1 - \alpha$ than are those of the traditional interval. (See Problem 1.16.)

Hints and comments: (b) Square and use the quadratic formula to solve for π. When $1 - \alpha = 95\%$, one often uses $\kappa = 1.96 \approx 2$ and thus $\tilde{p} \approx \frac{X+2}{n+4}$. (c) The difference between E and E^* is of little practical importance unless $\tilde{p}$ is near 0 or 1. For a more extensive discussion, see [BCD01].

1.18   For a discrete random variable X, the expected value (if it exists) is defined as $\mu = \mathrm{E}(X) = \sum_k kP\{X = k\}$, where the sum is taken over all possible values of k. Also, if X takes only nonnegative integer values, then one can show that $\mu = \sum_k P\{X > k\}$. In particular, if $X \sim \mathrm{BINOM}(n, \pi)$, then one can show that $\mu = \mathrm{E}(X) = n\pi$.

Also, the mode (if it exists) of a discrete random variable X is defined as the unique value k such that $P\{X = k\}$ is greatest. In particular, if X is binomial, then one can show that its mode is $\lfloor (n + 1)\pi \rfloor$; that is, the greatest integer in $(n+1)\pi$. Except that if $(n+1)\pi$ is an integer, then there is a "double mode": values $k = (n + 1)\pi$ and $(n + 1)\pi - 1$ have the same probability.

Run the following program for $n = 6$ and $\pi = 1/5$ (as shown); for $n = 7$ and $\pi = 1/2$; and for $n = 18$ and $\pi = 1/3$. Explain the code and interpret the answers in terms of the facts stated above about binomial random variables. (If necessary, use ?dbinom to get explanations of dbinom, pbinom, and rbinom.)

```
n = 6;  pp = 1/5;   k = 0:n
pdf = dbinom(k, n, pp);   sum(k*pdf)
cdf = pbinom(k, n, pp);   sum(1 - cdf)
mean(rbinom(100000, n, pp))
n*pp;   round(cbind(k, pdf, cumsum(pdf), cdf), 4)
k[pdf==max(pdf)];   floor((n+1)*pp)
```

1.19   *Average lengths of confidence intervals.* Problem 1.16 shows that, for most values of π, Agresti-Coull confidence intervals have better coverage probabilities than do traditional intervals based on formula (1.2). It is only reasonable to wonder whether this improved coverage comes at the expense of greater average length. For given n and π, the length of a confidence interval

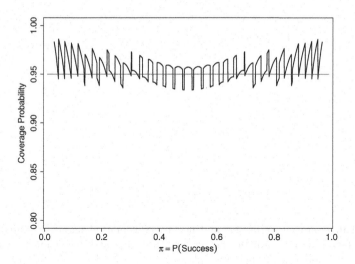

Figure 1.8. Coverage probabilities for Agresti-Coull 95% binomial confidence intervals when $n = 30$. By "adding two Successes and two Failures" to the data, coverage probabilities are much improved over those for traditional intervals, shown in Figure 1.6. See Problems 1.16 and 1.17 for a discussion of this adjustment.

is a random variable because the margin of error depends on the number of Successes observed. The program below illustrates the computation and finds the expected length.

```
n = 30;  pp = .2                        # binomial parameters
alpha = .05;  kappa = qnorm(1-alpha/2)  # level is 1 - alpha
adj = 0                      # 0 for traditional; 2 for Agresti-Coull
x = 0:n;  sp = (x + adj)/(n + 2*adj)
CI.len = 2*kappa*sqrt(sp*(1 - sp)/(n + 2*adj))
Prob = dbinom(x, n, pp);  Prod = CI.len*Prob
round(cbind(x, CI.len, Prob, Prod), 4)   # displays computation
sum(Prod)                                # expected length
```

a) Explain each statement in this program, and state the length of each named vector. (Consider a constant as a vector of length 1.)

b) Run the program as it is to find the average length of intervals based on (1.2) when $\pi = 0.1, 0.2$, and 0.5. Then use adj = 2 to do the same for Agresti-Coull intervals.

c) Figure 1.9 was made by looping through about 200 values of π. Use it to verify your answers in part (b). Compare the lengths of the two kinds of confidence intervals and explain.

d) **Write a program** to make a plot similar to Figure 1.9. Use the program of Example 1.5 as a rough guide to the structure. You can use plot for the first curve and lines to overlay the second curve.

Note: This program includes the entire length of any CI extending outside $(0, 1)$.

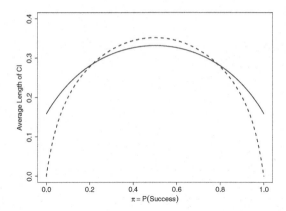

Figure 1.9. Average lengths of confidence intervals plotted against π. Agresti-Coull intervals (solid line) are a little shorter for π near $1/2$ and longer for π near 0 or 1 than are traditional confidence intervals based on (1.2). See Problem 1.19.

1.20 *Bayesian intervals.* Here is a confidence interval based on a Bayesian method and using the beta family of distributions. If x successes are observed in n binomial trials, we use the distribution $\mathsf{BETA}(x+1, n-x+1)$. An interval with nominal coverage probability $1 - \alpha$ is formed by cutting off probability $\alpha/2$ from each side of this beta distribution. For example, its 0.025 and 0.975 quantiles are the lower and upper limits of a 95% interval, respectively. In the R program of Example 1.6, replace the lines for lcl and ucl with the code below and run the program. Compare the coverage results for this Bayesian interval with results for the 95% confidence interval based on formula (1.2), which are shown in Figure 1.6. (Also, if you did Problem 1.16, compare it with the results for the Agresti-Coull interval.)

```
lcl = qbeta(alpha/2, x + 1, n - x + 1)
ucl = qbeta(1 - alpha/2, x + 1, n - x + 1)
```

Notes: The mean of this beta distribution is $(x+1)/(n+2)$, but this value need not lie exactly at the center of the resulting interval. If 30 trials result in 20 successes, then the traditional interval is $(0.4980, 0.8354)$ and the Agresti-Coull interval is $(0.4864, 0.8077)$. The mean of the beta distribution is 0.65625, and a 95% Bayesian interval is $(0.4863, 0.8077)$, obtained in R with qbeta(c(.025, .975), 21, 11). Bayesian intervals for π never extend outside $(0, 1)$. (These Bayesian intervals are based on a uniform prior distribution. Strictly speaking, the interpretation of such "Bayesian probability intervals" is somewhat different than for confidence intervals, but we ignore this distinction for now, pending a more complete discussion of Bayesian inference in Chapter 8.)

2

Generating Random Numbers

2.1 Introductory Comments on Random Numbers

Much of this book deals with simulation methods for probability models, also called Monte Carlo methods. We have seen a few introductory examples in Chapter 1. Even for some models that are easy to specify in a theoretical form, it may be difficult or impossible to "do the math" necessary to obtain the numerical results required in practice. Because of recent advances in computer hardware and software, simulation methods now offer feasible solutions to some of these troublesome computational problems.

In a simulation, the goal is to repeat an easily performed procedure thousands of times as a practical substitute for a direct computation that may not be feasible. Probability simulations are based on "random" numbers, which can be manipulated to produce observations from a probability model of interest.

How do we obtain such random numbers? One idea is to generate them with some physical device [KvG92]. Here are some examples:

- Successive tosses of a fair coin yield 0s (tails) or 1s (heads) at random.
- Rolling a fair die repeatedly produces a sequence of numbers 1 through 6.
- The faces of an icosahedron die (a regular 20-sided polyhedron) can be inscribed with two copies of the numbers 0 through 9; five rolls yield a randomly chosen number from 0 (for 00000) to 99999.
- A deck of 52 cards can be shuffled and cut to give a card interpreted as a randomly chosen digit from 1 to 52 (assuming more thorough shuffling than is usual in a friendly game of bridge.)
- Noise from an electronic circuit can be sampled at sufficiently long intervals to give cumulative counts, the last digits of which are randomly generated from among digits from 0 to 9. (In the early days of simulation, this method was used to make a book containing a million random digits [RAN55].)

Such mechanical procedures, carefully done, can give numbers that appear to be random: successive outcomes that are independent and equally likely.

E.A. Suess and B.E. Trumbo, *Introduction to Probability Simulation and Gibbs Sampling with R*, Use R!, DOI 10.1007/978-0-387-68765-0_2, © Springer Science+Business Media, LLC 2010

However, modern simulations often require hundreds of thousands or millions of random numbers, and so these physical methods are usually much too slow to be of practical use. The main goal of this chapter is to show how a computer can be programmed to generate "random numbers" quickly and simply enough to support the simulations we do throughout this book.

Specifically, in the next two sections, we show one method of generating numbers that are intended to be randomly sampled from the standard uniform distribution UNIF(0, 1). We also show some of the tests that are commonly used to verify whether these numbers behave as intended. In later sections of this chapter, we show some ways to transform random variables that are independent and identically distributed as UNIF(0, 1) into random observations from a variety of other families of distributions, such as binomial, Poisson, exponential, and normal.

2.2 Linear Congruential Generators

Because computers are programmed to carry out set arithmetic instructions, you may wonder how it is possible to get random numbers from a computer. Strictly speaking, it is not possible. However, with ingenuity and care, programs can be devised to produce **pseudorandom** numbers; that is, numbers that behave, for practical purposes, as if they were random. In this section, we hope you will understand how it is possible for the "random" procedures programmed into R to achieve an excellent and useful illusion of randomness.

A very common method of generating pseudorandom numbers on computers is called a **linear congruential generator**, a phrase we often shorten to "generator" in this chapter. First, we state and illustrate the general formula for such a generator. Then, in the next section, we discuss whether the results can be trusted to behave as random in practice.

A linear congruential generator produces integers r_i iteratively according to the mathematical formula

$$r_{i+1} = ar_i + b \ (\text{mod } d), \tag{2.1}$$

for integers $a > 0$, $b \geq 0$, and $d > 0$, where $i = 1, 2, 3 \ldots$. The notation *mod*, short for *modulo*, means that r_{i+1} is the remainder when $ar_i + b$ is divided by d. For example, we say that 3 is equal or "congruent" to 8 mod 5. Here, a is called the **multiplier**, b the **increment**, and d the **modulus** of the generator.

The generation process is started with a positive integer **seed** $s = r_1 < d$, often taken unpredictably from a computer clock. For suitably chosen a, b, and d, formula (2.1) can shuffle integers from among $0, 1, 2, \ldots, d - 1$ in ways that are not random but that pass many of the tests that randomly shuffled integers should pass. If $b = 0$, then the generator is called **multiplicative** and r_i cannot take the value 0.

Because the process described in formula (2.1) requires only simple computations, numbers can be generated fast enough to support practical computer simulations. Also, programmers like to use linear congruential generators because they can be made to give predictable results during debugging. From a generator specified by specific constants a, b, and d, you always get the same sequence of numbers if you start with the same seed.

Following standard terminology, we sometimes refer to the results from a linear congruential generator as "random numbers." But do not be confused by this terminology. First, the word *random* is more appropriately applied to the process by which a number is produced than to the number itself. There is no way to judge whether the number 0.785398 is random without seeing many other numbers that came from the same process. Second, *random* is shorthand for the more awkward *pseudorandom*.

Example 2.1. As a "toy" illustration, suppose we want to shuffle the 52 cards in a standard deck. For convenience, number them $1, 2, 3, \ldots, 52$. Then consider the generator with $a = 20$, $b = 0$, $d = 53$, and $s = 21$. To obtain its first 60 numbers, we treat R simply as a programming language, using the following code.

```
d = 53                # modulus
a = 20                # multiplier
b = 0                 # shift
s = 21                # seed
m = 60                # length of run (counting seed as #1)
r = numeric(m)        # initialize vector for random integers

r[1] = s              # set seed
for (i in 1:(m-1)) r[i+1] = (a * r[i] + b) %% d
                      # generates random integers

r                     # list of random integers generated
```

The results are:

```
> r                        # list of random integers generated
 [1] 21 49 26 43 12 28 30 17 22 16  2 40  5
[14] 47 39 38 18 42 45 52 33 24  3  7 34 44
[27] 32  4 27 10 41 25 23 36 31 37 51 13 48
[40]  6 14 15 35 11  8  1 20 29 50 46 19  9
[53] 21 49 26 43 12 28 30 17
```

Here are the details of the first few steps: To get from the seed $s = r_1 = 21$ to the first generated number r_2, compute $20(21) + 0 = 420$ and then find the remainder $r_2 = 49$ upon division by 53. Similarly, $r_3 = 20(49) + 0 = 671 = 26 \pmod{53}$. At step 53, we obtain $r_{53} = 21$ again, and the process repeats.

This generator runs through all 52 numbers before it repeats. We say that the **period** of this generator is 52 and that it has **full period** because only 52 values are possible for a multiplicative generator with modulus 53. As an illustration that not all generators have full period, you can easily verify for

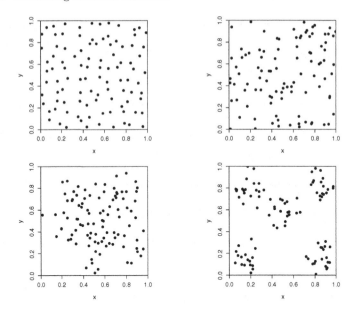

Figure 2.1. Randomness may be difficult to judge by eye. In which of these four plots are the 100 points placed at random within the square? Each point should have an equal chance of falling anywhere in the square, regardless of where other points might have fallen. See Problem 2.8.

yourself that, with $a = 23$, the sequence of results is $r_i = 21$, $r_2 = 36$, $r_3 = 2$, $r_4 = 47$, and then $r_5 = 21$ again, so that the period is only 4. $\Diamond$

The output of linear congruential generators is governed by intricate number-theoretic rules (some known and some apparently unknown), and we do not attempt a systematic discussion here. However, we see in Problem 2.3 that in a multiplicative generator with $d = 53$, the values $a = 15$ and $a = 30$ are among the multipliers that also give periods smaller than 52.

With $a = 20$, we have found a way to "shuffle" the 52-card deck. However, even in a simple application like shuffling cards for games of computer solitaire, this generator would not be a useful way to provide an unanticipated arrangement for each game. If we change the seed, we start with a different card, but from there on the order is the same as shown above. For example, if the seed is 5, then the sequence continues with 47, 39, 38, 18, and so on.

Carefully choosing values of a to get generators of full period, we could permute the cards into some fundamentally different orders. For example, with $a = 12$ and $s = 21$, we get a sequence that begins $21, 40, 3, 36, 8$, and so on. Even so, using a multiplicative generator that can assume only 52 possible values is not a promising approach for shuffling a deck of 52 cards. Of the $52! \approx 8 \times 10^{67}$ possible arrangements of 52 cards, multiplicative congruential generators with $d = 53$ can show fewer than a thousand.

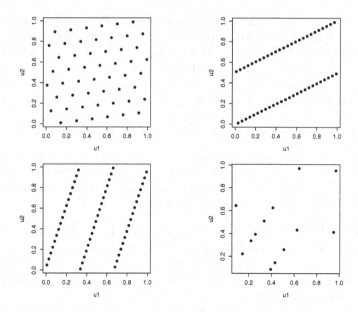

Figure 2.2. Grids of four multiplicative generators of modulus 53. The generator of Example 2.1 with multiplier $a = 20$ gives about the finest possible grid (upper left), but changing to $a = 27$ or 3 produces a grid with points lying in two or three widely separated parallel lines, and $a = 28$ results in less than the full period 52.

In the next section, we discuss some desirable properties of pseudorandom numbers and ways to test whether these properties are satisfied.

2.3 Validating Desirable Properties of a Generator

It is customary to rescale the values r_i output by a congruential generator to fit into the interval $(0, 1)$. For example, if a generator with modulus d takes values $0, 1, \ldots, d - 1$, then we can use $u_i = (r_i + 0.5)/d$.

Desirable properties of the u_i are that they have a large period, uniform distribution, and independent structure. In particular, we seek a generator with at least the following properties:

- A large modulus and full period. In practice, the period should be much larger than the number m of random values needed for the simulation at hand.
- A histogram of values u_i from the generator that is consistent with a random sample from $\mathsf{UNIF}(0, 1)$. Various statistical tests can be performed to see whether this is so.
- Pairwise independence. As a check for independence and uniform distribution, a plot of the $m - 1$ pairs (u_i, u_{i+1}), for $i = 1, 2, \ldots m - 1$, should

"fill" the unit square in a manner consistent with a uniform distribution over the unit square. Ideally, there would be no noticeable pattern of any kind. (Figure 2.1 illustrates that intuition may not always be a good guide in judging whether this is so.) However, on sufficiently close inspection, points from a congruential generator always show a grid pattern. We hope the grid is so fine as to be undetectable for m of the required size. We hope the grid lines, if visible, are very narrowly spaced in both directions. (Figure 2.2 illustrates obtrusive grid patterns for generators with $d = 53$.)

There are some rules for avoiding bad generators, but none for guaranteeing good ones. Each proposed generator has to be subjected to many tests before it can be used with reasonable confidence. Now we investigate a few generators, illustrating the principles just listed.

Example 2.2. Here we consider a generator that is "pretty good," but not really good enough for serious modern simulations. Let $a = 1093$, $b = 18\,257$, $d = 86\,436$, and $s = 7$. In elementary software applications, this generator has the advantage that the arithmetic involved never produces a number larger than 10^8 (see [PTVF92]).

This generator has full period 86 436. As shown in the top two panels of Figure 2.3, the histogram based on $m = 1000$ values u_i looks reasonably close to uniform, and no grid pattern is apparent in the corresponding bivariate plot. See Problem 2.4 for a formal test of fit to $\mathsf{UNIF}(0,1)$. The code for making these two panels is shown below.

```
# Initialize
a = 1093;  b = 18257;  d = 86436;  s = 7
m = 1000;  r = numeric(m);  r[1] = s

# Generate
for (i in 1:(m-1)) {r[i+1] = (a*r[i] + b) %% d}
u = (r + 1/2)/d                        # values fit in (0,1)

# Display Results
par(mfrow=c(1,2), pty="s")             # 2 square panels in a plot
   hist(u, breaks=10, col="wheat")     # left panel
      abline(h=m/10, lty="dashed")
   u1 = u[1:(m-1)]; u2 = u[2:m]        # right panel
   plot (u1, u2, pch=19)
par(mfrow=c(1,1), pty="m")             # return to default
```

If we generate $m = 50\,000$ values, then we obtain the two panels at the bottom of Figure 2.3. Fifty thousand is more than half the period of the generator, but not more than might be needed in a practical simulation. Initially, the histogram (lower left) looks promising because all the bars are of very nearly the same height. However, perhaps surprisingly, this is a serious flaw because the heights of the bars are much less variable than would be expected

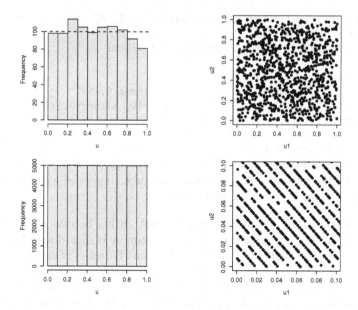

Figure 2.3. The "pretty good" generator of Example 2.2. The top two panels show 1000 generated values with unremarkable results. By contrast, the bottom panels show 50 000 values. Here the histogram bars are surprisingly regular, and an enlargement of a small corner of the 2-dimensional plot shows a clear grid pattern.

from truly random data. The result is "too good to be true": there is about one chance in 100 million of getting such a nearly perfect result.

Moreover, the 2-dimensional plot (lower right) shows that the grid is widely spaced in one direction. Here, to make a clear figure for publication, we have shown a magnified view of a small part of the entire 2-dimensional plot—specifically, the corner $(0, 0.1) \times (0, 0.1)$. ◇

We have seen that 2-dimensional plots can play an important role in testing a generator. The following example shows the importance of looking in higher dimensions as well.

Example 2.3. Some years ago, IBM introduced a generator called RANDU for some of its mainframe computers. Essentially a multiplicative generator with $a = 65\,539$ and $d = 2^{31}$, it came to be very widely used. But over time it acquired a bad reputation. First, it was found to yield wrong Monte Carlo answers in some computations where the right answers were known by other means. Later, it was found to concentrate its values on a very few planes in 3-dimensional space. The program below makes Figure 2.4.

```
a = 65539;   b = 0;   d = 2^31;   s = 10
m = 20000;   r = numeric(m);   r[1] = s
```

```
for (i in 1:(m-1)) {r[i+1] = (a*r[i] + b) %% d}
u = (r - 1/2)/(d - 1)
u1 = u[1:(m-2)]; u2 = u[2:(m-1)]; u3 = u[3:m]

par(mfrow=c(1,2), pty="s")
  plot(u1, u2, pch=19, xlim=c(0,.1), ylim=c(0,.1))
  plot(u1[u3 < .01], u2[u3 < .01], pch=19, xlim=c(0,1), ylim=c(0,1))
par(mfrow=c(1,1), pty="m")
```

In the left panel of Figure 2.4, we see no evidence of a grid pattern in a magnified view of part of the 2-dimensional plot. However, a view of the thin veneer from the entire front face of the unit cube shows the widely separated planes of the grid in 3-dimensional space. ◇

Subsequent to this discovery about RANDU, it has become standard practice to verify the grid structure of a generator in higher dimensions. Also, nowadays candidate generators are tested against batteries of simulation problems that have known answers and that have caused problems for previously disparaged generators. Problem 2.6 shows an example in which getting the wrong answer to one simple problem is enough to show that a particular generator is flawed.

In recent years, there has been tremendous progress in devising ever better random number generators. Currently, the R function runif uses a generator called the **Mersenne twister** which combines ideas of a linear congruential generator with other more advanced concepts to produce results that you can rely upon to be essentially independent and identically distributed as UNIF$(0, 1)$. Specifically, runif(10) produces a vector of ten observations from this distribution, without the need to write a loop.

The Mersenne twister has period $2^{19937} - 1 \approx 4.32 \times 10^{6001}$, and it has been tested for good behavior in up to 623 consecutive dimensions. These distinct generated values are mapped into the roughly 4.31 billion numbers that can be expressed within the precision of R. (See information provided at ?.Random.seed for some technical information about random number generation in R. Also, see the bottom panels of Figure 2.4, which show runif is at least better behaved in three dimensions than RANDU.)

Some commercial statistical software packages (such as SAS, S-Plus, Minitab, and so on) also use excellent generators. However, publishers of software intended mainly for nonstatistical customers have not always had the financial motivation or felt the corporate responsibility to use state-of-the-art random number generators. For example, a generator that may be adequate for "randomizing" the behavior of computer solitaire and other computer games may not be appropriate for large-scale business simulations or for simulations of the kind we show in this book. (For some specific critiques of generators in commercial software, see [LEc01] and [PTVF92].)

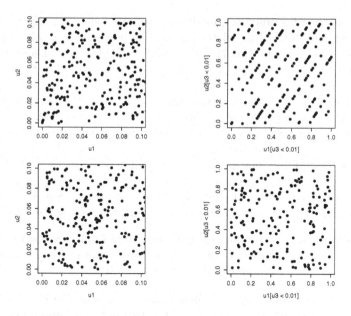

Figure 2.4. Plots in two and three dimensions. In a magnified corner of a 2-d plot (upper left), the RANDU generator of Example 2.3 seems alright. But a thin section from the front face of the unit cube (upper right) shows that in 3-d all points lie in very few planes. The second row shows no problem in analogous plots from `runif` in R. Each generator produced 20 000 values; each plot shows about 200 points.

2.4 Transformations of Uniform Random Variables

All of the random procedures in R, including the function `sample` used in the last chapter, are based on `runif`. In this section, we show how uniform random variables can be used to sample from a finite population or transformed into other random variables, such as binomial, exponential, and normal.

We begin with a few examples. These examples are elementary, but they illustrate some important ideas.

- If $U \sim \text{UNIF}(0, 1)$, then $2U + 5 \sim \text{UNIF}(5, 7)$, so we can generate a vector of ten observations from the latter distribution with `2*runif(10) + 5`. (See Problem 2.9.) Alternatively, we can use additional parameters of `runif` to show the endpoints of the interval in which we want to generate uniformly distributed observations: `runif(10, 5, 7)`. The code `runif(10)` is short for `runif(10, 0, 1)`.

- To simulate a roll of two fair dice, we can use `sample(1:6, 2, rep=T)`. This is equivalent to `ceiling(runif(2, 0, 6))`. Also, to draw one card at random from a standard deck, we can use `ceiling(runif(1, 0, 52))`. Sampling multiple cards without replacement is a little more complicated because at each draw we must keep track of the cards drawn previously, but this record keeping is already programmed into the `sample` function. So `sample(1:52, 5)` simulates fair dealing of a five-card poker hand.

- Suppose we want to simulate the number X of heads in four tosses of a fair coin. That is, $X \sim \text{BINOM}(4, 1/2)$. One method is to generate a vector of four uniform random variables with `runif(4)` and then compute `sum(runif(4) > .5)`, where the expression in parentheses is a logical vector. When taking the sum, R interprets `TRUE` as 1 and `FALSE` as 0. This method is considered wasteful because it requires four calls to the random number generator. In Example 2.5, we show a better way that requires only one call. Functions for simulating random samples from the binomial and several other commonly used probability distributions are available in R, and you can depend on these random functions being very efficiently programmed. So it is preferable to use `rbinom(1, 4, 0.5)`.

A real function of a random variable is another random variable. Random variables with a wide variety of distributions can be obtained by transforming a standard uniform random variable $U \sim \text{UNIF}(0, 1)$.

Example 2.4. The Square of a Uniform Random Variable. Let $U \sim \text{UNIF}(0, 1)$. We seek the distribution of $X = U^2$. The **support** of a continuous random variable is the part of the real line for which its density function is positive. So the support of U is the unit interval $(0, 1)$. Because the square of a number in the unit interval is again in the unit interval, it seems clear that the support of X is also $(0, 1)$.

However, it also seems clear that X cannot be uniformly distributed. For example, $P\{0 < U \leq 0.1\} = P\{0 < X \leq 0.01\} = 0.1$, so outcomes that put U into an interval of length 0.1 squeeze X into an interval of length 0.01. In contrast, $P\{0.9 < U \leq 1\} = P\{0.81 < X \leq 1\} = 0.1$, so outcomes that put U in an interval of length 0.1 stretch X over an interval of length 0.19. Figure 2.5 shows the result when 10 000 values of U are simulated (left panel) and transformed to values $X = U^2$ (right). Each interval in both histograms contains about 1000 simulated values. The following program makes such graphs.

```
# set.seed(1234)
m = 10000
u = runif(m);   x = u^2
xx = seq(0, 1, by=.001)
cut.u = (0:10)/10;   cut.x = cut.u^2

par(mfrow=c(1,2))
  hist(u, breaks=cut.u, prob=T, ylim=c(0,10))
    lines(xx, dunif(xx), col="blue")
  hist(x, breaks=cut.x, prob=T, ylim=c(0,10))
    lines(xx, .5*xx^-.5, col="blue")
par(mfrow=c(1,1))
```

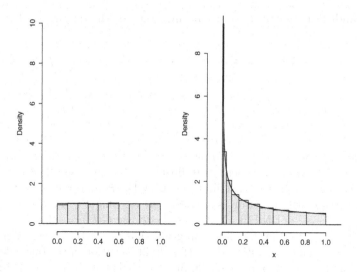

Figure 2.5. Square of a uniform random variable. The histogram at the left shows 10 000 simulated observations from UNIF(0, 1). Upon squaring, these observations make the histogram at the right, in which each bar represents about 1000 observations. It approximates the density function of BETA(0.5, 1). See Example 2.4.

The density function of X can be derived using an argument based on its cumulative distribution function (CDF). Recall that the CDF of U is given by $F_U(u) = P\{U \le u\} = u$, for $0 < u < 1$. Then, for $0 < x < 1$, we have

$$F_X(x) = P\{X \le x\} = P\{U^2 \le x\} = P\{U \le x^{1/2}\} = x^{1/2}. \qquad (2.2)$$

Taking the derivative of this CDF to get the density function of X, we obtain $f_X(x) = 0.5x^{-0.5}$, which is plotted along with the histogram of simulated values of X in Figure 2.5. This is the density function of BETA(0.5, 1). ◇

In Example 2.4, we have seen how to find the density function of a transformed standard uniform random variable. Now suppose we want to simulate random samples from a particular distribution. Then the question is how to find the transformation that will change a standard uniform distribution into the desired one. In Example 2.4, notice that the CDF of $X \sim$ BETA(0.5, 1) is $F_X(x) = x^{1/2}$, for $0 < x < 1$, that the quantile function (inverse CDF) of X is $x = F_X^{-1}(u) = u^2$, and that $X = F_X^{-1}(U) = U^2$ is the transformation we used to generate $X \sim$ BETA(0.5, 1).

This result for BETA(0.5, 1) is not a coincidence. In general, if F_Y is the CDF of a continuous random variable, then $F_Y^{-1}(U)$ has the distribution of Y. This is called the **quantile transformation method** of simulating a distribution. In order for this method to be useful, it is usually necessary for F_Y to be expressed in closed form so that we can find its inverse, the quantile function F_Y^{-1}. A few additional examples of this method are listed below, where U

has a standard uniform distribution, and some of the details are left to the problems.

- A random variable $X \sim \mathsf{BETA}(\alpha, 1)$ has density function $f_X(x) = \alpha x^{\alpha-1}$, CDF $F_X(x) = x^\alpha$, and quantile function $x = F_X^{-1}(u) = u^{1/\alpha}$, where $0 < x, u < 1$. Thus, X can be simulated as $X = U^{1/\alpha}$. To get a vector of ten independent observations from such a distribution, we could use `qbeta(runif(10), alpha, 1)`, where the constant `alpha` is suitably defined. However, to sample from any beta distribution, it is best to use the R function `rbeta` because it is programmed to use efficient methods for all choices of α and β. Problem 2.10 illustrates the case $\alpha = 2$, $\beta = 1$.
- If X has an exponential distribution with CDF $F_X(x) = 1 - e^{-x}$, for $x > 0$, then the quantile function is $x = F_X^{-1}(u) = -\log(1 - u)$, for $0 < u < 1$. Because $1 - U \sim \mathsf{UNIF}(0, 1)$, it is easier to use $G(U) = -\log(U)$ to simulate X. See Problem 2.11 for a program that generates observations from an exponential distribution using this slight variation of the quantile transformation method. Here again, in practice, it is best to use the R function `rexp`, with appropriate parameters, to simulate samples from any exponential distribution. Notice that R parameterizes the exponential family of distributions according to the rate (reciprocal mean) rather than the mean.
- Some commonly used distributions have CDFs that cannot be expressed in closed form. Examples are normal distributions and some members of the beta and gamma families. Even for these, R provides quantile functions that are accurate to many decimal places, but they require computation by approximate numerical methods. Thus, although one could get two independent standard normal random variables from `qnorm(runif(2))`, it is simpler and faster—and maybe more accurate—to use `rnorm(2)`.

Properly interpreted, the quantile transformation method also works for simulating discrete distributions. The next example illustrates this method for a binomial random variable.

Example 2.5. Suppose we want to simulate an observation $X \sim \mathsf{BINOM}(5, 0.6)$. In particular, $P\{X = 3\} = F_X(3) - F_X(2) = 0.66304 - 0.31744 = 0.34560$. The left panel of Figure 2.6 shows the CDF of this distribution, and the length of the heavy vertical line above $x = 3$ corresponds to this probability. What mechanism of simulation would lead to the outcome $X = 3$? The quantile function F_X^{-1} is plotted in the right panel of the figure. Here, the length of the heavy horizontal line at $F_X^{-1}(u) = 3$ is 0.34560. If $U \sim \mathsf{UNIF}(0, 1)$, then $P\{X = 3\} = F_U(0.66304) - F_U(0.31744) = 0.34560$. So we can simulate the event $\{X = 3\}$ as $\{0.31744 < U \le 0.66304\}$.

More generally, this shows that we can use the quantile transformation method to simulate an observation X: `qbinom(runif(1), 5, 0.6)`. This is an efficient method because it uses only one simulated value of U for each simulated value of X produced. Consider the following experiment.

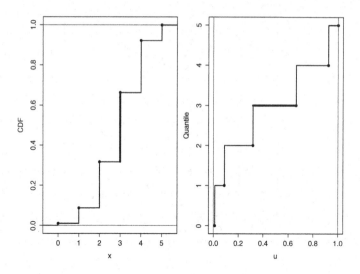

Figure 2.6. Binomial CDF and quantile function. The CDF of $X \sim \mathsf{BINOM}(5, 0.6)$ is shown in the left panel. Its inverse is the quantile function (right). The length of the heavy line segment in each plot represents $P\{X = 3\}$. See Example 2.5.

```
> set.seed(1234); rbinom(10, 5, 0.4)
 [1] 1 2 2 2 3 2 0 1 2 2
> set.seed(1234); qbinom(runif(10), 5, 0.4)
 [1] 1 2 2 2 3 2 0 1 2 2
> set.seed(1234); rbinom(10, 5, 0.6)
 [1] 4 3 3 3 2 3 5 4 3 3
> set.seed(1234); 5 - qbinom(runif(10), 5, 0.4)
 [1] 4 3 3 3 2 3 5 4 3 3
> set.seed(1234); qbinom(runif(10), 5, 0.6)
 [1] 2 3 3 3 4 3 0 2 4 3
```

This suggests that the random function `rbinom` in R uses a method equivalent to the quantile transformation when the success probability $\pi \leq 0.5$ but that a slight variation is used for larger π. $\Diamond$

Note: For a discrete random variable, interpretation of a quantile function as the inverse of the cumulative distribution function must be done with care. For example, `qbinom(0.5, 5, 0.6)` returns 3, as shown in the right panel in Figure 2.6. But no cumulative probability in $\mathsf{BINOM}(5, 0.6)$ equals 0.5, so `pbinom(3, 5, 0.6)` returns 0.66304, not 0.5. Also, `pbinom(3.5, 5, 0.6)` returns 0.66304, which is consistent with the left panel in the figure, but `qbinom(0.66304, 5, 0.6)` returns 3, not 3.5. Exact inverse relationships hold only for the values shown by heavy dots in Figure 2.6. Nevertheless, in R definitions of `qbinom` and quantile functions of other discrete distributions are compatible with the quantile transformation method.

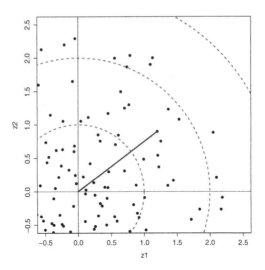

Figure 2.7. Normal errors for hits on a target. Vertical and horizontal errors are standard normal. The point roughly at $(1.2, 0.9)$ is $R \approx 1.5$ units away from the origin. The line from the origin (bull's eye) to this point makes an angle of $\Theta \approx 37$ degrees with the positive horizontal axis. See Example 2.6 and Problem 2.7.

2.5 Transformations Involving Normal Random Variables

In this section, we explore some important relationships among normal, exponential, and uniform random variables. One goal—accomplished near the end of the section—is to illustrate one very common method of generating normal random variables from uniform ones. However, in our first example, we take it for granted that the R function `rnorm` generates independent standard normal variates.

Example 2.6. Target Practice with Normal Errors. Suppose we aim an arrow at a target, as shown in Figure 2.7. The goal is to hit the bull's eye, located at the origin, but shots are random.

We assume that errors in the horizontal and vertical directions are independent standard normal random variables, Z_1 and Z_2, respectively, so that any one shot hits at the point (Z_1, Z_2). Thus the squared distance from the origin is $T = R^2 = Z_1^2 + Z_2^2$.

It is shown in books on probability theory and mathematical statistics that $T \sim \mathsf{CHISQ}(\nu = 2) = \mathsf{EXP}(1/2)$. This distribution has density function $f_t(t) = 0.5e^{-0.5t}$, for $t > 0$, and hence $\mathrm{E}(T) = 2$. (See Problems 2.12 and 2.13.) However, without any formal proof, the following program allows us to see in Figure 2.8 the excellent agreement between this density function and the histogram of 40 000 values of T as simulated from Z_1 and Z_2.

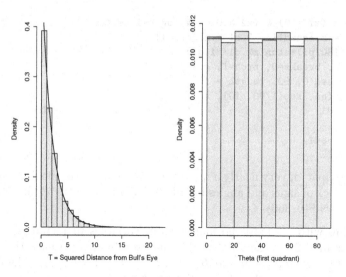

Figure 2.8. Based on polar coordinates for hits on a target. The histogram at the left shows 40 000 simulated values of $R^2 \sim$ EXP(2), the squared distance from the origin. For the roughly 10 000 simulated hits in the first quadrant, the histogram at the right shows $\Theta \sim$ UNIF(0, 10) degrees. See Example 2.6.

```
# set.seed(1212)          # this seed for exact results shown
m = 40000;  z1 = rnorm(m);  z2 = rnorm(m)
t = z1^2 + z2^2;  r = sqrt(t)

hist(t, breaks=30, prob=T, col="wheat")
   tt = seq(0, max(t), length=100);  dens = 0.5*exp(-0.5*tt)
   lines(tt, dens, col="blue")
mean(t);  mean(r);  mean(r < 2)

> mean(t);  mean(r);  mean(r < 2)
[1] 2.005963
[1] 1.254416
[1] 0.86445
```

The first numerical result is consistent with the known value $E(T) = 2$. From the second numerical result, we see that on average the arrow misses the bull's eye by about 1.25 units (not $\sqrt{2}$). From the last result, we see that about 86% of the hits are within 2 units of the bull's eye. (Use `pchisq(4, 2)` for the exact value of $P\{R < 2\} = P\{T < 4\}$.)

Also of interest is the angle Θ of the hit as measured counterclockwise from the positive horizontal axis. If the hit is in the first quadrant, then $\Theta = \arctan(Z_2/Z_1)$. Restricting attention to (Z_1, Z_2) in the first quadrant for simplicity, the following additional code shows that $\Theta \sim$ UNIF(0, 90), measured in degrees.

```
quad1 = (z1 > 0) & (z2 > 0)    # logical vector
z1.q1 = z1[quad1]; z2.q1 = z2[quad1]
th = (180/pi)*atan(z2.q1/z1.q1)
hist(th, breaks=9, prob=T, col="wheat")
  aa = seq(0, 90, length = 10)
  lines(aa, rep(1/90, 10), col="blue")
sum(quad1)                      # number of hits in quadrant 1

> sum(quad1)
[1] 10032
```

As anticipated, very nearly a quarter of the hits were in the first quadrant. Keeping track of signs and angles for all 40 000 hits, one gets results consistent with $\Theta \sim \mathsf{UNIF}(0, 360)$. $\Diamond$

Starting with standard normal random variables in rectangular coordinates, Example 2.6 suggests that in polar coordinates we have the random variables $\Theta \sim \mathsf{UNIF}(0, 360)$ and R, with $R^2 = T \sim \mathsf{EXP}(1/2)$. (The distance R itself is said to have a **Rayleigh distribution**.) Moreover, we already know that the distributions of Θ and T are easily simulated as transformations of standard uniform distributions.

Reversing this procedure, we can start with a pair of standard uniform random variates (U_1, U_2) to simulate the location of a random hit in polar coordinates and then transform to rectangular coordinates to obtain two standard normal random variates (Z_1, Z_2) that express the location of the same hit. Specifically, two independent observations from $\mathsf{NORM}(0, 1)$ can be simulated from two independent observations from $\mathsf{UNIF}(0, 1)$ according to the equations

$$Z_1 = \sqrt{-2\log(U_1)}\,\cos(2\pi U_2) \text{ and } Z_2 = \sqrt{-2\log(U_1)}\,\sin(2\pi U_2).$$

This transformation from (U_1, U_2) to (Z_1, Z_2) is accurate and efficient. It is known as the **Box-Muller method** of generating standard normal random variables. Problem 2.14 illustrates its use. This method and variations of it are widely used in statistical software to simulate random observations from $\mathsf{NORM}(0, 1)$.

We hope the examples in this chapter have helped you to understand how modern probability simulations can be based on a carefully constructed random number generator of uniform random variables that may then be transformed to imitate output from various other probability models and distributions. The book by James Gentle [Gen98] includes more detailed and advanced discussions of the topics we have illustrated at a relatively superficial level in this chapter, and it provides over 20 pages of references to original papers and yet more extensive discussions.

In the chapters that follow, we use a variety of random functions in R, including `sample`, `rnorm`, `rbeta`, `rgamma`, `rbinom`, and `rpois`. All of these are efficiently programmed and based on the excellent generator implemented for `runif`.

2.6 Problems

Problems for Section 2.2 (Congruential Generators)

2.1   Before congruential generators became widely used, various mathematical formulas were tried in an effort to generate useful pseudorandom numbers. The following unsuccessful procedure from the mid-1940s has been ascribed to the famous mathematician John von Neumann, a pioneer in attempts to simulate probability models.

Start with a seed number r_1 having an even number of digits, square it, pad the square with initial zeros at the left (if necessary) so that it has twice as many digits as the seed, and take the center half of the digits in the result as the pseudorandom number r_2. Then square this number and repeat the same process to get the next number in the sequence, and so on.

To see why this method was never widely used, start with $r_1 = 23$ as the seed. Then $23^2 = 529$; pad to get 0529 and $r_2 = 52$. Show that the next number is $r_3 = 70$. Continue with values $r_i, i = 3, 4 \ldots$ until you discover a difficulty. Also try starting with 19 as the seed.

2.2   The digits of transcendental numbers such as $\pi = 3.14159\ldots$ pass many tests for randomness. A search of the Web using the search phrase `pi digits` retrieves the URLs of many sites that list the first n digits of π for very large n. We put the first 500 digits from one site into the vector v and then used `summary(as.factor(v))` to get the number of times each digit appeared:

```
> summary(as.factor(v))
 0  1  2  3  4  5  6  7  8  9
45 59 54 50 53 50 48 36 53 52
```

a) Always be cautious of information from unfamiliar websites, but for now assume this information is correct. Use it to simulate 500 tosses of a coin, taking even digits to represent Heads and odd digits to represent Tails. Is this simulation consistent with 500 independent tosses of a fair coin?

b) Repeat part (a) letting numbers 0 through 4 represent Heads and numbers 5 through 9 represent Tails.

c) Why do you suppose digits of π are not often used for simulations?

Hint: (a, b) One possible approach is to find a 95% confidence interval for $P(\text{Heads})$ and interpret the result.

2.3   Example 2.1 illustrates one congruential generator with $b = 0$ and $d = 53$. The program there shows the first $m = 60$ numbers generated. Modify the program, making the changes indicated in each part below, using `length(unique(r))` to find the number of distinct numbers produced, and using the additional code below to make a 2-dimensional plot. Each part requires two runs of such a modified program. Summarize findings, commenting on differences within and among parts.

```
u = (r - 1/2)/(d-1)
u1 = u[0:(m-1)];   u2 = u[2:m]
plot(u1, u2, pch=19)
```

a) Use $a = 23$, first with $s = 21$ and then with $s = 5$.
b) Use $s = 21$, first with $a = 15$ and then with $a = 18$.
c) Use $a = 22$ and then $a = 26$, each with a seed of your choice.

Problems for Section 2.3 (Validating Generators)

2.4   *A Chi-squared test for Example 2.2.* Sometimes it is difficult to judge by eye whether the evenness of the bars of a histogram is consistent with a uniform distribution. The chi-squared goodness-of-fit statistic allows us to quantify the evenness and formally test the null hypothesis that results agree with $\mathsf{UNIF}(0, 1)$. If the null hypothesis is true, then each u_i is equally likely to fall into any one of the h bins of the histogram, so that the *expected* number of values in each bin is $E = m/h$. Let N_j denote the *observed* number of values in the jth of the h bins. The chi-squared statistic is

$$Q = \sum_{j=1}^{h} \frac{(N_j - E)^2}{E}.$$

If the null hypothesis is true and E is large, as here, then Q is very nearly distributed as $\mathsf{CHISQ}(h - 1)$, the chi-squared distribution with $h - 1$ degrees of freedom. Accordingly, $\mathrm{E}(Q) = h - 1$. For our example, $h = 10$, so values of Q "near" 9 are consistent with uniform observations. Specifically, if Q falls outside the interval $[2.7, 19]$, then we suspect the generator is behaving badly. The values 2.7 and 19 are quantiles 0.025 and 0.975, respectively, of $\mathsf{CHISQ}(9)$.

In some applications of the chi-squared test, we would reject the null hypothesis only if Q is too large, indicating some large values of $|Ni - E|$. But when we are validating a generator we are also suspicious if results are "too perfect" to seem random. (One similarly suspicious situation occurs if a fair coin is supposedly tossed 8000 times independently and exactly 4000 Heads are reported. Another is shown in the upper left panel of Figure 2.1.)

a) Run the part of the program of Example 2.2 that initializes variables and the part that generates corresponding values of u_i. Instead of the part that prints a histogram and 2-dimensional plot, use the code below, in which the parameter `plot=F` suppresses plotting and the suffix `$counts` retrieves the vector of 10 counts. What is the result, and how do you interpret it?

```
# Compute chi-squared statistic
h = 10;  E = m/h;   cut = (0:h)/h
N = hist(u, breaks=cut, plot=F)$counts
Q = sum((N - E)^2/E); Q
```

b) Repeat part (a), but with $m = 50\,000$ iterations.

c) Repeat part (a) again, but now with $m = 1000$ and $b = 252$. In this case, also make the histogram and the 2-dimensional plot of the results and comment. Do you suppose the generator with increment $b = 252$ is useful? (Problem 2.6 below investigates this generator further.)

d) Repeat part (a) with the original values of a, b, d, and s, but change to $m = 5000$ and add the step u = u^0.9 before computing the chi-squared statistic. (We still have $0 < u_i < 1$.) Also, make and comment on the histogram.

e) Find and interpret the chi-squared goodness-of-fit statistic for the 10 counts given in the statement of Problem 2.2.

Answers: In (a)–(e), $Q \approx 7$, 0.1, 0.2, 46, and 7, respectively. Report additional decimal places, and provide interpretation.

2.5   When beginning work on [Tru89], Trumbo obtained some obviously incorrect results from the generator included in Applesoft BASIC on the Apple II computer. The intended generator would have been mediocre even if programmed correctly, but it had a disastrous bug in the machine-level programming that led to periods of only a few dozen for some seeds [Spa83]. A cure (proposed in a magazine for computer enthusiasts [HRG83]) was to import the generator $r_{i+1} = 8192r_i$ (mod 67 099 547). This generator has full period, matched the capabilities of the Apple II, and seemed to give accurate results for the limited simulation work at hand.

a) Modify the program of Example 2.3 to make plots for this generator analogous to those in Figure 2.4. Use u = (r + 1/2)/d.

b) Perform chi-square goodness-of-fit tests as in Problem 2.4, based on 1000, and then 100 000 simulated uniform observations from this generator.

Comment: (b) Not a bad generator. Q varies with seed.

2.6   Consider $m = 50\,000$ values $u_i = (r + .5)/d$ from the generator with $a = 1093$, $b = 252$, $d = 86\,436$, and $s = 6$. We try using this generator to simulate many tosses of a fair coin.

a) For a particular $n \leq m$, you can use the code sum(u[1:n] < .5)/n to simulate the proportion of heads in the first n tosses. If the values u_i are uniform in the interval $(0, 1)$, then each of the n comparisons inside the parentheses has probability one-half of being TRUE, and thus contributing 1 to the sum. Evaluate this for $n = 10\,000$, $20\,000$, $30\,000$, $40\,000$, and $50\,000$. For each n, the 95% margin of error is about $n^{-1/2}$. Show that all of your values are within this margin of the true value $P\{\text{Head}\} = 0.5$. So, you might be tempted to conclude that the generator is working satisfactorily. But notice that all of these proportions are above 0.5—and by similar amounts. Is this a random coincidence or a pattern? (See part (c).)

b) This generator has serious problems. First, how many distinct values do you get among m? Use length(unique(r)). So, this generator repeats a few values many times in $m = 50\,000$ iterations. Second, the period depends heavily on the seed s. Report results for $s = 2$, 8 and 17.

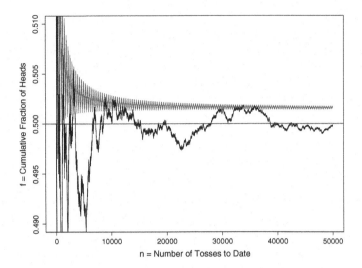

Figure 2.9. A generator with a short period. The upper "sawtooth" trace results from 50 000 tosses of a supposedly fair coin simulated with the defective generator of Problem 2.6. It converges far too rapidly, and to a value above $1/2$. The truly random "meandering" trace from `runif` converges more slowly, but to $1/2$.

c) Based on this generator, the code below makes a plot of the proportion of heads in n tosses for all values $n = 1, 2 \ldots m$. For comparison, it does the same for values from `runif`, which are known to simulate $\mathsf{UNIF}(0, 1)$ accurately. Explain the code, run the program (which makes Figure 2.9), and comment on the results. In particular, what do you suppose would happen towards the right-hand edge of the graph if there were millions of iterations m? (You will learn more about such plots in Chapter 3.)

```
a = 1093;  b = 252;  d = 86436;  s = 6
m = 50000;  n = 1:m
r = numeric(m);  r[1] = s
for (i in 1:(m-1)) {r[i+1] = (a*r[i] + b) %% d}

u = (r + 1/2)/d;  f = cumsum(u < .5)/n
plot(n, f, type="l", ylim=c(.49,.51), col="red")  # 'ell', not 1
    abline(h=.5, col="green")

set.seed(1237)  # Use this seed for exact graph shown in figure.
g = cumsum(sample(0:1, m, repl=T))/n;  lines(n, g)
```

Note: This generator "never had a chance." It breaks one of the number-theoretic rules for linear congruential generators, that b and d not have factors in common. [AS64]

2.7   In R, the statement `runif(m)` makes a vector of m simulated observations from the distribution UNIF$(0, 1)$. Notice that no explicit loop is required to generate this vector.

```
m = 10^6;  u = runif(m)        # generate one million from UNIF(0, 1)
u1 = u[1:(m-2)]; u2 = u[2:(m-1)]; u3 = u[3:m]  # 3 dimensions

par(mfrow=c(1,2), pty="s")    # 2 square panels per graph
   plot(u1, u2, pch=".", xlim=c(0,.1), ylim=c(0,.1))
   plot(u1[u3<.01], u2[u3<.01], pch=".", xlim=c(0,1), ylim=c(0,1))
par(mfrow=c(1,1), pty="m")    # restore default plotting
```

a) Run the program and comment on the results. Approximately how many points are printed in each graph?
b) Perform chi-square goodness-of-fit tests as in Problem 2.4 based on these one million simulated uniform observations.

2.8   The code used to make the two plots in the top row of Figure 2.1 (p26) is shown below. The function `runif` is used in the left panel to "jitter" (randomly displace) two plotted points slightly above and to the right of each of the 100 grid points in the unit square. The same function is used more simply in the right panel to put 200 points at random into the unit square.

```
set.seed(121);  n = 100
par(mfrow=c(1,2), pty="s")      # 2 square panels per graph
# Left Panel
   s = rep(0:9, each=10)/10      # grid points
   t = rep(0:9, times=10)/10
   x = s + runif(n, .01, .09)    # jittered grid points
   y = t + runif(n, .01, .09)
   plot(x, y, pch=19, xaxs="i", yaxs="i", xlim=0:1, ylim=0:1)
      #abline(h = seq(.1, .9, by=.1), col="green")  # grid lines
      #abline(v = seq(.1, .9, by=.1), col="green")

# Right Panel
   x=runif(n);  y = runif(n)     # random points in unit square
   plot(x, y, pch=19, xaxs="i", yaxs="i", xlim=0:1, ylim=0:1)
par(mfrow=c(1,1), pty="m")      # restore default plotting
```

a) Run the program (without the grid lines) to make the top row of Figure 2.1 for yourself. Then remove the # symbols at the start of the two `abline` statements so that grid lines will print to show the 100 cells of your left panel. See Figure 2.12 (p47).
b) Repeat part (a) several times without the seed statement (thus getting a different seed on each run) and without the grid lines to see a variety of examples of versions of Figure 2.1. Comment on the degree of change in the appearance of each with the different seeds.
c) What do you get from a single plot with `plot(s, t)`?

d) If 100 points are placed at random into the unit square, what is the probability that none of the 100 cells of this problem are left empty? (Give your answer in exponential notation with four significant digits.)

Note: Consider nesting habits of birds in a marsh. From left to right in Figure 2.1, the first plot shows *territorial* behavior that tends to avoid close neighbors. The second shows random nesting in which birds choose nesting sites entirely independently of other birds. The third shows a strong preference for nesting near the center of the square. The last shows *social* behavior with a tendency to build nests in clusters.

Problems for Section 2.4 (Transforming Uniform Distributions)

2.9   *(Theoretical)* Let $U \sim \mathsf{UNIF}(0, 1)$. In each part below, modify equation (2.2) to derive the cumulative distribution function of X, and then take derivatives to find the density function.

a) Show that $X = (b - a)U + a \sim \mathsf{UNIF}(a, b)$, for real numbers a and b with $a < b$. Specify the support of X.
b) What is the distribution of $X = 1 - U$? [Hints: Multiplying an inequality by a negative number changes its sense (direction). $P(A^c) = 1 - P(A)$. A continuous distribution assigns probability 0 to a single point.]

2.10   In Example 2.4, we used the random R function `runif` to sample from the distribution $\mathsf{BETA}(0.5, 1)$. Here we wish to sample from $\mathsf{BETA}(2, 1)$.

a) Write the density function, cumulative distribution function, and quantile function of $\mathsf{BETA}(2, 1)$. According to the quantile transformation method, explain how to use $U \sim \mathsf{UNIF}(0, 1)$ to sample from $\mathsf{BETA}(2, 1)$.
b) Modify equation (2.2) as appropriate to this situation.
c) Modify the program of Example 2.4 to illustrate the method of part (a), Of course, you will need to change the code for x and `cut.x` and the code used to plot the density function of $\mathsf{BETA}(2, 1)$. Also, change the code to simulate a sample of 100 000 observations, and use 20 bars in each of the histograms. Finally, we suggest changing the `ylim` parameters so that the vertical axes of the histograms include the interval $(0, 2)$. See Figure 2.10.

2.11   The program below simulates 10 000 values of $X \sim \mathsf{EXP}(\lambda = 1)$, using the quantile transformation method. That is, $X = -\log(U)/\lambda$, where $U \sim \mathsf{UNIF}(0, 1)$. A histogram of results is shown in Figure 2.11.

```
# set.seed(1212)
m = 10000;   lam = 1
u = runif(m);   x = -log(u)/lam

cut1 = seq(0, 1, by=.1)                    # for hist of u, not plotted
cut2 = -log(cut1)/lam;  cut2[1] = max(x); cut2 = sort(cut2)
hist(x, breaks=cut2, ylim=c(0,lam), prob=T, col="wheat")
  xx = seq(0, max(x), by = .01)
  lines(xx, lam*exp(-lam*xx), col="blue")
```

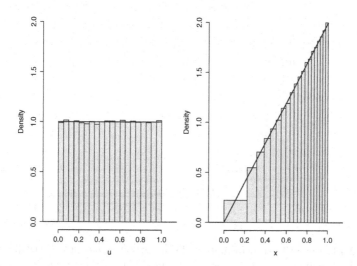

Figure 2.10. Simulating $X \sim$ BETA$(2, 1)$. The left histogram shows 100 000 simulated values $U \sim$ UNIF$(0, 1)$. Because the quantile function of BETA$(2, 1)$ is $x = \sqrt{u}$, for $0 < u, x < 1$, the values $X = \sqrt{U} \sim$ BETA$(2, 1)$. Histogram intervals for X are images of respective intervals for U. See Problem 2.10.

```
mean(x);   1/lam                      # simulated and exact mean
median(x);  qexp(.5, lam)             # simulated and exact median
hist(u, breaks=cut1, plot=F)$counts   # interval counts for UNIF
hist(x, breaks=cut2, plot=F)$counts   # interval counts for EXP
```

a) If $X \sim$ EXP(λ), then E$(X) = 1/\lambda$ (see Problem 2.12). Find the median of this distribution by setting $F_X(x) = 1 - e^{-\lambda x} = 1/2$ and solving for x. How accurately does your simulated sample of size 10 000 estimate the population mean and median of EXP(1)? [The answer for $\lambda = 1$ is qexp(.5).]

b) The last two lines of the program (counts from unplotted histograms) provide counts for each interval of the realizations of U and X, respectively. Report the 10 counts in each case. Explain why their order gets reversed when transforming from uniform to exponential. What is the support of X? Which values in the support $(0, 1)$ of U correspond to the largest values of X? Also, explain how cut2 is computed and why.

c) In Figure 2.11, each histogram bar represents about 1000 values of X, so that the bars have approximately *equal area*. Make a different histogram of these values of X using breaks=10 to get about 10 intervals of *equal width* along the horizontal axis. (For most purposes, intervals of equal width are easier to interpret.) Also, an alternate method to overlay the density curve, use dexp(xx, lam) instead of lam*exp(-lam*xx).

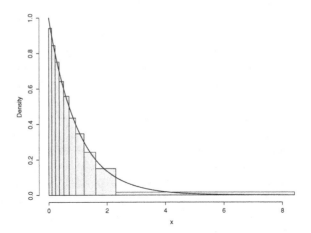

Figure 2.11. Quantile transformation. From 10 000 values of $U \sim \mathsf{UNIF}(0,1)$, we obtain 10 000 values of $X = -\log(U) \sim \mathsf{EXP}(1)$. Each histogram bar represents about 1000 values of X. See Problem 2.11.

d) Run the program again with `lam = 1/2`. Describe and explain the results of this change. (Notice that $\mathsf{EXP}(1/2) = \mathsf{CHISQ}(2) = \mathsf{GAMMA}(1, 1/2)$. See Problems 2.12 and 2.13.)

Problems for Section 2.5 (Normal and Related Distributions)

2.12   *Distributions in Example 2.6 (Theoretical).* The density function for the gamma family of distributions $\mathsf{GAMMA}(\alpha, \lambda)$ is $f_T(t) = \frac{\lambda^\alpha}{\Gamma(\alpha)} t^{\alpha-1} e^{-\lambda t}$, for $t > 0$. Here $\alpha > 0$ is the shape parameter and $\lambda > 0$ is the rate parameter. Two important subfamilies are the exponential $\mathsf{EXP}(\lambda)$ with $\alpha = 1$ and the chi-squared $\mathsf{CHISQ}(\nu)$ with $\nu = 2\alpha$ (called *degrees of freedom*) and $\lambda = 1/2$.

a) Show that the density function of $\mathsf{CHISQ}(2)$ shown in the example is consistent with the information provided above. Recall that $\Gamma(\alpha) = (\alpha - 1)!$, for integer $\alpha > 0$.
b) For $T \sim \mathsf{CHISQ}(2)$, show that $\mathrm{E}(T) = 2$. (Use the density function and integration by parts, or use the moment generating function in part (c).)
c) The moment generating function of $X \sim \mathsf{CHISQ}(\nu)$, for $s < 1/2$, is

$$m_X(s) = E(e^{sX}) = (1 - 2s)^{-\nu/2}.$$

If $Z \sim \mathsf{NORM}(0,1)$, with density function $\varphi(z) = \frac{1}{\sqrt{2\pi}} e^{-z^2/2}$, then the moment generating function of Z^2 is

$$m(s) = m_{Z^2}(s) = \int_{-\infty}^{\infty} \exp(sz^2)\varphi(z)\,dz = 2\int_{0}^{\infty} \exp(sz^2)\varphi(z)\,dz.$$

Show that this simplifies to $m(s) = (1 - 2s)^{-1/2}$, so that $Z^2 \sim \mathsf{CHISQ}(1)$. Recall that $\Gamma(1/2) = \sqrt{\pi}$.

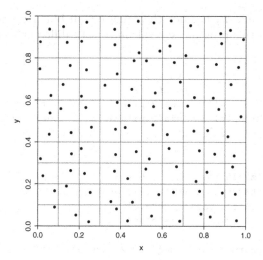

Figure 2.12. Illustrating nonrandomness. In the top left panel of Figure 2.1 (p26), each of 100 grid squares contains exactly one point. The points are too regularly spaced to be "random." Specifically, the probability is very small that independently and uniformly distributed points would give such a result; see Problem 2.8 (p43).

d) If X and Y are independent random variables with moment generating functions $m_X(s)$ and $m_Y(s)$, respectively, then $m_{X+Y}(s) = m_X(s)m_Y(s)$. Use this property of moment generating functions to show that, if Z_i are independently NORM$(0, 1)$, then $Z_1^2 + Z_2^2 + \ldots + Z_\nu^2 \sim$ CHISQ(ν).

2.13 *Simulations for chi-squared random variables.* The first block of code in Example 2.6 illustrates that the sum of squares of two standard normal random variables is distributed as CHISQ(2). (Problem 2.12 provides formal proof.) Modify the code in the example to do each part below. For simplicity, when plotting the required density functions, use `dens = dchisq(tt, df)` for `df` suitably defined.

a) If $Z \sim$ NORM$(0, 1)$, then illustrate by simulation that $Z^2 \sim$ CHISQ(1) and that $\mathrm{E}(Z^2) = 1$.

b) If Z_1, Z_2, and Z_3 are independently NORM$(0, 1)$, then illustrate by simulation that $T = Z_1^2 + Z_2^2 + Z_3^2 \sim$ CHISQ(3) and that $\mathrm{E}(T) = 3$.

2.14 *Illustrating the Box-Muller method.* Use the program below to implement the Box-Muller method of simulating a random sample from a standard normal distribution. Does the histogram of simulated values seem to agree with the standard normal density curve? What do you conclude from the chi-squared goodness-of-fit statistic? (This statistic, based on 10 bins, has the same approximate chi-squared distribution as in Problem 2.4, but here the expected counts E_i are not the same for all bins.) Before drawing firm conclusions, run this program several times with different seeds.

```
# set.seed(1234)
m = 2*50000; z = numeric(m)
u1 = runif(m/2);   u2 = runif(m/2)
z1 = sqrt(-2*log(u1)) * cos(2*pi*u2)   # half of normal variates
z2 = sqrt(-2*log(u1)) * sin(2*pi*u2)   #     other half
z[seq(1, m, by = 2)] = z1              # interleave
z[seq(2, m, by = 2)] = z2              #     two halves

cut = c(min(z)-.5, seq(-2, 2, by=.5), max(z)+.5)
hist(z, breaks=cut, ylim=c(0,.4), prob=T)
   zz = seq(min(z), max(z), by=.01)
   lines(zz, dnorm(zz), col="blue")

E = m*diff(pnorm(c(-Inf, seq(-2, 2, by=.5), Inf))); E
N = hist(z, breaks=cut, plot=F)$counts; N
Q = sum(((N-E)^2)/E);  Q;  qchisq(c(.025,.975), 9)
```

2.15 *Summing uniforms to simulate a standard normal.* The Box-Muller transformation requires the evaluation of logarithms and trigonometric functions. Some years ago, when these transcendental computations were very time-intensive (compared with addition and subtraction), the following method of simulating a standard normal random variable Z from 12 independent random variables $U_i \sim$ UNIF$(0,1)$ was commonly used: $Z = U_1+U_2+\ldots+U_{12}-6$.

However, with current hardware, transcendental operations are relatively fast, so this method is now deprecated—partly because it makes 12 calls to the random number generator for each standard normal variate generated. (You may discover other reasons as you work this problem.)

a) Using the fact that a standard uniform random variable has mean $1/2$ and variance $1/12$, show that Z as defined above has $E(Z) = 0$ and $V(Z) = 1$. (The assumed near normality of such a random variable Z is based on the Central Limit Theorem, which works reasonably well here—even though only 12 random variables are summed.)

b) For the random variable Z of part (a), evaluate $P\{-6 < Z < 6\}$. Theoretically, how does this result differ for a random variable Z that is precisely distributed as standard normal?

c) The following program implements the method of part (a) to simulate 100,000 (nearly) standard normal observations by making a $100\,000 \times 12$ matrix DTA, summing its rows, and subtracting 6 from each result.

```
# set.seed(1234)
m = 100000;  n = 12
u = runif(m*n)
UNI = matrix(u, nrow=m)
z = rowSums(UNI) - 6
```

```
cut = c(min(z)-.5, seq(-2, 2, by=.5), max(z)+.5)
hist(z, breaks=cut, ylim=c(0,.4), prob=T)
  zz = seq(min(z), max(z), by=.01)
  lines(zz, dnorm(zz), col="blue")

E = m*diff(pnorm(c(-Inf, seq(-2, 2, by=.5), Inf))); E
N = hist(z, breaks=cut, plot=F)$counts; N
Q = sum(((N-E)^2)/E);   Q;   qchisq(c(.025,.975), 9)
```

Does the histogram of simulated values seem to agree with the standard normal density curve? What do you conclude from the chi-squared goodness-of-fit test? (This statistic, based on 10 bins, has the same approximate chi-squared distribution as in Problem 2.4, but here the expected counts E_i are not the same for all bins.) Before drawing firm conclusions, run this program several times with different seeds. Also, make a few runs with $m = 10\,000$ iterations.

2.16   *Random triangles (Project).* If three points are chosen at random from a standard bivariate normal distribution ($\mu_1 = \mu_2 = \rho = 0$, $\sigma_1 = \sigma_2 = 1$), then the probability that they are vertices of an obtuse triangle is $3/4$. Use simulation to illustrate this result. Perhaps explore higher dimensions. (See [Por94] for a proof and for a history of this problem tracing back to Lewis Carroll.)

3

Monte Carlo Integration and Limit Theorems

In Chapter 1, we did a few simulations by sampling from finite populations. In Chapter 2, we discussed (pseudo)random numbers and the simulation of some familiar discrete and continuous distributions. In this chapter, we investigate how simulation is used to approximate integrals and what some fundamental limit theorems of probability theory have to say about the accuracy of these approximations. Section 3.1 sets the stage with elementary examples that illustrate some methods of integration.

3.1 Computing Probabilities of Intervals

Many practical applications of statistics and probability require the evaluation of the probability $P\{a < X \le b\}$ that a continuous random variable X lies in a particular interval (a, b). Sometimes a simple integration gives the answer, and sometimes numerical methods of various degrees of sophistication are required. Here are some examples.

- Trains run on a strict schedule leaving Central Station on the hour and the half hour. So a rider who arrives at a random time will have to wait for W minutes, where $W \sim \text{UNIF}(0, 30)$, before the next train leaves. What is the probability that her wait does not exceed 10 minutes? Calculating the probability $P\{W \le 10\} = P\{0 < W \le 10\} = 1/3$ involves only simple arithmetic. (See Problem 3.1.)
- An electronic component chosen at random from a particular population has an exponentially distributed lifetime T with mean $E(T) = 2$ years and thus density function $f_T(t) = (1/2)e^{-t/2}$, for $t > 0$. The probability that it survives for a year is $P\{T > 1\} = 1 - \int_0^1 f_T(t)\, dt = e^{-1/2} = 0.6065$. Nowadays, a cheap calculator can evaluate $e^{-1/2}$, but 50 years ago one would have consulted a table of exponentials. Values in such tables—and from calculators and computer software—are typically based on analytic relationships such as series expansions. (See Problems 3.1 and 3.2.)

E.A. Suess and B.E. Trumbo, *Introduction to Probability Simulation and Gibbs Sampling with R,* Use R!, DOI 10.1007/978-0-387-68765-0_3, © Springer Science+Business Media, LLC 2010

- A process produces batches of a protein. The nominal yield is 100, and we hope to obtain yields between 90 and 110. If the yield Y of a randomly chosen batch is normally distributed with mean $\mu = 100$ and standard deviation $\sigma = 10$, then $P\{90 < Y \leq 110\} = P\{-1 < Z \leq 1\} \approx 0.6827$, where Z is a standard normal random variable.

Because the cumulative distribution function (CDF) Φ of Z cannot be expressed in closed form, this is a more difficult computation than those above. In practice, we might get the answer from a table of the standard normal CDF or use software (on a calculator or computer) that is written to do this kind of computation. For example, in R we can use the statement pnorm(1) - pnorm(-1) to evaluate $P\{-1 < Z \leq 1\} = \Phi(1) - \Phi(-1) = 0.6826895$. Similarly, $P\{0 < Z \leq 1\} = 0.3413447$. Used in this way, the function pnorm acts like a standard normal CDF table built into R.

Tables and software for Φ are widely available, so you might not have had the opportunity to think about methods for computing its values. As an introduction to some of the computational methods available, we look briefly at several ways to evaluate $J = P\{a < Z \leq b\} = \Phi(b) - \Phi(a) = \int_a^b \varphi(z)\, dz$, where $a < b$ and $\varphi(z) = \frac{1}{\sqrt{2\pi}} e^{-z^2/2}$ is the standard normal density function. Each of these methods can be used with a wide variety of other distributions, for which tables and software may *not* be readily available.

Example 3.1. Riemann Approximation. The Riemann integral J is defined as the limit of the sum of the areas of increasingly many rectangles. The heights of these rectangles depend on φ, and the union of their bases is the required interval. As the number of rectangles increases, their widths shrink to 0. In Figure 3.1, there are $m = 5$ rectangles. For any practical purpose at hand, we can use a large enough number m of such rectangles that the sum of their areas is a sufficiently accurate approximation to J. Because it is based on the definition of the Riemann integral, this method of numerical integration in terms of areas of rectangles is sometimes called **Riemann approximation**.

The R script below implements this method to integrate φ over an interval $(a, b]$, with $a < b$. For $m = 5000$, $a = 0$, and $b = 1$, the 5000 rectangles have bases centered at values in the vector g: 0.0001, 0.0003, ..., 0.9997, 0.9999. The areas of these rectangles, elements of the vector w*h, are summed to give the desired approximation.

```
m = 5000;  a = 0;  b = 1            # constants
w = (b - a)/m                       # width of each rectangle
g = seq(a + w/2, b - w/2, length=m) # m  "grid" points
const = 1/sqrt(2 * pi)              # const. for density function
h = const * exp(-g^2 / 2)           # vector of m heights
sum(w * h)                          # total area (approx. prob.)

> sum(w * h)
[1] 0.3413447
```

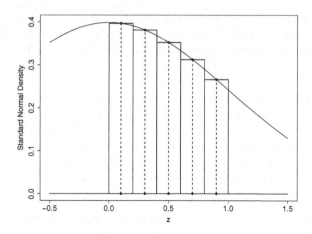

Figure 3.1. Approximating the area under a standard normal curve. The Riemann approximation of Example 3.1 is illustrated for $m = 5$ rectangles with bases centered at grid points 0.1, 0.3, 0.5, 0.7, and 0.9. Heights are represented by dotted lines. The total area within the five rectangles is 0.3417, whereas $\int_0^1 \varphi(z)\, dz = 0.3413$.

The result is exactly the same as we obtained above for $P\{0 < Z \leq 1\}$ by using pnorm. In Problem 3.3, we see that we can get five-place accuracy with a much smaller number m of rectangles. ◇

The method of Example 3.1 is simple and effective. The program is very brief, it implements a familiar theoretical idea, it can give more accurate results than are generally needed in practice, and it can be easily modified for use with the density function of any continuous random variable. Embellishments that use trapezoids or strips with polynomial-shaped tops—instead of simple rectangles—can sometimes considerably decrease the number m of iterations needed, but today's computers run so fast that the extra programming effort may not be worthwhile in simple applications.

In Riemann approximation, the bases of the rectangles lie on a grid of equally spaced points. But it can be difficult to construct such a regular grid for some advanced applications, particularly those requiring integration over multidimensional regions. Perhaps surprisingly, it turns out that we can use points that are scattered at random rather than equally spaced on a grid.

Example 3.2. Monte Carlo Integration. Here we randomly select points u according to the uniform distribution on $(0, 1)$. Because points are randomly selected, this method is called **Monte Carlo integration**—after a southern European resort town long known for its casinos.

Except for the way the points of evaluation are chosen, this method is very similar to the one in Example 3.1. Heights of the density function φ are determined at each of m randomly chosen points, and each height is multiplied by the *average* horizontal distance $w = 1/m$ corresponding to each random point. The sum of these products approximates the desired integral. As we see later in this chapter, the Law of Large Numbers guarantees that this method works, provided the number of random points is large enough. Problem 3.14 gives a graphical illustration that the approximation gets better as the number of randomly chosen points increases.

The following R script implements the Monte Carlo method for evaluating $P\{0 < Z \le 1\}$. As we have seen in Chapter 2, the R function runif generates a specified number of random observations uniformly distributed in the interval $(0, 1)$. Also, in R the standard normal density is called dnorm. In Example 3.1, the two lines of code used to make the vector h could have been written more compactly as h = dnorm(g). We use the function dnorm below.

```
# set.seed(12)
m = 500000                    # number or random points
a = 0;  b = 1                 # interval endpoints
w = (b - a)/m
u = a + (b - a) * runif(m)    # vector of m random points
h = dnorm(u)                  # hts. of density above rand. points
sum(w * h)                    # approximate probability

>  sum(w * h)
[1] 0.3413249
```

When the set.seed statement is "commented out" as shown, different random points will be selected each time the program is run, so the exact result will differ somewhat from run to run. It can be shown that, when we use $m = 500\,000$ points to evaluate $P\{0 < Z \le 1\}$, the simulation error rarely exceeds 0.00015. (See Problem 3.5.) If we were to use only $m = 5000$, the 95% margin of error becomes 0.0015, which is still small enough for many practical purposes. ◇

Example 3.3. The Acceptance-Rejection Method. Another simulation method that can be used to find $P\{0 < Z \le 1\}$ is to surround the desired area by the rectangle with diagonal vertices at $(0,0)$ and $(1, 0.4)$, put a large number of points into this rectangle at random, and find the fraction of points that falls in the area beneath the density curve—the "accepted" points (see Figure 3.2). R code to implement this method for our specific problem is as follows.

```
# set.seed(12)
m = 500000
u = runif(m, 0, 1)
h = runif(m, 0, 0.4)
frac.acc = mean(h < dnorm(u))
0.4*frac.acc

> 0.4*frac.acc
[1] 0.341056
```

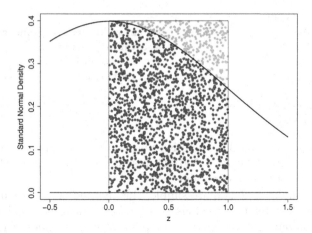

Figure 3.2. In the acceptance-rejection method of Example 3.3, we put 500 000 points at random into a rectangle that surrounds the area for $P\{0 < Z \leq 1\}$. The fraction falling below the density curve, the (darker) "accepted points," is used to approximate this probability. For clarity, the figure shows only the first 2000 points.

The elements of the logical vector h < dnorm(u) take values TRUE and FALSE. When we find the mean of this vector, TRUE is interpreted as 1 and FALSE as 0, so that the result is the proportion of the $m = 500\,000$ points that lie within the desired region. In the last line of the program, the fraction of these accepted points is multiplied by the area of the rectangle to obtain the approximate value $P\{0 < Z \leq 1\} \approx 0.341056$. It can be shown that about 95% of the runs of this program will give a value within ± 0.0004 of the correct answer.

This method is a relatively inefficient one for evaluating simple normal probabilities. However, like Monte Carlo integration, it can be useful in more complicated situations. ◊

The methods of the preceding three examples all require us to know the *density function* of the standard normal distribution. The method illustrated in the next example takes a fundamentally different approach.

Example 3.4. Random Sampling Method. To evaluate $J = P\{a < Z \leq b\}$, we can take a large random sample from a standard normal distribution and then see what proportion of the observations falls between a and b. It is easy for us to generate standard normal observations at random because (as we saw in Chapter 2) the R function rnorm is programmed to do this specific task.

The script below uses this function to generate the vector z. The logical vector z > a & z <= b has m elements, each of which takes the value TRUE, interpreted as 1, when and only when the corresponding element of the vector z lies in $(a, b]$.

```
# set.seed(1212)
m = 500000                          # number of observations sampled
a = 0;  b = 1                       # interval endpoints
z = rnorm(m)                        # m random obs. from std. normal
mean(z > a & z <= b)                # proportion of obs. in (c, d)

> mean(z > a & z <= b)
 [1] 0.34146
```

As in Examples 3.2 and 3.3, there is some random variation from one simulation to the next. With $m = 500\,000$, results for $P\{0 < Z \leq 1\}$ are likely
within ± 0.0013 of the correct value.

Although the R code is quite simple, this method actually requires more
computation than for the previous two methods. In some applications, the
amount of computation required to sample values of a random variable X of
interest may be time-consuming even on modern computers. Also, for the task
of evaluating $P\{0 < Z \leq 1\}$, the sampling method has the largest margin of
error. Consequently, you may wonder why the random sampling method is
ever used. The answer is that there are many important applications in which
the density function of a distribution is not known but in which a way can be
found to sample at random from the distribution. This is a major theme of
Chapters 4 and beyond. For now, see the simple example of Problem 3.10. ◇

So far, we have concentrated attention on integrals that provide probabilities of intervals, but the methods just described are often used to find
expected values of continuous random variables. For example, up to a constant, the integral $\int_0^1 x^2 \, dx$ can be interpreted as one of two expectations or
as a probability. (This is Problem 3.15; see other examples in Problems 3.7
and 3.8.) Also, we can use simulation methods to evaluate integrals that are
not directly related to probability (as in Problems 3.6 and 3.24 through 3.26).

As is true for many computer-intensive computations in probability and
statistics, the examples of this section are based on limiting processes.
It is important to distinguish between two kinds of limiting processes—
deterministic and random. Examples of deterministic convergence are the
convergence of $(1 + 1/m)^m$ to $e = 2.718282$ as $m \to \infty$ and the convergence in the definition of a Riemann integral. Here each term in the sequence
has a definite value. In contrast, the Monte Carlo methods we have just seen
are based on convergence of sequences of random variables. We explore this
topic in more detail in the next two sections.

3.2 Convergence—The Law of Large Numbers

Why does simulation work? Why is it ever possible to generate a lot of "fake"
data on a computer and find out something useful from the results? Even
if simulation does give accurate results in some cases, how can we know if

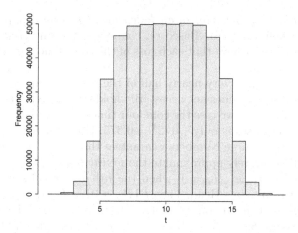

Figure 3.3. Histogram of the sum T of independent random variables U and V, where $U \sim$ UNIF$(0, 10)$ and $V \sim$ NORM$(5, 1)$. In Problem 3.10, without knowing the density function of T, we take a sample of size 500 000 from its distribution to evaluate $P\{T > 15\}$. Can you roughly guess the answer by looking at the histogram?

a particular simulation is useful? In this section, we start to answer these questions, first with some general guidelines.

- *Relevance.* Simulated observations must be carefully generated according to rules that we have reason to believe are relevant. For example, the rules may be based on a combination of past experience and current data from which we want to make a decision, or they may be based on a hypothetical probability model that we want to explore and better understand.
- *Stability.* Limit theorems of probability theory guarantee that, if a simulation is properly designed and run for enough iterations, it will give a stable result instead of useless random noise. Two of these theorems are the Law of Large Numbers and the Central Limit Theorem, which we illustrate later in this section and in the next.
- *Diagnostics.* A variety of numerical and graphical methods can be used to see whether a simulation has stabilized. By running a simulation several times, we can distinguish the random quirks of any one run ("noise") from the valid results shared by all runs ("signal"). We can look at the numbers from the first few steps of a simulation and verify them by hand to make sure our programming is correct. Also, by running simulations of similar models where some or all of the results can be found analytically, we can verify that we have programmed the correct model and that the random number generator we used is of adequate quality.

We begin this section by simulating the repeated tossing of a fair coin. This simple model illustrates the role of probability laws in simulation, and it allows us to compare a simulated value with a known result.

Example 3.5. Limiting Behavior of an Independent Process. Suppose we have a coin that we believe is fair. Based on common sense and past experience tossing coins, we also believe that each toss of the coin is independent of other tosses.

Intuitively, we expect very nearly half of the tosses of this coin will result in Heads. Of course, this does not necessarily hold true for very small numbers of tosses. For example, if we toss the coin four times, the probability of getting two Heads in four tosses is only $\binom{4}{2}/2^4 = 0.375$. However, for large numbers of tosses, the Law of Large Numbers guarantees that the *proportion* of Heads we observe will accurately approximate the *probability* of Heads.

More formally, we let $H_n = 0$ when the nth toss results in a Tail and $H_n = 1$ when it results in a Head. Then $P\{H_n = 0\} = P\{H_n = 1\} = 1/2$, and the H_n are independent random variables. After the nth toss, the distribution of the Heads count $S_n = H_1 + H_2 + \cdots + H_n$ is BINOM$(n, 1/2)$. Also, we can use this distribution to evaluate probabilities for $Y_n = \bar{H}_n = S_n/n$, the proportion of heads observed so far.

The Law of Large Numbers involves choosing a positive "tolerable error" ϵ as close to 0 as we please. For any $\epsilon > 0$, as the number of tosses goes to infinity, it becomes a sure thing that the proportion Y_n of Heads is within $\pm\epsilon$ of $1/2$. In symbols, defining $P_n = P\{|Y_n - 1/2| < \epsilon\}$, we have

$$\lim_{n\to\infty} P_n = \lim_{n\to\infty} P\{|Y_n - 1/2| < \epsilon\} = 1.$$

We say Y_n **converges in probability** to $1/2$ and write $Y_n \xrightarrow{p} 1/2$.

To illustrate, let us choose $\epsilon = 0.02$. For $n = 4$ tosses, we know that $P\{S_4 = 2\} = P\{Y_4 = 1/2\} = 0.375$, and no other possible values of Y_4 come within ϵ of $1/2$, so $P_4 = P\{|Y_4 - 1/2| < \epsilon\} = 0.375$. Further computations with the binomial distribution reveal that $P_{100} = 0.236$, $P_{1000} = 0.783$, $P_{5000} = 0.995$, and $P_{10\,000} > 0.999$. (Also, see Problem 3.16 and Figure 3.12.)

Here is the crucial point in making a formal mathematical statement about the "limiting behavior" of random variables. We cannot know the exact values of the *random variables* H_n or Y_n. But we can know the exact values of the *probabilities* P_n, and we formulate a limit theorem about the random variables in terms of these probabilities.

As a demonstration, we now simulate $m = 10\,000$ coin tosses according to the probability model stated above. After each simulated toss, we plot the proportion Y_n of Heads obtained so far against the number n of tosses so far. This gives a **trace** of the process. The Law of Large Numbers says we should see a trace that gets very close to $1/2$ as n increases.

The R code for such a simulation is shown below. A graph based on one run of this program is shown in Figure 3.4.

The randomly generated Bernoulli random variable H_n has equal probabilities of taking the values 0 and 1. The vector h contains ten thousand 0s and 1s. The nth element of the vector y is the mean of the first n elements of h, that is, the proportion of 1s (Heads) in the first n tosses.

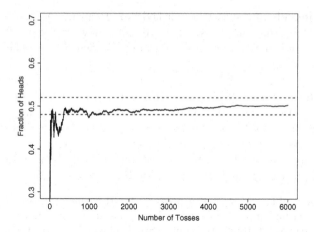

Figure 3.4. This trace of the proportions of Heads after each toss of a fair coin results from one run of the program in Example 3.5. (For clarity in print, this is a magnified view of part of the plot specified there.) The dashed lines at 0.48 and 0.52 illustrate the Law of Large Numbers with $\epsilon = 0.02$.

```
#Initialize:
# set.seed(1212)
m = 10000                # total number of tosses
n = 1:m                  # vector: n = 1, 2, ..., m; Toss number

#Simulate and Plot:
h = rbinom(m, 1, 1/2)    # vector: H = 0 or 1 each with prob. 1/2
y = cumsum(h)/n          # vector: Proportion of heads
plot (n, y, type="l", ylim=c(0,1))    # Plot trace

#Verify:
Show = cbind(n,h,y)      # matrix: 3 vectors as cols.: n, h, y
Show[1:10, ]             # print first 10 rows of Show
Show[(m-4):m, ]          # print last 5 rows of Show

> Show[1:10, ]                    > Show[(m-4):m, ]
        n h        y                       n h        y
 [1,]   1 0 0.0000000             [1,]   9996 1 0.49990
 [2,]   2 0 0.0000000             [2,]   9997 0 0.49985
 [3,]   3 1 0.3333333             [3,]   9998 0 0.49980
 [4,]   4 0 0.2500000             [4,]   9999 1 0.49985
 [5,]   5 1 0.4000000             [5,]  10000 1 0.49990
 [6,]   6 0 0.3333333
 [7,]   7 0 0.2857143
 [8,]   8 0 0.2500000
 [9,]   9 1 0.3333333
[10,]  10 0 0.3000000
```

The commands in the section marked `Verify` print the first ten and last five values of n, H_n, and Y_n so that we can follow in detail what is going on. Output from one run of this program is shown above. The final value $Y_{10\,000} = 0.49990$ is very near $1/2$.

Towards the right of the graph, our simulated values of Y_n have begun to cluster tightly around $1/2$ in a way consistent with the Law of Large Numbers. Computations with the binomial distribution show that the interval 0.5 ± 0.01 is 95% sure to include $Y_{10\,000}$. In particular, our (especially lucky) simulated value 0.49990 falls inside this interval. As judged by this information, randomly generated coin tosses, obtained quickly by a simple program, are behaving just as if they were tosses by hand of a real coin. ◇

> Note: We could have written a program for the simulation above with a loop, simulating one coin toss and updating the cumulative number of heads on each passage through the loop. (See Problem 3.18.) But R executes such "explicit" loops in a relatively inefficient way, and it is best to avoid them when possible. Because the H_n in the example above are independent random variables, we were able to "vectorize" the program to avoid writing a loop. Of course, in executing our vectorized code, R performed several loops to store the random variables in vectors and perform arithmetic on these vectors. But R executes these implicit loops more efficiently within its data-handling structure. Generally speaking, if an R program requires a large number m of iterations, then arithmetic on an m-vector is faster than running through an explicit loop m times.

We have just stated and illustrated the Law of Large Numbers for coin tossing. The same principle holds more generally for sequences of random variables from a wide variety of discrete and continuous distributions. Here is a more general statement.

The Law of Large Numbers. Let $X_1, X_2, \ldots$ be a sequence of independent random variables each with mean μ and finite standard deviation σ, and let the "running average" $\bar{X}_n$ be the sample mean of the first n random variables. Then, for any $\epsilon > 0$,

$$\lim_{n \to \infty} P\{|\bar{X}_n - \mu| < \epsilon\} = 1.$$

Notice that in the coin-toss example the H_i correspond to the X_i, and Y_n to $\bar{X}_n$, in the general statement above. Thus, in that example $\mu = \mathrm{E}(X_i) = 1/2$ and $\sigma = \mathrm{SD}(X_i) = 1/2$.

One crucial assumption of the Law of Large Numbers is that the X_i must have finite variance. One well-known distribution that does not is the "heavy-tailed" Student's t distribution with 1 degree of freedom, also called a **Cauchy distribution**. In the program of Example 3.5, if we replace the specification of h with `h = rt(m, 1)` and remove the `ylim` restrictions on the vertical axis, then we get the nonconvergent sequence of running averages shown in Figure 3.5. One can prove that the mean of n independent random variables with this Cauchy distribution again has this same Cauchy distribution, so no "stability" is gained by averaging.

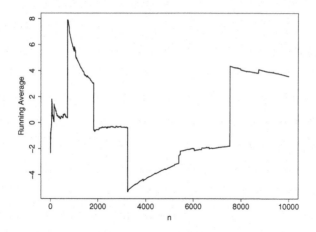

Figure 3.5. The running averages of Cauchy random variables do not converge in probability to a constant. The Law of Large Numbers does not apply because the Cauchy random variables do not have a finite standard deviation.

Another important assumption in the coin-toss example above is that the tosses are independent. Now we look at a simple process where the outcome at each step depends on the previous one.

Example 3.6. Limiting Behavior of a Dependent Process. On an imaginary tropical island, days can be classified as either Sunny (coded as 0) or Rainy (1). Once either a rainy or a sunny weather pattern has begun, it tends to continue for a while. In particular, if one day is Sunny, then the probability that the next day will also be Sunny is 0.97; and if one day is Rainy, then the probability that the next day will also be rainy is 0.94.

More formally, let the random variable W_n, which takes only the values 0 and 1, describe the weather on the nth day. The W_n are not independent random variables. In particular, $P\{W_{n+1} = 1 | W_n = 0\} = 0.03$, whereas $P\{W_{n+1} = 1 | W_n = 1\} = 0.94$. However, we will see in Chapter 6 that the probabilities $P\{W_n = 1 | W_1 = 0\}$ and $P\{W_n = 1 | W_1 = 1\}$ become equal as $n \to \infty$. We will also see that the proportion $Y_n = \bar{W}_n = (1/n) \sum_{i=1}^{n} W_i$ of rainy days obeys a Law of Large Numbers, converging in probability to $1/3$. Thus, over the long run, it rains on a third of the days.

Using the R code below, we simulate the weather for $m = 10\,000$ days (about 27 years). Then we can make a plot that shows the behavior of the "average" weather over this period. The result $Y_{10\,000} = 0.3099$ suggests that it may rain on about $1/3$ of the days, but this is not a very satisfying match with the known theoretical value. The convergence in this dependent model is not as fast as it was in the independent coin-toss model of Example 3.5, so $10\,000$ days is not really long enough to qualify as "the long run" in this case. (See Figure 3.6.)

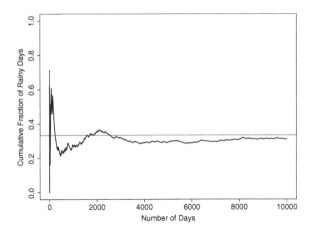

Figure 3.6. Counting sunny days as 0s and rainy days as 1s, the running averages of Example 3.6 converge very slowly to the known limit 1/3 (horizontal reference line). The slow convergence is due to the dependence of each day's weather on the weather of the previous day, in a pattern with highly positive correlation.

Because the weather on each day depends on the weather on the previous day, we could find no straightforward way to vectorize this program entirely, although we did keep the `for`-loop as simple as possible.

```
#Initialize:
# set.seed(1237)
m = 10000;  n = 1:m;  alpha = 0.03;  beta = 0.06
w = numeric(m);  w[1] = 0

#Simulate:
for (i in 2:m)
{
    if (w[i-1]==0)  w[i] = rbinom(1, 1, alpha)
    else            w[i] = rbinom(1, 1, 1 - beta)
}
y = cumsum(w)/n

#Results:
y[m]
plot(y, type="l",  ylim=c(0,1))
```

The dependent structure is reflected in the two plots of Figure 3.7. The upper panel, made with the additional code `plot(w[1:500], type="l")`, is a plot of the first 500 values W_i against n, with line segments connecting successive values. It shows the tendency for the weather to "stick" either at Rainy or Sunny for many days in a row. The lower panel plots the **autocorrelation**

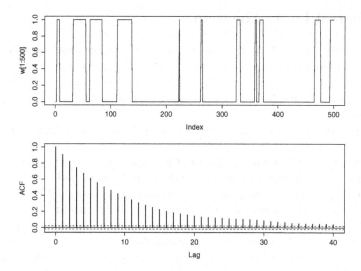

Figure 3.7. The trace of the first 500 values of W (upper panel) shows long periods of Rainy and of Sunny weather. The autocorrelation function (lower) shows a very highly positive autocorrelation for small lags, slowly decreasing for larger lags.

function, made with `acf(w)`. It shows the very high sample correlations between the sequences (W_n) and (W_{n-g}) for small **lags** g. For larger lags, the correlations decrease but are still significantly positive out as far as $g = 40$. The weather tomorrow depends very strongly on the weather today, but the weather a month from now is noticeably less dependent on today's weather.

Although our simulation did not stabilize within $m = 10\,000$ iterations, this slow convergence is a property of the particular *process* of this example, not a fault of our method of *simulation*. In this regard, simulation has accurately reflected the model we proposed but did not provide a very good approximation for the proportion of rainy days encountered over the long run.

Taken together, the three graphs in Figures 3.6 and 3.7 are very useful in detecting the slow convergence and diagnosing the reason for it. We use such plots repeatedly throughout this book. (See Problem 3.20 for two better-behaved weather models.) ◊

The Law of Large Numbers applies to most models we simulate in this book, independent or not, guaranteeing that they will stabilize "over the long run." But, in practice, we have just seen that the long run may be very long indeed. Especially for a process with positively correlated stages, one needs to verify whether a particular simulation has stabilized enough to give useful results.

3.3 Central Limit Theorem

We have seen that the Law of Large Numbers gives assurance that simulations based on independent models converge eventually, but it does not provide a good way to assess the accuracy of results after a given number m of iterations. Accordingly, we turn now to a brief discussion of a result that can provide useful assessments.

If we sum or average many observations from a population, the result tends to be normally distributed. Later in this section, we state this idea more precisely as the **Central Limit Theorem**. As we have already seen in Chapter 1, one consequence of this tendency is that, for sufficiently large n, $X_n \sim \mathsf{BINOM}(n, \pi)$ is approximately normal. The Central Limit Theorem applies because we can consider the binomial random variable X_n as the sum of n independent Bernoulli random variables V_i, each taking the value 1 with probability π and the value 0 with probability $1 - \pi$.

Three of the simulations we did earlier in this chapter provide elementary examples of this tendency for binomial random variables with large n to be approximately normal.

- In Example 3.5, the random variable $S_n \sim \mathsf{BINOM}(n, 1/2)$ is the number of Heads seen in the first $n = 100$ tosses of a coin, so that $\mathrm{E}(S_n) = n\pi = 50$ and $\mathrm{SD}(S_n) = \sqrt{n\pi(1-\pi)} = 5$. There, with $Y_n = S_n/n$, we claim that $P_n = P\{|Y_n - 1/2| < 0.02\} = 0.236$. However,

$$
\begin{aligned}
P_{100} &= P\{0.48 < Y_{100} < 0.52\} = P\{48 < S_{100} \le 51\} \\
&= P\{48.5 < S_{100} < 51.5\} \approx P\{48.5 < Z_{100} < 51.5\},
\end{aligned}
$$

where $Z_{100} \sim \mathsf{NORM}(50, 5)$. Thus P_{100} can be evaluated exactly in terms of S_{100}, and approximately in terms of Z_{100}, as follows:

```
> pbinom(51,100,.5) - pbinom(48,100,.5)
[1] 0.2356466
> pnorm(51.5,50,5) - pnorm(48.5,50,5)
[1] 0.2358228
```

The two answers agree to three places. Because of the good approximation of binomial probabilities by normal ones, for large m we feel comfortable taking S_m to be approximately $\mathsf{NORM}(m/2, \sqrt{m}/2)$ and $Y_m = S_m/m$ to be approximately $\mathsf{NORM}(1/2, 1/2\sqrt{m})$. Thus, in a simulation run with $m = 10\,000$, we have good reason to claim, as we did after the program in Section 3.5, to be 95% sure that $Y_{10\,000}$ will be within $\pm 1/\sqrt{m}$ or ± 0.01 of the correct answer.

- In Example 3.4, we used the sampling method to evaluate $P\{0 < Z \le 1\}$, finding that the proportion 0.34146 of our $m = 500\,000$ random samples from a standard normal distribution fell in the unit interval. Of course, in this problem, we know the exact answer is 0.3413. But in a practical simulation problem we would not know the answer, so it seems reasonable to use a standard 95% binomial confidence interval to assess the margin of error of the simulated result 0.34146. The estimated margin of error is $1.96 \, [0.34146(1 - 0.34146)/m]^{1/2} = 0.0013$, which is what we claimed at the end of that simulation. (Because m is so large, there is no point in using the Agresti-Coull adjustment discussed in Chapter 1.)

- The acceptance-rejection method of Example 3.3 gave 0.341056 as the simulated value of $P\{0 < Z \leq 1\}$. Thus, of the $m = 500\,000$ points used, the proportion of accepted points was about $0.341056/0.40 = 0.85264$. So a 95% confidence interval for the true proportion of acceptable points is $0.85264 \pm 1.96 \, [0.85264(1-0.85264)/m]^{1/2}$ or 0.85264 ± 0.00098. Multiplying both the estimate and the margin of error by 0.40 gives the corresponding confidence interval for the answer: 0.341056 ± 0.00039. It was on this basis that we concluded it is reasonable to expect the acceptance-rejection method with this number of iterations to give answers with about a 0.0004 margin of error.

In the next section, we look at the margins of error for estimates of $P\{0 < Z \leq 1\}$ using Riemann and Monte Carlo estimation. Although the three examples we have just considered all involve sums of Bernoulli random variables, the Central Limit Theorem works for random samples from a wide variety of discrete and continuous distributions. Here is a general statement.

The Central Limit Theorem. Let $X_1, X_2, \ldots$ be a sequence of independent, identically distributed random variables each with mean μ and finite standard deviation σ, and let $S_n = \sum_{i=1}^{n} X_i$ be the sum of the first n random variables and $\bar{X}_n = S_n/n$ be their mean. Further, let

$$Z_n = \frac{S_n - n\mu}{\sqrt{n}\sigma} = \frac{\bar{X}_n - \mu}{\sigma/\sqrt{n}}.$$

Then $\lim_{n \to \infty} F_{Z_n}(z) = \lim_{n \to \infty} P\{Z_n \leq z\} = \Phi(z)$ for any real z. We say that Z_n **converges in distribution** to standard normal. In symbols, we write this as $Z_n \overset{d}{\to} Z \sim \text{NORM}(0, 1)$. For $\overset{d}{\to}$, one reads *converges in distribution to.*

It is worthwhile to point out an essential difference between the Law of Large Numbers and the Central Limit Theorem. For simplicity, let the X_i have $\mu = \text{E}(X_i) = 0$ and $\sigma = \text{SD}(X_i) = 1$, and set $S_n = X_1 + \cdots + X_n$. Then the former theorem states that $\bar{X}_n = S_n/n \overset{p}{\to} \mu = 0$ and the latter states that $Z_n = S_n/\sqrt{n} \overset{d}{\to} \text{NORM}(0, 1)$. It matters greatly whether we divide S_n by n or by $\sqrt{n}$. Dividing by n in the Law of Large Numbers yields convergence to a constant value. Dividing by $\sqrt{n}$ in the Central Limit Theorem permits relatively more variation around $\mu = 0$ and results in convergence to a distribution. Partly because of this distinction, the Central Limit Theorem often provides useful information for much smaller values of n (sometimes useful for n as small as 10) than does the Law of Large Numbers (often useful only for n in the hundreds or thousands).

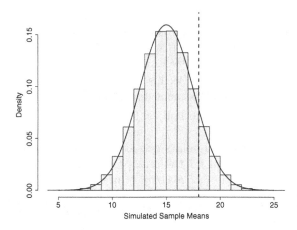

Figure 3.8. Illustrating the Central Limit Theorem. The density of NORM(15, 2.5) closely fits the histogram of sample means of 500 000 simulated samples of size 12 from UNIF(0, 30). The area under the curve to the right of the dashed line is 0.1151, while the area of the corresponding histogram bars is 0.1165. (See Example 3.7.)

Example 3.7. If the distribution of the X_i is relatively short-tailed and not far from symmetrical, the Central Limit Theorem is sometimes of practical use as an approximation for surprisingly small values of n. For example, let the X_i be a random sample of size $n = 12$ from UNIF(0, 30) so that $\mu = E(X_i) = 15$ and $\sigma = SD(X_i) = 30/\sqrt{12}$, as with the waiting times for trains at the beginning of Section 3.1. Then $E(\bar{X}_{12}) = 15$ and $SD(\bar{X}_{12}) = 30/12 = 2.5$, so the probability that the average waiting time $\bar{X}_{12}$ over 12 trips exceeds 18 minutes is approximated as

$$P\{\bar{X}_{12} > 18\} = P\left\{\frac{\bar{X}_{12} - \mu}{\sigma/\sqrt{n}} > \frac{18 - 15}{2.5} = 1.2\right\} \approx 1 - \Phi(1.2) = 0.1151.$$

In contrast, the following R program solves this problem using the sampling method with the actual distribution of $\bar{X}_{12}$ instead of using the normal approximation. (We use the sampling method because finding the density function of the actual distribution would be messy.) Each of the $m = 500\,000$ rows of the matrix DTA contains one simulated sample of size $n = 12$.

```
# set.seed(1212)
m = 500000;   n = 12
x = runif(m*n, 0, 30);   DTA = matrix(x, m)
x.bar = rowMeans(DTA);   mean(x.bar > 18)

> mean(x.bar > 18)
[1] 0.11655
```

The result 0.1165 ± 0.001 agrees with the normal approximation to two decimal places. The margin of error is approximated as $2[0.1165(1 - 0.1165)/m]^{1/2}$. (See Figure 3.8.)

If the distribution of the random variables being averaged is markedly skewed, as for the electronic components with exponentially distributed lifetimes in Section 3.1, then $n = 12$ is too small a sample size for the normal approximation to be useful. (See Problem 3.13.) $\Diamond$

Now we revisit Monte Carlo integration to elaborate on its justification in terms of the Law of Large Numbers and methods of estimating the accuracy of results obtained using this method of integration.

3.4 A Closer Look at Monte Carlo Integration

In Section 3.1, we used Monte Carlo integration to evaluate $P\{0 < Z \le 1\} = P\{0 \le Z \le 1\}$, where Z is standard normal. Now, more generally, suppose we want to integrate a bounded, piecewise-continuous function $h(x)$ over a finite interval $[a, b]$, with $a < b$. Then, for sufficiently large m, the integral $J = \int_a^b h(u)\, du$ is approximated by

$$A_m = \frac{b - a}{m} \sum_{i=1}^{m} h(U_i) = \frac{1}{m} \sum_{i=1}^{m}(b - a)h(U_i) = \frac{1}{m} \sum_{i=1}^{m} Y_i = \bar{Y}_m,$$

where U_i are sampled at random from $\mathsf{UNIF}(a, b)$, and $Y_i = (b - a)h(U_i)$.

Because the A_m are appropriate averages, the Law of Large Numbers guarantees that the $A_m \xrightarrow{P} J$ as $m \to \infty$. The following steps show that the Law of Large Numbers applies:

- The density function of each U_i is $f_U(u) = \frac{1}{b-a}I_{[a,b]}(u)$, which takes the value $1/(b - a)$ on $[a, b]$ and 0 elsewhere.
- The Y_i are independent and identically distributed random variables, each Y_i with mean J, because

$$\mathrm{E}(Y) = \mathrm{E}[(b - a)h(U)] = \int_a^b (b - a)h(u)f_U(u)\, du = \int_a^b h(u)\, du = J.$$

- $A_m = \bar{Y}_m$ is the sample mean of the $Y_i = (b - a)h(U_i)$, $i = 1, 2, ..., m$.
- Thus the Law of Large Numbers states that $A_m = \bar{Y}_m \xrightarrow{P} \mathrm{E}(Y) = J$.

In addition, for large m, the Central Limit Theorem states that $A_m = \bar{Y}_m$ is approximately normally distributed with mean $\mathrm{E}(Y)$ and standard deviation $\mathrm{SD}(\bar{Y}_m) = \mathrm{SD}(Y)/\sqrt{m}$. This gives an idea of the largest likely error in approximating J by A_m; in only 5% of simulations will the absolute error exceed $2\mathrm{SD}(\bar{Y}_m)$. Both the Law of Large Numbers and the Central Limit Theorem require the existence of $\mathrm{SD}(Y)$, but this is assured by our restrictions that h is bounded and piecewise continuous on the finite interval $[a, b]$. (See Problem 3.24 for examples where the integrand is not bounded.)

Example 3.8. Monte Carlo Margin of Error. We show results of the Riemann
and Monte Carlo methods of integration for $J = \int_0^{1.5} x^2\, dx = 1.125$. Recall
that the Riemann approximation is based on summing rectangles:

$$R_m = \frac{b-a}{m} \sum_{i=1}^{m} h(x_i) = \sum_{i=1}^{m} \frac{b-a}{m}\, h(x_i),$$

where the x_i are m grid points evenly spaced in $[a, b]$ and $R_m \to J$. The
following program implements both methods.

```
m = 100000
a = 0;   b = 3/2;   w = (b - a)/m
x = seq(a + w/2, b-w/2, length=m)
h = x^2;   rect.areas = w*h
sum(rect.areas)                         # Riemann

# set.seed(1212)
u = runif(m, a, b)
h = u^2;   y = (b - a)*h
mean(y)                                 # Monte Carlo
2*sd(y)/sqrt(m)                         # MC margin of error

> sum(rect.areas)                       # Riemann
[1] 1.125
> mean(y)                               # Monte Carlo
[1] 1.123621
> 2*sd(y)/sqrt(m)                       # MC margin of error
[1] 0.006358837
```

Riemann approximation gives the exact answer and Monte Carlo integration comes reasonably close. The estimated margin of error of the latter is
about 0.006, so here we can expect one- or (most often) two-place accuracy
from Monte Carlo integration with $m = 100\,000$. ◇

   In general, for reasonably smooth functions h on a finite interval of the real
line, the Riemann approximation gives better results than the Monte Carlo
approximation. However, we can make no comprehensive comparisons because
different factors affect the accuracy of each method. We have seen that the
error of the Monte Carlo method decreases as $1/\sqrt{m}$ with increasing m and
depends on $\mathrm{SD}(Y) = (b-a)\mathrm{SD}[h(U)]$. The error of the Riemann procedure is
increased by the roughness or "wiggliness" of h; if h is relatively smooth, its
error decreases as $1/m$.
   As we mentioned in Section 3.1, the importance of Monte Carlo integration
lies mainly in the evaluation of multiple integrals. The Monte Carlo method
often works well in higher dimensions, while the Riemann method usually does
not. Roughly speaking, the number m of grid points needed for an accurate
approximation with the Riemann method increases as the power of the number

of dimensions, whereas the required number m of random points in the Monte Carlo method depends on the variance of h (which may be large when the dimensionality is large) but not explicitly on the number of dimensions. (For more detail on the last two paragraphs, see [Liu01], Chapter 1.)

> The term *Monte Carlo* is used in slightly different ways by different authors. Most broadly, it can be used to refer to almost any kind of simulation. In this chapter we have used it to refer to the most basic kind of integration by simulation, involving points uniformly distributed in an interval of interest. All simulations are ultimately based on pseudorandom numbers that can be taken as $UNIF(0, 1)$. Other random variables are obtained by transformation. Hence it is theoretically possible to consider many kinds of simulations as transformations of integration problems based on uniform values in a unit interval, square, cube, or hypercube. As an especially straightforward example, each sample in the simulation of Example 3.7 arises from a uniform distribution on the 12-dimensional unit hypercube. (See [KG80], p233.)

Because many problems of practical interest require integration in higher dimensions, we now show an example in two dimensions, which is extended to three dimensions in the problems.

Example 3.9. A Bivariate Normal Probability. In the examples of Section 3.1, a deterministic Riemann approximation with rectangles performed better than Monte Carlo integration. Now we consider an integral over a two-dimensional region, for which the deterministic method loses some of its advantage.

Let Z_1 and Z_2 be independent standard normal random variables, and let φ denote the standard normal density. We wish to evaluate

$$J = P\{Z_1 > 0, Z_2 > 0, Z_1 + Z_2 < 1\},$$

which corresponds to the volume under a bivariate standard normal density surface and above the triangle with vertices at $(0,0), (0,1)$, and $(1,0)$. One can show that $J = 0.06773$, so we are able to judge the accuracy of both the deterministic and the Monte Carlo methods. (See Problem 3.27(c).)

We use $m = 10\,000$ for both approximations. Thus, in the Riemann approximation, we make a grid of 100×100 points within the unit square and consider volumes of solid rectangular "posts" that have square bases centered at grid points and approximate the height of the normal surface. We use only grid points inside the triangular region of integration (but some of the square bases can extend outside the triangle).

```
m = 10000
g = round(sqrt(m))             # no. of grid pts on each axis
x1 = rep((1:g - 1/2)/g, times=g)   # these two lines give
x2 = rep((1:g - 1/2)/g, each=g)    #    coordinates of grid points
hx = dnorm(x1)*dnorm(x2)
sum(hx[x1 + x2 < 1])/g^2       # Riemann approximation
(pnorm(sqrt(1/2)) - 0.5)^2     # exact value of J
```

```
> sum(hx[x1 + x2 < 1])/g^2           # Riemann approximation
[1] 0.06715779
> (pnorm(sqrt(1/2)) - 0.5)^2         # exact value of J
[1] 0.06773003
```

Note: The rep function, used above to make the coordinates of the grid points, may be new to you. Here are two simple examples illustrating its use: The expression rep(1:3, times=3) returns the vector $(1, 2, 3, 1, 2, 3, 1, 2, 3)$, and rep(1:3, each=3) returns $(1, 1, 1, 2, 2, 2, 3, 3, 3)$.

The approximation $J \approx 0.06716$ agrees to two decimal places with the exact answer $J = 0.06773$. If we change < to <= in the line of code that sums volumes of posts, then the result changes to $J \approx 0.06830$. With this change, all of the grid squares along the hypotenuse of the triangle, excluded in the original program, are included. Either choice—inclusion or exclusion— is defensible. For the case $m = 100$, the top two panels of Figure 3.9 show the square bases of the posts and illustrate exclusion (left) and inclusion of posts corresponding to such boundary points.

In the Monte Carlo procedure for evaluating J, we need to sample from a uniform distribution on the triangle. We do this by sampling $m = 10\,000$ points from the uniform distribution on the unit square and rejecting those that do not fall within the triangle. Because the area of the triangle is $1/2$, about half of the points will be accepted, so that the actual number of sampled points m' will be about 5000. Accordingly, in order to obtain the Monte Carlo approximation to J, we must multiply by $1/2$ the average value $\bar{Y}_{m'}$ for the sampled points. (The last two lines of the block of code below show this multiplication, which is analogous to multiplying by the length $b - a$ when the one-dimensional region of integration is the interval $[a, b]$.)

```
# set.seed(1237)
u1 = runif(m)                         # these two lines give a random
u2 = runif(m)                         #   point in the unit square
hu = dnorm(u1)*dnorm(u2)
hu.acc = hu[u1 + u2 < 1]              # heights above accepted points
m.prime = length(hu.acc); m.prime    # no. of points in triangle
(1/2)*mean(hu.acc)                    # Monte Carlo result
2*(1/2)*sd(hu.acc)/sqrt(m.prime)      # aprx. Marg. of Err. = 2SD(A)

>  m.prime = length(hu.acc); m.prime # no. of points in triangle
[1] 5066
>  (1/2)*mean(hu.acc)                 # Monte Carlo result
[1] 0.06761627
>  2*(1/2)*sd(hu.acc)/sqrt(m.prime)   # aprx. Marg. of Err. = 2SD(A)
[1] 0.0001972224
```

This run of the Monte Carlo method has an effective run size of $m' = 5066$ accepted points. (The lower left panel of Figure 3.9 shows 47 accepted points

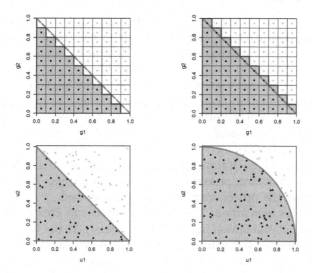

Figure 3.9. Integrating over a triangle. Graphs in the upper row show square bases of "posts" used in Riemann approximation, excluding (left) and including boundary points. At the lower left, in Monte Carlo integration, 47 accepted random points out of 100 fall within the region of integration. Example 3.9 uses 10 000 points instead of 100. (Analogously, the graph at the lower right illustrates Monte Carlo integration over the quarter of the unit circle in the first quadrant, see Problem 3.27(b).)

in a run with $m = 100$.) We see that we can expect, as achieved in this run, about three-place accuracy. Integrating over this 2-dimensional region, we have achieved better results with Monte Carlo integration than with deterministic Riemann approximation.

A fundamental difficulty of Riemann approximation in higher dimensions is illustrated here. In Riemann approximation over two dimensions, errors typically arise not only because flat-topped posts above grid squares do not exactly fit the normal surface, but also because the bases of the posts cannot exactly fit the region of integration. In higher dimensions, hard-to-fit edges and surfaces can proliferate. The random points of the Monte Carlo method lack regularity, but they have the great advantage of always lying precisely within the region of integration. In Problems 3.27 through 3.29, we consider related examples of multidimensional integration. ◇

3.5 When Is Simulation Appropriate?

One harsh traditional view is that simulation is inelegant and to be avoided whenever an analytic solution is possible. Now that we have run a few simulations for various purposes, we are in a position to consider when simulation may be appropriate and when it is not.

- *Simulation used to test a procedure.* If our goal is to evaluate $\int_0^1 x^2\,dx$, then simulation is clearly not the way to go because it is a trivial calculus exercise to obtain the exact answer. But in Example 3.8, we used the Monte Carlo method to approximate this integral and learn how to find the degree of accuracy we could expect from the method. Once we know Monte Carlo integration works well enough when the integrand is x^2, we are willing to try it for other integrands that do not have explicit indefinite integrals, for example $x^{-2}\sin^2 x$ in Problem 3.26. So the principle of testing a simulation procedure on cases where the answer is known is not just pedagogical. Such tests are important in program development to see whether a simulation procedure works for a particular kind of problem—and whether the procedure is properly programmed.

- *Analysis plus simulation.* Traditional analysis can greatly simplify some problems, but simulation may still be necessary at the last stage to get a numerical answer. In Example 3.9, we used analytic methods to express the double integral over a triangular region in terms of the univariate standard normal cumulative distribution function Φ. But then the integral Φ must be evaluated—either by Riemann approximation or some simulation method. For many analogous instances in higher dimensions, a Monte Carlo method would be the clear choice for obtaining a numerical answer.

- *Simulation because analysis is "impossible."* Sometimes an analytic solution is not possible, even as an initial step. Here we must understand that *possible* is a slippery word. In practice, what is possible often depends on how much you know or how much time you have to ponder the problem before a solution is due. If you know about ergodic Markov chains, an analytic solution to Problem 3.20 comes quickly (see Chapter 6). If not, a simulation as in Example 3.6 would serve well. Also, there are many problems where no known analytic solution exists, or even where it is known that no analytic solution exists. In particular, the expected value in Problem 3.8 seems an unlikely candidate for analytic solution.

- *Simulation for exploration or convenience.* With recent advances in congruential generators, computer hardware, and statistical software, simulations of important problems can be programmed easily and very long runs can be done in seconds. In practice, simulations are often used to explore ideas or obtain approximate answers. Perhaps "elegant" analytic solutions come later when one knows it is worthwhile to look for them, or perhaps analytic solutions would be desirable but remain elusive. Perhaps the line of inquiry collapses once we see disappointing results in several simulations. Also, both of us have used simulations to check analytic results, sometimes discovering errors in the analysis.

As you approach a practical problem where numerical results are required, you should try to use the combination of analysis and simulation that meets your needs for generality, accuracy, speed, and convenience. Often analytic solutions generalize more easily than simulation methods, but sometimes the reverse is true.

If the model you are computing is based on approximations and simplifying assumptions, a relatively tiny additional error introduced by simulation may be irrelevant. There may be practical situations in which a good approximation by noon today is worth more than the exact answer sometime next week. Of course, when publishing methods or software for use by others, one is expected to put the criteria of generality and accuracy above one's personal convenience.

3.6 Problems

Problems for Section 3.1 (Probabilities of Intervals)

3.1   Computation of simple integrals in Section 3.1.

a) If $W \sim \text{UNIF}(0, 30)$, sketch the density function of W and the area that represents $P\{W \leq 10\}$. In general, if $X \sim \text{UNIF}(\alpha, \beta)$, write the formula for $P\{c < X < d\}$, where $\alpha < c < d < \beta$.

b) If $T \sim \text{EXP}(0.5)$, sketch the exponential distribution with rate $\lambda = 0.5$ and mean $\mu = 2$. Write $P\{T > 1\}$ as an integral, and use calculus to evaluate it.

3.2   We explore two ways to evaluate $e^{-1/2} \approx 0.61$, correct to two decimal places, using only addition, subtraction, multiplication, and division—the fundamental operations available to the makers of tables 50 years ago. On most modern computers, the evaluation of e^x is a chip-based function.

a) Consider the Taylor (Maclauren) expansion $e^x = \sum_{k=0}^{\infty} x^k/k!$. Use the first few terms of this infinite series to approximate $e^{-1/2}$. How many terms are required to get two-place accuracy? Explain.

b) Use the relationship $e^x = \lim_{n \to \infty} (1 + x/n)^n$. Notice that this is the limit of an increasing sequence. What is the smallest value of k such that $n = 2^k$ gives two-place accuracy for $e^{-1/2}$?

c) Run the following R script. For each listed value of x, say whether the method of part (a) or part (b) provides the better approximation of e^x.

```
x = seq(-2, 2, by=.25)
taylor.7 = 1 + x + x^2/2 + x^3/6 + x^4/24 + x^5/120 + x^6/720
seq.1024  = (1 + x/1024)^1024
exact = exp(x)
round(cbind(x, taylor.7, seq.1024, exact), 4)
```

3.3   Change the values of the constants in the program of Example 3.1 as indicated.

a) For $a = 0$ and $b = 1$, try each of the values $m = 10, 20, 50$, and 500. Among these values of m, what is the smallest that gives five-place accuracy for $P\{0 < Z < 1\}$?

b) For $m = 5000$, modify this program to find $P\{1.2 < Z \le 2.5\}$. Compare your answer with the exact value obtained using the R function pnorm.

3.4   Modify the program of Example 3.1 to find $P\{X \le 1\}$ for X exponentially distributed with mean 2. The density function is $f(x) = \frac{1}{2}e^{-x/2}$, for $x > 0$. Run the program, and compare the result with the exact value obtained using calculus and a calculator.

3.5   Run the program of Example 3.2 several times (omitting set.seed) to evaluate $P\{0 < Z \le 1\}$. Do any of your answers have errors that exceed the claimed margin of error 0.00015? Also, changing constants as necessary, make several runs of this program to evaluate $P\{0.5 < Z \le 2\}$. Compare your results with the exact value.

3.6   Use Monte Carlo integration with $m = 100\,000$ to find the area of the first quadrant of the unit circle, which has area $\pi/4$. Thus obtain a simulated value of $\pi = 3.141593$. How many places of accuracy do you get?

3.7   Here we consider two very similar random variables. In each part below we wish to evaluate $P\{X \le 1/2\}$ and $E(X)$. Notice that part (a) can be done by straightforward analytic integration but part (b) cannot.

a) Let X be a random variable distributed as BETA$(3, 2)$ with density function $f(x) = 12x^2(1-x)$, for $0 < x < 1$, and 0 elsewhere. Use the numerical integration method of Example 3.1 to evaluate the specified quantities. Compare the results with exact values obtained using calculus.

b) Let X be a random variable distributed as BETA$(2.9, 2.1)$ with density function $f(x) = \frac{\Gamma(5)}{\Gamma(2.9)\Gamma(2.1)}\, x^{1.9}(1-x)^{1.1}$, for $0 < x < 1$, and 0 elsewhere. Use the method of Example 3.1.

c) Use the Monte Carlo integration method of Example 3.2 for both of the previous parts. Compare results.

Hints and answers: (a) From integral calculus, $P\{X \le 1/2\} = 5/16$ and $E(X) = 3/5$. (Show your work.) For the numerical integration, modify the lines of the program of Example 3.1 that compute the density function. Also, let $a = 0$ and let $b = 1/2$ for the probability. For the expectation, let $b = 1$ use h = 12*g^3*(1-g) or h = g*dbeta(g, 3, 2). Why? (b) The constant factor of $f(x)$ can be evaluated in R as gamma(5)/(gamma(2.9)*gamma(2.1)), which returns 12.55032. Accurate answers are 0.3481386 (from pbeta(.5, 2.9, 2.1)) and 29/50.

3.8   The yield of a batch of protein produced by a biotech company is $X \sim$ NORM$(100, 10)$. The dollar value of such a batch is $V = 20 - |X - 100|$ as long as the yield X is between 80 and 120, but the batch is worthless otherwise. (Issues of quality and purity arise if the yield of a batch is much different from 100.) Find the expected monetary value $E(V)$ of such a batch.

Hint: In the program of Example 3.1, let $a = 80$, $b = 120$. Also, for the line of code defining h, substitute h = (20 - abs(g - 100))*dnorm(g, 100, 10). Provide the answer (between 12.0 and 12.5) correct to two decimal places.

3.9 Suppose you do not know the value of $\sqrt{2}$. You can use simulation to approximate it as follows. Let $X = U^2$, where $U \sim \text{UNIF}(0,1)$. Then show that $2P\{0 < X \le 1/2\} = \sqrt{2}$, and use the sampling method with large m to approximate $\sqrt{2}$.

Note: Of the methods in Section 3.1, only sampling is useful. You could find the density function of X, but it involves $\sqrt{2}$, which you are pretending not to know.

3.10 A computer processes a particular kind of instruction in two steps. The time U (in μs) for the first step is uniformly distributed on $(0, 10)$. Independently, the additional time V for the second step is normally distributed with mean $5\,\mu s$ and standard deviation $1\,\mu s$. Represent the total processing time as $T = U + V$ and evaluate $P\{T > 15\}$. Explain each step in the suggested R code below. Interpret the results. Why do you suppose we choose the method of Example 3.4 here—in preference to those of Examples 3.1–3.3? (The histogram is shown in Figure 3.3, p57.)

```
m = 500000;   u = 10*runif(m);   v = 5 + rnorm(m)
t = u + v;   mean(t > 15);   mean(t);   sd(t);   hist(t)
```

Comments: The mean of the $500\,000$ observations of T is the balance point of the histogram. How accurately does this mean simulate $E(T) = E(U)+E(V) = 10$? Also, compare simulated and exact $\text{SD}(T)$. The histogram facilitates a rough guess of the value $P\{T > 15\}$. Of the $m = 500\,000$ sampled values, it seems that approximately $20\,000$ (or 4%) exceed 15. Your value from the program should be more precise.

3.11 *The acceptance-rejection method for sampling from a distribution.* Example 3.3 illustrates how the acceptance-rejection (AR) method can be used to approximate the probability of an interval. A generalization of this idea is sometimes useful in sampling from a distribution, especially when the quantile transformation method is infeasible. Suppose the random variable X has the density function f_X with support S (that is, $f_X(x) > 0$ exactly when $x \in S$). Also, suppose we can find an "envelope" $Bb(x) \ge f_X(x)$, for all x in S, where B is a known constant and $b(x)$ has a finite integral over S.

Then, to sample a value at random from X, we sample a "candidate" value y at random from a density $g(x)$ that is proportional to $b(x)$, accepting the candidate value as a random value of X with probability $f_X(y)/Bb(y)$. (Rejected candidate values are ignored.) Generally speaking, this method works best when the envelope function is a reasonably good fit to the target density so that the acceptance rate is relatively high.

a) As a trivial example, suppose we want to sample from $X \sim \text{BETA}(3, 1)$ without using the R function **rbeta**. Its density function is $f_X(x) = 3x^2$, on $S = (0, 1)$. Here we can choose $Bb(x) = 3x \ge f_X(x)$, for x in $(0, 1)$, and $b(x)$ is proportional to the density $2x$ of BETA$(2, 1)$. Explain how the following R code implements the AR method to simulate X. (See Problem 2.10 and Figure 2.10, p45.)

```
m = 20000;  u = runif(m);  y = sqrt(u)
acc = rbinom(m, 1, y);  x = y[acc==T]
mean(x);  sd(x);  mean(x < 1/2);  mean(acc)
hist(x, prob=T, ylim=c(0,3), col="wheat")
lines(c(0,1), c(0, 3), lty="dashed", lwd=2, col="darkgreen")
xx = seq(0, 1, len=1000)
lines(xx, dbeta(xx, 3, 1), lwd=2, col="blue")
```

The top panel of Figure 3.10 illustrates this method. We say this is a trivial example because we could easily use the quantile transformation method to sample from BETA$(3, 1)$.)

b) As a more serious example, consider sampling from $X \sim$ BETA$(1.5, 1.5)$, for which the quantile function is not so easily found. Here the density is $f_X(x) = (8/\pi)x^{0.5}(1 - x)^{0.5}$ on $(0, 1)$. The mode occurs at $x = 1/2$ with $F_X(1/2) = 4/\pi$, so we can use $Bb(x) = 4/\pi$. Modify the program of part (a) to implement the AR method for simulating values of X, beginning with the following two lines. Annotate and explain your code. Make a figure similar to the bottom panel of Figure 3.2. For verification, note that $E(X) = 1/2$, $SD(X) = 1/4$, and $F_X(1/2) = 1/2$. What is the acceptance rate?

```
m = 40000;  y = runif(m)
acc = rbinom(m, 1, dbeta(y, 1.5, 1.5)/(4/pi));  x = y[acc==T]
```

c) Repeat part (b) for $X \sim$ BETA$(1.4, 1.6)$. As necessary, use the R function gamma to evaluate the necessary values of the Γ-function. The function rbeta implements very efficient algorithms for sampling from beta distributions. Compare your results from the AR method in this part with results from rbeta.

Answers: (a) Compare with exact values $E(X) = 3/4$, $SD(X) = (3/80)^{1/2} = 0.1936$, and $F_X(1/2) = P\{X \le 1/2\} = 1/8$.

Problems for Section 3.2 (Law of Large Numbers)

3.12   In Example 3.5, interpret the output for the run shown in the example as follows. First, verify using hand computations the values given for $Y_1, Y_2, \ldots, Y_5$. Then, say exactly how many Heads were obtained in the first 9996 simulated tosses and how many Heads were obtained in all 10 000 tosses.

3.13   Run the program of Example 3.5 several times (omitting set.seed). Did you get any values of $Y_{10\,000}$ outside the 95% interval $(0.49, 0.51)$ claimed there? Looking at the traces from your various runs, would you say that the runs are more alike for the first 1000 values of n or the last 1000 values?

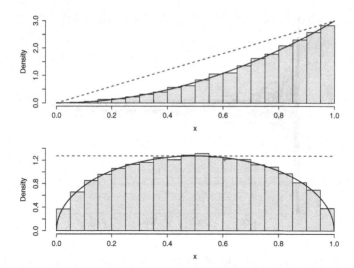

Figure 3.10. Using the acceptance-rejection method to simulate samples from two beta distributions. In each plot, the dotted line shows our choice of envelope. The solid lines show the density functions of the target distributions BETA(3, 1) (at the top) and BETA(1.5, 1.5). See Problem 3.11.

3.14   By making minor changes in the program of Example 3.2 (as below), it is possible to illustrate the convergence of the approximation to $J = 0.341345$ as the number n of randomly chosen points increases to $m = 5000$. Explain what each statement in the code does. Make several runs of the program. How variable are the results for very small values of n, and how variable are they for values of n near $m = 5000$? (Figure 3.11 shows superimposed traces for 20 runs.)

```
m = 5000;  n = 1:m
u = runif(m)
h = dnorm(u)
j = cumsum(h)/n
plot(n, j, type="l", ylim=c(0.32, 0.36))
abline(h=0.3413, col="blue");  j[m]
```

Note: The plotting parameter `ylim` establishes a relatively small vertical range for the plotting window on each run, making it easier to assess variability within and among runs.

3.15   Consider random variables $X_1 \sim$ BETA(1, 1), $X_2 \sim$ BETA(2, 1), and $X_3 \sim$ BETA(3, 1). Then, for appropriate constants, K_i, $i = 1, 2, 3$, the integral $\int_0^1 x^2 \, dx = 1/3$ can be considered as each of the following: $K_1 \mathrm{E}(X_1^2)$, $K_2 \mathrm{E}(X_2)$, and $K_3 P\{0 < X_3 \leq 1/2\}$. Evaluate K_1, K_2, and K_3.

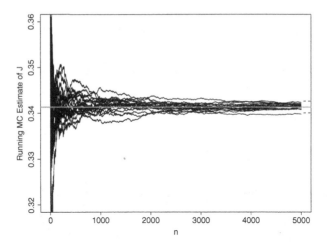

Figure 3.11. Traces from 20 runs of the program in Problem 3.14 overlaid on one plot. Horizontal dotted lines at the right show the interval inside which 95% of runs should end. In this typical set of 20, one run ends just outside the interval.

3.16   In Example 3.5, let $\epsilon = 1/10$ and define $P_n = P\{|Y_n - 1/2| < \epsilon\} = P\{1/2 - \epsilon < Y_n < 1/2 + \epsilon\}$. In R, the function `pbinom` is the cumulative distribution function of a binomial random variable.

a) In the R Console window, execute
   `n = 1:100; pbinom(ceiling(n*0.6)-1, n, 0.5) - pbinom(n*0.4, n, 0.5)`
   Explain how this provides values of P_n, for $n = 1, 2, \ldots 100$. (Notice that the argument `n` in the function `pbinom` is a vector, so 100 results are generated by the second statement.) Also, report the five values $P_{20}, P_{40}, P_{60}, P_{80}$, and P_{100}, correct to six decimal places, and compare results with Figure 3.12.
b) By hand, verify the R results for $P_1, \ldots, P_6$.
c) With $\epsilon = 1/50$, evaluate the fifty values $P_{100}, P_{200}, \ldots, P_{5000}$. (Use the expression `n = seq(100, 5000, by=100)`, and modify the parameters of `pbinom` appropriately.)

3.17   Modify the program of Example 3.5 so that there are only $n = 100$ tosses of the coin. This allows you to see more detail in the plot. Compare the behavior of a fair coin with that of a coin heavily biased in favor of Heads, $P(\text{Heads}) = 0.9$, using the code `h = rbinom(m, 1, 0.9)`. Make several runs for each type of coin. Some specific points for discussion are: Why are there long upslopes and short downslopes in the paths for the biased coin but not for the fair coin? Which simulations seem to converge faster—fair or biased? Do the autocorrelation plots `acf(h)` differ between fair and biased coins?

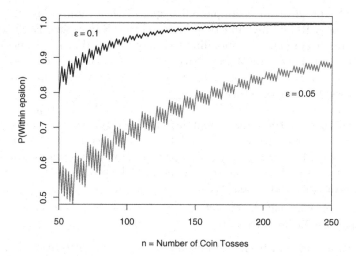

Figure 3.12. Convergence rates for the Law of Large numbers. The rate at which $P_n = P\{|Y_n - 1/2| < \epsilon\}$ converges to 1 in Example 3.5 depends on the value of ϵ. The upper plot shows P_{250} very near 1, with $\epsilon = 0.1$ as in Problem 3.16. But if $\epsilon = 0.05$, then convergence of P_n to 1 is much slower (lower plot).

3.18   A version of the program in Example 3.5 with an explicit loop would substitute one of the two blocks of code below for the lines of the original program that make the vectors h and y.

```
# First block: One operation inside loop
h = numeric(m)
for (i in 1:m)   {h[i] = rbinom(1, 1, 1/2)}
y = cumsum(h)/n

# Second block: More operations inside loop
y = numeric(m);   h = numeric(m)
for (i in 1:m)   {
  if (i==1)
    {b = rbinom(1, 1, 1/2); h[i] = y[i] = b}
  else
    {b = rbinom(1, 1, 1/2); h[i] = b;
        y[i] = ((i - 1)*y[i - 1] + b)/i}  }
```

Modify the program with one of these blocks, use $m = 500\,000$ iterations, and compare the running time with that of the original "vectorized" program. To get the running time of a program accurate to about a second, use as the first line t1 = Sys.time() and as the last line t2 = Sys.time(); t2 - t1.

Note: On computers available as this is being written, the explicit loops in the substitute blocks take noticeably longer to execute than the original vectorized code.

3.19   The program in Example 3.6 begins with a Sunny day. Eventually, there will be a Rainy day, and then later another Sunny day. Each return to Sun (0) after Rain (1), corresponding to a day n with $W_{n-1} = 1$ and $W_n = 0$, signals the end of one Sun–Rain "weather cycle" and the beginning of another. (In the early part of the plot of Y_n, you can probably see some "valleys" or "dips" caused by such cycles.)

If we align the vectors $(W_1, \ldots, W_{9999})$ and $(W_2, \ldots, W_{10\,000})$, looking to see where 0 in the former matches 1 in the latter, we can count the complete weather cycles in our simulation. The R code to make this count is `length(w[w[1:(m-1)]==0 & w[2:m]==1])`. Type this line in the Console window after a simulation run—or append it to the program. How many cycles do you count with `set.seed(1237)`?

Hint: One can show that the *theoretical* cycle length is 50 days. Compare this with the top panel of Figure 3.7 (p63).

3.20   Branching out from Example 3.6, we discuss two additional imaginary islands. Call the island of the example Island E.

a) The weather on Island A changes more readily than on Island E. Specifically, $P\{W_{n+1} = 0 | W_n = 0\} = 3/4$ and $P\{W_{n+1} = 1 | W_n = 1\} = 1/2$. Modify the program of Example 3.6 accordingly, and make several runs. Does Y_n appear to converge to $1/3$? Does Y_n appear to stabilize to its limit more quickly or less quickly than for Island E?

b) On Island B, $P\{W_{n+1} = 0 | W_n = 0\} = 2/3$ and $P\{W_{n+1} = 1 | W_n = 1\} = 1/3$. Modify the program, make several runs, and discuss as in part (a), but now comparing all three islands. In what fundamental way is Island B different from Islands E and A?

c) Make `acf` plots for Islands A and B, and compare them with the corresponding plot in the bottom panel of Figure 3.7 (p63).

Note: We know of no real place where weather patterns are as extremely persistent as on Island E. The two models in this problem are both more realistic.)

3.21   *Proof of the Weak Law of Large Numbers (Theoretical).* Turn the methods suggested below (or others) into carefully written proofs. Verify the examples. (Below we assume continuous random variables. Similar arguments, with sums for integrals, would work for the discrete case.)

a) *Markov's Inequality.* Let W be a random variable that takes only positive values and has a finite expected value $E(W) = \int_0^\infty x f_W(w) \, dw$. Then, for any $a > 0$, $P\{W \geq a\} \leq E(W)/a$.

Method of proof: Break the integral into two nonnegative parts, over the intervals $(0, a)$ and (a, ∞). Then $E(W)$ cannot be less than the second integral, which in turn cannot be less than $aP\{W \geq a\} = a \int_a^\infty f_W(w) \, dw$. Example: Let $W \sim \text{UNIF}(0, 1)$. Then, for $0 < a < 1$, $E(W)/a = 1/2a$ and $P\{W \geq a\} = 1 - P\{W < a\} = 1 - a$. Is $1 - a < 1/2a$?

b) *Chebyshev's Inequality.* Let X be a random variable with $E(X) = \mu$ and $V(X) = \sigma^2 < \infty$. Then, for any $k > 0$, $P\{|X - \mu| \geq k\sigma\} \leq 1/k^2$.

Method of proof: In Markov's Inequality, let $W = (X - \mu)^2 \geq 0$ so that $E(W) = V(X)$, and let $a = k^2\sigma^2$.

Example: If Z is standard normal, then $P\{|Z| \geq 2\} < 1/4$. Explain briefly how this illustrates Chebyshev's Inequality.

c) *WLLN.* Let $Y_1, Y_2, \ldots, Y_n$ be independent, identically distributed random variables each with mean μ and variance $\sigma^2 < \infty$. Further denote by $\bar{Y}_n$ the sample mean of the Y_i. Then, for any $\epsilon > 0$, $\lim_{n\to\infty} P\{|\bar{Y}_n - \mu| < \epsilon\} = 1$.

Method of proof: In Chebyshev's Inequality, let $X = \bar{Y}_n$, which has $V(\bar{Y}_n) = \sigma^2/n$, and let $k = n\epsilon/\sigma$. Then use the complement rule and let $n \to \infty$.

Note: What we have referred to in this section as the Law of Large Numbers is usually called the Weak Law of Large Numbers (WLLN), because a stronger result can be proved with more advanced mathematical methods than we are using in this book. The same assumptions imply that $P\{\bar{Y}_n \to \mu\} = 1$. This is called the Strong Law of Large Numbers. The proof is more advanced because one must consider the joint distribution of all Y_n in order to evaluate the probability.

Problems for Section 3.3 (Central Limit Theorem)

3.22   In Example 3.5, we have $S_n \sim \mathsf{BINOM}(n, 1/2)$. Thus $E(S_n) = n/2$ and $V(S_n) = n/4$. Find the mean and variance of Y_n. According to the Central Limit Theorem, Y_n is very nearly normal for large n. Assuming $Y_{10\,000}$ to be normal, find $P\{|Y_{10\,000} - 1/2| \geq 0.01\}$. Also find the margin of error in estimating $P\{\text{Heads}\}$ using $Y_{10\,000}$.

3.23   In Example 3.7, we see that the mean of 12 observations from a uniform population is nearly normal. In contrast, the electronic components of Section 3.1 have exponentially distributed lifetimes with mean 2 years (rate $1/2$ per year). Because the exponential distribution is strongly skewed, convergence in the Central Limit Theorem is relatively slow. Suppose you want to know the probability that the average lifetime $\bar{T}$ of 12 randomly chosen components of this kind exceeds 3.

a) Show that $E(\bar{T}) = 2$ and $SD(\bar{T}) = \sqrt{1/3}$. Use the normal distribution with this mean and standard deviation to obtain an (inadequate) estimate of $P\{\bar{T} > 3\}$. Here the Central Limit Theorem does not provide useful estimates when n is as small as 12.

b) Modify the program of Example 3.7, using `x = rexp(m*n, rate=1/2)`, to simulate $P\{\bar{T} > 3\}$. One can show that $\bar{T} \sim \mathsf{GAMMA}(12, 6)$ precisely. Compare the results of your simulation with your answer to part (a) and with the exact result obtained using `1 - pgamma(3, 12, rate=6)`.

c) Compare your results from parts (a) and (b) with Figure 3.13 and numerical values given in its caption.

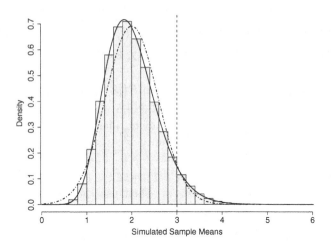

Figure 3.13. The histogram shows the simulated distribution of the mean $\bar{T}$ of 12 observations from $EXP(1/2)$. The normal approximation is poor (broken curve), giving $P\{\bar{T} > 3\} \approx 0.042$. The simulation and the exact gamma distribution (solid) give $P\{\bar{T} > 3\} = 0.055$. (See Problem 3.23 and compare with Figure 3.8.)

Problems for Section 3.4 (Closer Look at Monte Carlo)

In these problems, $\pi = 3.14159$.

3.24 Here are some modifications of Example 3.8 with consequences that may not be apparent at first. Consider $J = \int_0^1 x^d \, dx$.

a) For $d = 1/2$ and $m = 100\,000$, compare the Riemann approximation with the Monte Carlo approximation. Modify the method in Example 3.8 appropriately, perhaps writing a program that incorporates d = 1/2 and h = x^d, to facilitate easy changes in the parts that follow. Find $V(Y)$.

b) What assumption of Section 3.4 fails for $d = -1/2$? What is the value of J? Of $V(Y)$? Try running the two approximations. How do you explain the unexpectedly good behavior of the Monte Carlo simulation?

c) Repeat part(b), but with $d = -1$. Comment.

3.25 This problem shows how the rapid oscillation of a function can affect the accuracy of a Riemann approximation.

a) Let $h(x) = |\sin k\pi x|$ and k be a positive integer. Then use calculus to show that $\int_0^1 h(x) \, dx = 2/\pi = 0.6366$. Use the code below to plot h on $[0,1]$ for $k = 4$.

```
k = 4
x = seq(0,1, by = 0.01);   h = abs(sin(k*pi*x))
plot(x, h, type="l")
```

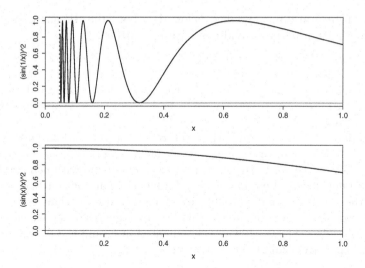

Figure 3.14. Two functions used in Problem 3.26. The top panel shows a plot of $\sin^2(1/x)$ in the interval $(0.05, 1]$; to the left of $x = 0.05$, the oscillation becomes too rapid for clear plotting. The bottom panel shows a plot of $x^{-2}\sin^2 x$ on $(0, 1]$.

b) Modify the program of Example 3.8 to integrate h. Use $k = 5000$ throughout, and make separate runs with $m = 2500$, 5000, $10\,000$, $15\,000$, and $20\,000$. Compare the accuracy of the resulting Riemann and Monte Carlo approximations, and explain the behavior of the Riemann approximation.

c) Use calculus to show that $V(Y) = V(h(U)) = 1/2 - 4/\pi^2 = 0.0947$. How accurately is this value approximated by simulation? If $m = 10\,000$, find the margin of error for the Monte Carlo approximation in part (b) based on $SD(Y)$ and the Central Limit Theorem. Are your results consistent with this margin of error?

3.26   The integral $J = \int_0^1 \sin^2(1/x)\,dx$ cannot be evaluated analytically, but advanced analytic methods yield $\int_0^\infty \sin^2(1/x)\,dx = \pi/2$.

a) Assuming this result, show that $J = \pi/2 - \int_0^1 x^{-2}\sin^2 x\,dx$. Use R to plot both integrands on $(0, 1)$, obtaining results as in Figure 3.14.

b) Use both Riemann and Monte Carlo approximations to evaluate J as originally defined. Then evaluate J using the equation in part (a). Try both methods with $m = 100$, 1000, 1001, and $10\,000$ iterations. What do you believe is the best answer? Comment on differences between methods and between equations.

Note: Based on a problem in [Liu01], Chapter 2.

3.27   Modify the program of Example 3.9 to approximate the volume beneath the bivariate standard normal density surface and above two additional regions of integration as specified below. Use both the Riemann and Monte Carlo methods in parts (a) and (b), with $m = 10\,000$.

a) Evaluate $P\{0 < Z_1 \le 1,\ 0 < Z_2 \le 1\}$. Because Z_1 and Z_2 are independent standard normal random variables, we know that this probability is $0.341345^2 = 0.116516$. For each method, say whether it would have been better to use $m = 10\,000$ points to find $P\{0 < Z \le 1\}$ and then square the answer.

b) Evaluate $P\{Z_1^2 + Z_2^2 < 1\}$. Here the region of integration does not have area 1, so remember to multiply by an appropriate constant. Because $Z_1^2 + Z_2^2 \sim \mathsf{CHISQ}(2)$, the exact answer can be found with `pchisq(1, 2)`.

c) The joint density function of (Z_1, Z_2) has circular contour lines centered at the origin, so that probabilities of regions do not change if they are rotated about the origin. Use this fact to argue that the exact value of $P\{Z_1 + Z_2 < 1\}$, which was approximated in Example 3.9, can be found with `(pnorm(1/sqrt(2)) - 0.5)^2`.

3.28   Here we extend the idea of Example 3.9 to three dimensions. Suppose three items are drawn at random from a population of items with weights (in grams) distributed as $\mathsf{NORM}(100, 10)$.

a) Using the R function `pnorm` (cumulative distribution function of a standard normal), find the probability that the sum of the three weights is less than 310 g. Also find the probability that the minimum weight of these three items exceeds 100 g.

b) Using both Riemann and Monte Carlo methods, approximate the probability that (simultaneously) the minimum of the weights exceeds 100 g and their sum is less than 310 g.

   The suggested procedure is to (i) express this problem in terms of three standard normal random variables, and (ii) modify appropriate parts of the program in Example 3.9 to approximate the required integral over a triangular cone (of area 1/6) in the unit cube $(0, 1)^3$ using $m = g^3 = 25^3$. Here is R code to make the three grid vectors needed for the Riemann approximation:

```
x1 = rep((1:g - 1/2)/g, each=g^2)
x2 = rep(rep((1:g - 1/2)/g, each=g), times=g)
x3 = rep((1:g - 1/2)/g, times=g^2)
```

Answers: (a) `pnorm(1/sqrt(3))` for the sum. (b) Approximately 0.009. To understand the three lines of code provided, experiment with a five-line program. Use `g = 3` before these three lines and `cbind(x1, x2, x3)` after.

3.29   For $d = 2$, 3, and 4, use Monte Carlo approximation with $m = 100\,000$ to find the probability that a d-variate (independent) standard normal distribution places in the part of the d-dimensional unit ball with all coordinates positive. To do this, imitate the program in Example 3.9, and use the fact that the entire 4-dimensional unit ball has hypervolume $\pi^2/2$. Compare the results with appropriate values computed using the chi-squared distribution with d degrees of freedom; use the R function `pchisq`.

Note: In case you want to explore higher dimensions, the general formula for the hypervolume of the unit ball in d dimensions is $\pi^{d/2}/\, \Gamma((d+2)/2)$; for a derivation see [CJ89], p459. Properties of higher dimensional spaces may seem strange to you. What happens to the hypervolume of the unit ball as d increases? What happens to the probability assigned to the (entire) unit ball by the d-variate standard normal distribution? What happens to the hypervolume of the smallest hypercube that contains it? There is "a lot of room" in higher dimensional space.

4

Sampling from Applied Probability Models

In Chapter 3, we used the sampling method to find probabilities and expectations involving a random variable with a distribution that is easy to describe but with a density function that is not explicitly known. In this chapter, we explore additional applications of the sampling method. The examples are chosen because of their practical importance or theoretical interest. Sometimes, analytic methods can be used to get exact results for special cases, thus providing some confidence in the validity of more general simulation results. Also, in an elementary way, some of the examples and problems show how simulation can be useful in research. At least they have illustrated this to us personally because we have gained insights from simulation in these settings that we might never have gained by analytic means.

4.1 Models Based on Exponential Distributions

The exponential family of distributions is often used to model waiting times for random events. In order to establish our point of view and notation, we begin with a brief review of exponential distributions and their properties.

Suppose the number N_t of random events that occur in the time interval $(0, t]$, where $t > 0$, has the distribution $\mathsf{POIS}(\lambda t)$. Thus the probability of seeing none of these random events by time $t > 0$ is $P\{N_t = 0\} = e^{-\lambda t}$. Another way to specify that there are no events in the interval $(0, t]$ is to let X be the waiting time, starting at $t = 0$, until we see the first event. Then

$$P\{X > t\} = 1 - F_X(t) = P\{N_t = 0\} = e^{-\lambda t}, \tag{4.1}$$

for $t > 0$. Here, according to our usual notation, F_X is the cumulative distribution function of X. Then, by differentiation, the density function of X is $f_X(t) = F_X'(t) = \lambda e^{-\lambda t}$, for $t > 0$ (and 0 elsewhere). We say that X has an exponential distribution with rate λ; in symbols, $X \sim \mathsf{EXP}(\lambda)$.

E.A. Suess and B.E. Trumbo, *Introduction to Probability Simulation and Gibbs Sampling with R*, 87
Use R!, DOI 10.1007/978-0-387-68765-0_4, © Springer Science+Business Media, LLC 2010

Starting with the distribution of a discrete random variable $N_t \sim \mathsf{POIS}(\lambda t)$ that counts the number of random events in an interval $(0, t]$, we have found the distribution of the continuous random variable X that takes values $t > 0$.

One can show that $\mathrm{E}(X) = \mathrm{SD}(X) = 1/\lambda$ (see Problem 4.2). Intuitively speaking, if the Poisson events occur at an average rate $\lambda = 1/2$ events per minute (one event every two minutes), then it seems reasonable that the average waiting time for such an event should be $1/\lambda = 2$ minutes.

Some texts parameterize the exponential family of distributions according to the mean $\mu = 1/\lambda$. But following the notation of R, we always use the rate λ as the parameter. For example, if $t = 1.5$ and $\lambda = 2$, then $N_{1.5} \sim \mathsf{POIS}(3)$ and $X \sim \mathsf{EXP}(2)$. Thus, $P\{N_{1.5} = 0\} = P\{X > 1.5\} = e^{-3} = 0.0498$. In R, this can be evaluated as `dpois(0, 3)` or as `1 - pexp(1.5, 2)`.

An important characteristic of an exponential random variable X is the **no-memory property**. In symbols,

$$P\{X > s + t | X > s\} = \frac{P\{X > s + t,\, X > s\}}{P\{X > s\}} = \frac{P\{X > s + t\}}{P\{X > s\}}$$
$$= \frac{\exp[-\lambda(s + t)]}{\exp(-\lambda s)} = e^{-\lambda t} = P\{X > t\}, \qquad (4.2)$$

where $s, t > 0$ (see Problem 4.3). This means that the probability of waiting at least another t time units until an event occurs is the same whether we start waiting at time 0 or at time s.

Real-life waiting times are approximately exponentially distributed when the no-memory property is approximately true. Suppose a bank teller takes an exponentially distributed length of time to serve a customer. You are next in line waiting to be served. You might hope that the fact you have already been waiting $s = 3$ minutes improves your chances of being served within the next $t = 1$ minute. But, if the teller's service times are truly exponential, it is irrelevant how long you have been waiting.

Suppose a type of device fails only because of "fatal" Poisson events. Then a device that has been in use for s units of time has the same additional life expectancy as a new one (in use for 0 units of time). The no-memory property is often expressed by saying that for such a device "used is as good as new."

Certainly then, an exponential distribution should not be used to model the lifetimes of people. Typically, an 80-year-old person does not have the same additional life expectancy as a 20-year-old. People can be killed by accident, but they also die by "wearing out." In contrast, an exponential distribution may be a good choice to model the lifetime of a computer chip that fails only by accident (such as a hit by a cosmic ray) rather than by wearing out. A computer chip is, after all, only a very complicated piece of sand for which wearing out is not much of an issue. Also, exponential lifetime models apply to some systems that are continually maintained by replacing parts before they wear out, so that failures are due almost exclusively to accidents. Up to a point, it might be possible to maintain an automobile in this way.

In practice, the exponential distribution is often used even when it is only approximately correct. This preference for exponential distributions is sometimes because of the simple mathematical form of the exponential density function, but perhaps more often because of the great convenience of not having to take past history into account.

Example 4.1. Three Bank Scenarios. Two tellers at a bank serve customers from a single queue. The time it takes one teller to serve a randomly selected customer is distributed as $\mathsf{EXP}(1/5)$ so that his average service time is 5 minutes. The other, more experienced teller's service times are distributed as $\mathsf{EXP}(1/4)$, averaging 4 minutes. Tellers work independently.

Scenario 1. Both tellers are busy and you are next in line. What is the probability it will be longer than 5 minutes before you begin to be served? Let X_1 and X_2 be the respective waiting times until the tellers are free to take their next customers. The time until one of them starts serving you is $V = \min(X_1, X_2)$. The following R code simulates $P\{V > 5\}$. The function pmin, for *parallel minimum*, compares two m-vectors elementwise and makes an m-vector of the minimums (see Problem 4.5).

```
# set.seed(1212)
m = 100000;  lam1 = 1/5;  lam2 = 1/4      # constants
x1 = rexp(m, lam1);  x2 = rexp(m, lam2)   # simulation
v = pmin(x1, x2)                          # m-Vector of minimums
mean(v > 5)                               # approximates P{V > 5}
hist(v, prob=T)
xx = seq(0, 15, by=.01);  lines(xx, dexp(xx, 9/20))

> mean(v > 5)
[1] 0.10572
```

This simulated result $P\{V > 5\} = 0.106 \pm 0.002$ is easily verified analytically: $P\{V > 5\} = P\{X_1 > 5,\, X_2 > 5\} = e^{-(1/5)5}e^{-(1/4)5} = e^{-9/4} = 0.1054$. A similar computation gives the following general result. If $X_1 \sim \mathsf{EXP}(\lambda_1)$ and $X_2 \sim \mathsf{EXP}(\lambda_2)$ are independent, then $V = \min(X_1, X_2) \sim \mathsf{EXP}(\lambda_1 + \lambda_2)$. In our case, $V \sim \mathsf{EXP}(9/20)$. The top panel in Figure 4.1 compares the histogram of the simulated distribution of V with the known density function of V.

Scenario 2. In order to finish your banking business, you need to be served by both tellers in sequence. Fortunately, it isn't a busy time at the bank and you don't need to wait to begin service with either teller. What is the probability it will take you more than 5 minutes to finish? With X_1 and X_2 distributed as in Scenario 1, the time it takes you to finish is $T = X_1 + X_2$. The following R code simulates $P\{T > 5\}$.

```
# set.seed(1212)
m = 100000;  lam1 = 1/5;  lam2 = 1/4
x1 = rexp(m, lam1);  x2 = rexp(m, lam2); t = x1 + x2
mean(t > 5); hist(t, prob=T)
```

```
> mean(t > 5)
[1] 0.69371
```

The histogram of the simulated distribution of T is shown in the middle panel of Figure 4.1. The required probability is approximately 0.694 ± 0.003, but here it is not easy to get an exact result by analytic means. (If both tellers served with rate $\lambda = 1/5$, then we would have $T \sim \text{GAMMA}(2, 1/5)$ and the probability would be 0.7358. If both rates were $\lambda = 1/4$, then the probability would be 0.6446. Our result is intermediate, which makes sense intuitively.)

Scenario 3. It is closing time, and no more customers will enter the bank. No one is in line, but both tellers are still busy serving customers. You have a date for coffee with one of the tellers after work. Neither is allowed to leave until both are finished. What is the probability that this takes more than 5 minutes? We need to find $P\{W > 5\}$, where $W = \max(X_1, X_2)$. The R code to simulate this probability and make the bottom histogram in Figure 4.1 is essentially the same as in Scenario 1, except for replacing pmin by pmax.

The maximum of two exponentials is not exponential. However, in this relatively simple case, it is not difficult to find the exact distribution (see Problem 4.7). Intuitively, W cannot be exponential because it clearly does matter how long you have been waiting. The longer you wait, the greater the chance one of the two tellers has finished—and then the prospects of finishing soon become better. The simulated probability from set.seed(1212) is 0.550 ± 0.003 and the exact value is 0.5490. The bottom panel of Figure 4.1 shows the simulated distribution of W along with the density function of W. (See Problem 4.7(d) for the derivation of the density of W). ◇

Situations similar to these three scenarios form the basis of queueing theory—the theory of waiting lines. Using exponential distributions with various rates to model both times between arrivals of customers and service times between their departures, we could construct a realistic system for the flow of customers through a bank or similar business where customers generally arrive at random times (without appointments).

Queueing theory is a field with many theoretical results and analytic approximations for reasonably simple structures, but in which simulation is widely used for more complex ones. Typical questions of interest are the average length of the queue, a customer's average time between arrival and departure, and the proportion of time servers are not busy serving customers. Similar questions arise in the design of telecommunications and broadband networks, in the design of computer hardware, and in understanding some biological systems. An extensive exploration of these questions would lead us too far from the central purpose of this book.

In our next example, we take a brief look at another application of the maximum of exponential distributions. If a certain type of device has too short a lifetime for a particular purpose, one may be able to connect several of them "in parallel" so that a system of these devices continues to function as

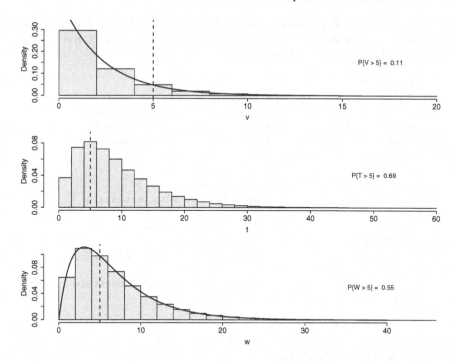

Figure 4.1. Three simulated waiting-time distributions of Example 4.1: minimum (top), sum (center), and maximum. Each is based on two independent random variables distributed as $\mathsf{EXP}(1/4)$ and $\mathsf{EXP}(1/5)$. For each, the probability of a wait exceeding 5 minutes is approximated. Analytic results for the sum are problematic.

long as any one of the component devices survives. The lifetime of the system is then the maximum of the lifetimes of its components.

Example 4.2. Parallel Redundancy in a Communications Satellite. An on-board computer controlling crucial functions of a communications satellite will fail if its CPU is disabled by cosmic radiation. Reliability of the CPU is important because it is not feasible to repair a disabled satellite.

In particular, suppose that the level of radiation is such that fatal events befall such a CPU on average once in 4 years, so that the rate of these events is $\lambda = 1/4$. The random lifetime of such a CPU is $X \sim \mathsf{EXP}(1/4)$ and its **reliability function** is

$$R_X(t) = P\{X > t\} = 1 - F_X(t) = e^{-t/4},$$

for $t > 0$. For example, its probability of surviving longer than 5 years is $R_X(5) = 0.2865$.

For greater reliability, suppose we connect five CPU chips in parallel. Then the lifetime of the resulting CPU system is $W = \max(X_1, X_2, \ldots, X_5)$, where the X_i are independently distributed as $\mathsf{EXP}(1/4)$. We want to evaluate $\mathrm{E}(W)$,

$P\{W > 5\}$, and the 90th percentile of the distribution of W. The R code below can be used to simulate these quantities. Each of the $m = 100\,000$ rows of the matrix DTA simulates one 5-chip system.

```
# set.seed(12)
m = 100000                  # iterations
n = 5;  lam = 1/4           # constants: system parameters
x = rexp(m*n, lam)          # vector of all simulated data
DTA = matrix(x, nrow=m)     # each row a simulated system
w = apply(DTA, 1, max)      # lifetimes of m systems
mean(w)                     # approximates E(W)
mean(w > 5)                 # approximates P{W > 5}
quantile(w, .9)             # approximates 90th percentile

> mean(w)                      # approximates E(W)
[1] 9.128704
> mean(w > 5)                  # approximates P{W > 5}
[1] 0.81456
> quantile(w, .9)              # approximates 90th percentile
      90%
15.51431
```

This run indicates that $E(W) \approx 9.129$, $P\{W > 5\} \approx 0.814$, and the 90th percentile is about 15.51 years.

In this example, we are able to get exact analytic results for the expectation and the probability: $E(W) = 137/15 = 9.1333$ and $P\{W > 5\} = 0.8151$. Also, once we have the approximate value 15.51, it is easy to verify that $P\{W \le 15.51\} = 0.9007$. So our simulation gives useful approximations. (See Problem 4.9 for some analytic results.)

The empirical cumulative distribution function (ECDF) of the simulated distribution of W is shown in Figure 4.2 along with the ECDF (dashed line) of a single component. At all points in time, the parallel system is less likely to have died than is a single component. The ECDFs are made with the *additional* code shown below. We truncated the graph at about 30 years even though the longest lifetime among our 100 000 simulated 5-component systems exceeded 60 years.

```
ecdf = (1:m)/m
w.sort = sort(w)        # sorted lifetimes for system simulated above
y = rexp(m, lam)        # simulation for single CPU
y.sort = sort(y)        # sorted lifetimes for single CPU
plot(w.sort, ecdf, type="l", xlim=c(0,30), xlab="Years")
lines(y.sort, ecdf, lty="dashed")
```

Recall that the ECDF takes a jump of size i/m at a data value, where i is the number of tied observations at that data value. Because the random variables simulated here are continuous, ties are nonexistent in theory and extremely rare in practice, so i is almost always 1. (We could have used the R

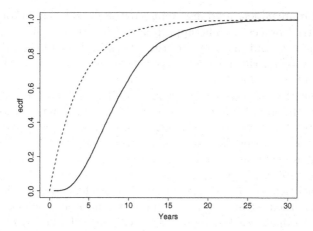

Figure 4.2. ECDFs of two simulated lifetime distributions. The solid plot corresponds to a system with five identical exponential devices in parallel; the dashed plot is for a single device. At each point in time the 5-component system is more reliable (that is, less likely to have failed). See Example 4.2.

expression `plot.ecdf(w)` to make a plot similar to the heavy line in Figure 4.2. It automatically makes values for the vertical axis and sorts the lifetime values, but gives less control over the appearance of the plot.) ◇

4.2 Range of a Normal Sample

Suppose we take a random sample from a population or process distributed as $\mathsf{NORM}(\mu, \sigma)$ to obtain n observations $Y_1, Y_2, \ldots, Y_n$. The sample variance $S^2 = \frac{1}{n-1} \sum_i (Y_i - \bar{Y})^2$ and sample standard deviation S are commonly used measures of the dispersion of the sample. These statistics are also used as estimates of the population variance σ^2 and standard deviation σ, respectively.

For normal data, $(n-1)S^2/\sigma^2 \sim \mathsf{CHISQ}(n-1)$, the chi-squared distribution with $n-1$ degrees of freedom, and also with mean $n-1$. Hence $\mathrm{E}(S^2) = \sigma^2$, and we say that S^2 is an unbiased estimator of σ^2. If L and U cut off probability 0.025 from the lower and upper tails of $\mathsf{CHISQ}(n-1)$, respectively, then a 95% confidence interval for σ^2 is $((n-1)S^2/U, (n-1)S^2/L)$. Take square roots of the endpoints of this interval to get a 95% confidence interval for σ (see Problem 4.15).

In some circumstances, it is customary or preferable to use the range $R = Y_{(n)} - Y_{(1)}$ as a measure of sample dispersion, where subscripts in parentheses indicate the **order statistics** $Y_{(1)} \le Y_{(2)} \le \ldots \le Y_{(n)}$. We multiply R by an appropriate constant K (less than 1) so that $\mathrm{E}(KR) = \sigma$. Thus $R_{\mathrm{unb}} = KR$ is a convenient unbiased estimator of σ. For small samples, such estimates of σ are widely used in making control charts to monitor the stability of industrial processes. The appropriate value of K depends strongly on the sample size n, and many engineering statistics books have tables of constants to be used for this purpose. Tables are necessary because the sampling distribution of R is difficult to handle analytically.

Observing that almost all of the probability of a normal *population* is contained in the interval $\mu \pm 3\sigma$, an interval of length 6σ, authors of some elementary textbooks recommend general use of $K = 1/6$ or $1/5$ to estimate σ (or approximate S) from R. However, this is not a workable idea because no one constant is useful across the wide variety of *sample* sizes encountered in practice (see Problem 4.14).

From statistical theory we know that S^2 must have the smallest variance among all unbiased estimators of σ^2, and so it is natural to wonder how much precision of estimation is lost by basing an estimate of σ^2 on R instead of S^2. In the next example we use a straightforward simulation to obtain information about the distribution of R and the suitability of an interval estimate based on R compared with one based on S^2.

Example 4.3. Let $Y_1, Y_2, \ldots, Y_{10}$ be a random sample from NORM(100, 10). The program below simulates the sample range R and for $m = 100\,000$ such samples in order to learn about the distribution of R.

```
# set.seed(1237)
m = 100000;   n = 10;   mu = 100;   sg = 10
x = rnorm(m*n, mu, sg);   DTA = matrix(x, m)
x.mx = apply(DTA, 1, max);   x.mn = apply(DTA, 1, min)
x.rg = x.mx - x.mn        # vector of m sample ranges
mean(x.rg);   sd(x.rg)
quantile(x.rg, c(.025,.975))
hist(x.rg, prob=T)

> mean(x.rg); sd(x.rg)
[1] 30.80583
[1] 7.960399
> quantile(x.rg, c(.025,.975))
    2.5%    97.5%
16.77872 47.88030
```

From this simulation, we have obtained approximate values $E(R) \approx 30.806$ and $SD(R) \approx 7.960$. These are in reasonably good agreement with exact values $E(R) = 30.8$ and $SD(R) = 7.97$, obtainable by a combination of advanced analytical methods and numerical integration.

In this example, we know the population standard deviation is $\sigma = 10$, so $E(R) = 3.08\sigma$, for $n = 10$. Thus we can estimate σ by $R/3.08 = 0.325R$. Also, based on the simulated quantiles, we expect that in about 95% of samples of size $n = 10$ the sample range R lies between 17 and 48. More generally, because we know $\sigma = 10$ here, we can write

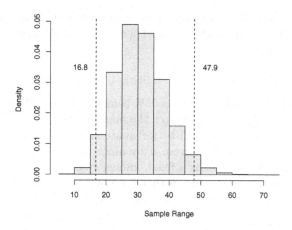

Figure 4.3. Histogram of the 100 000 simulated sample ranges, based on samples of size $n = 10$ from $\mathsf{NORM}(100, 10)$. Skewness becomes more pronounced as n increases. Vertical dashed lines show quantiles 0.025 and 0.975. See Example 4.3.

$$P\{1.7 < R/\sigma < 4.8\} = P\{R/4.8 < \sigma < R/1.7\} = 95\%,$$

so that $(R/4.8, R/1.7)$ is a 95% confidence interval for σ. This interval has expected length

$$\mathrm{E}(L_R) = \mathrm{E}(R)(1/1.7 - 1/4.8) = 3.08\sigma(0.38) = 1.17\sigma.$$

The usual 95% confidence interval for σ, based on $9S^2/\sigma^2 \sim \mathsf{CHISQ}(9)$, is $(0.688S, 1.826S)$. It has average length $\mathrm{E}(L_S) = 1.107\sigma$ (see Problem 4.15). On average, this confidence interval for σ is shorter than one based on R—but not by much when n is as small as 10. As n increases, estimates of σ based on R become less useful.

The histogram in Figure 4.3 shows that the distribution of R is slightly skewed to the right and so it is not normal. Because R is based on extremes, as n increases its distribution becomes even less like the normal. In contrast, the Central Limit Theorem says that the distribution of S^2, which is based on sums (of squares), converges to normal. Intuitively, it is clear that R must behave badly for very large samples from a normal distribution. A normal distribution has tails extending out towards $\pm\infty$. If we take a large enough sample, we are bound to get some very extreme values. ◊

4.3 Joint Distribution of the Sample Mean and Standard Deviation

Suppose random observations $Y_1, Y_2, \ldots, Y_n$ from a laboratory instrument have the distribution $N(\mu, \sigma_0)$. For practical purposes, μ is an unknown value determined by the specimen being analyzed and σ_0 is a known value that is a property of the instrument. Then $\bar{Y} \sim \mathsf{NORM}(\mu, \sigma_0/\sqrt{n})$, so that $Z = \frac{\bar{Y}-\mu}{\sigma_0/\sqrt{n}} \sim \mathsf{NORM}(0,1)$ and

$$P\{-1.96 \le Z \le 1.96\} = P\{\bar{Y} - 1.96\sigma_0/\sqrt{n} \le \mu \le \bar{Y} + 1.96\sigma_0/\sqrt{n}\}.$$

Thus we say that $\bar{Y} \pm 1.96\sigma_0/\sqrt{n}$ is a 95% confidence interval for μ. We call $\bar{Y}$ the point estimate of μ and $1.96\sigma_0/\sqrt{n}$ the margin of error. Intuitively, in 95% of such samples of size n, we will be fortunate enough to get data yielding a value of $\bar{Y}$ that does not differ from μ by more than the margin of error. In this case, the margin of error is a fixed, known value.

A more common circumstance is that both of the normal parameters μ and σ are unknown, and we estimate μ by $\bar{Y}$ and σ by $S = [\sum_i (Y_i - \bar{Y})^2/(n-1)]^{1/2}$. In this case, $T = \frac{\bar{Y}-\mu}{S/\sqrt{n}} \sim \mathsf{T}(n-1)$, Student's t distribution with $n-1$ degrees of freedom. If t^* cuts off 2.5% of the area from the upper tail of $\mathsf{T}(n-1)$, then

$$P\{-t^* < T < t^*\} = P\{\bar{Y} - t^*S/\sqrt{n} \le \mu \le \bar{Y} + t^*S/\sqrt{n}\}.$$

We say that $\bar{Y} \pm t^*S/\sqrt{n}$ is a 95% confidence interval for μ. An important distinction from the case where σ is known is that we now use S to estimate σ, so both the denominator of T and the margin of error are now random variables.

For normal data, $\bar{Y}$ and S are independent random variables. For many students, this fact does not match intuition—perhaps not least because $\bar{Y}$ appears explicitly in the definition of S. The independence of $\bar{Y}$ and S can be proved rigorously in several ways (for example, using linear algebra or moment generating functions), but these proofs have an abstract flavor that may be satisfying in a logical sense while still not overcoming intuitive misgivings.

In practice, it is not foolish to be suspicious about the independence of $\bar{Y}$ and S. This independence holds *only* for normal data. However, most applied statisticians know that precisely normal data are rare in practice. The simulation in the next example demonstrates the independence of $\bar{Y}$ and S for normal data. Later in this section, we consider what happens when the Y_i are not normally distributed. (Some of the examples are from [TSF01].)

Example 4.4. Independence of the Sample Mean and Standard Deviation for Normal Data. To be specific, consider a random sample of size $n = 5$ from $\mathsf{NORM}(200, 10)$. We simulate $m = 100\,000$ such samples and find $\bar{Y}$ and S for each. Thus we have m observations from a bivariate distribution. If $\bar{Y}$ and S really are independent, then the population correlation $\rho_{\bar{Y},S} = 0$, and we should find that the sample correlation $r_{\bar{Y},S} \approx 0$. Of course, it is possible that the random variables $\bar{Y}$ and S are associated (not independent) and yet have $\rho = 0$. But verifying that our simulated data have $r \approx 0$ seems a good first step in investigating independence.

A second step is to make a scatterplot of $\bar{Y}$ and S, looking for any patterns of association. In particular, going more directly to the definition of independence of two random variables, we can estimate some probabilities from our simulated distribution to see whether

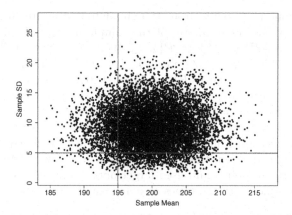

Figure 4.4. Scatterplot of the bivariate distribution of the sample mean and standard deviation. Each of the $10\,000$ points represents a simulated sample of five observations from $\mathrm{NORM}(200, 10)$. The two variables $\bar{Y}$ and S are independent. See Example 4.4.

$$P\{\bar{Y} \le a\}\, P\{S \le b\} = P\{\bar{Y} \le a,\, S \le b\}$$

is a believable equality. Clearly, we cannot investigate all possibilities (as would be required for an ironclad proof of independence), but seeing that this equality holds for several choices of a and b lends credence to independence.

Note: In the R code below, we use `rowSums` to find the vector of m standard deviations. For us, this ran faster (by about an order of magnitude) than code with `apply(DTA, 1, sd)` or `sd(t(DTA))`.

```
# set.seed(37)
m = 100000;  n = 5;   mu = 200;   sg = 10
y = rnorm(m*n, mu, sg);  DTA = matrix(y, nrow=m)
samp.mn = rowMeans(DTA)
samp.sd = sqrt(rowSums((DTA - samp.mn)^2)/(n-1))
cor(samp.mn, samp.sd)
A = mean(samp.mn <= 195); B = mean(samp.sd <= 5);  A;  B;  A*B
mean(samp.mn <= 195 & samp.sd <= 5)
plot(samp.mn, samp.sd, pch=".")

> cor(samp.mn, samp.sd)
[[1] 0.003977406
> A = mean(samp.mn <= 195);  B = mean(samp.sd <= 5);  A;  B;  A*B
[1] 0.13014
[1] 0.08922
[1] 0.01161109
> mean(samp.mn <= 195 & samp.sd <= 5)
[1] 0.01155
```

All the numerical results from the simulated bivariate distribution are consistent with the independence of the sample mean and variance: $\rho \approx 0$ and $P\{\bar{Y} \leq 195\}\, P\{S \leq 5\} = P\{\bar{Y} \leq 195,\ S \leq 5\} \approx 0.012$. For exact values, `pnorm(195, 200, 10/sqrt(5))` gives $P\{\bar{Y} \leq 195\} = 0.1318$, and `pchisq(4*5^2/10^2, 4)` gives $P\{S \leq 5\} = 0.0902$. Thus the simulated values for these probabilities are correct to two places.

Figure 4.4 shows a scatterplot of the simulated bivariate distribution, in which we see no evidence of association. Vertical and horizontal reference lines bound regions corresponding to the probabilities just discussed. For clarity in print, the figure is different than you will see on your screen. We have plotted only the first 10 000 points and used larger dots. From the code, you will get a more detailed plot with ten times as many smaller dots. ◇

In estimating the parameters μ and σ of a normal population from a random sample of size n, statistical theory shows that the two statistics $\bar{Y}$ and S are **sufficient statistics** for estimating μ and σ; that is, they contain all of the relevant information for estimation. Statisticians often view the n observations of a sample as a vector in n-dimensional space. Thus, by using the bivariate distribution of the sufficient statistics $\bar{Y}$ and S, we have reduced the dimensionality of the estimation problem from n to 2.

When we construct the 95% confidence interval $\bar{Y} \pm t^* S/\sqrt{n}$ for μ, we know that about 95% of the points in a plot such as Figure 4.4 will lead to a confidence interval that covers (includes) the value of μ and that about 5% will not. One might guess that the points that yield covering confidence intervals must be the ones nearest a vertical line through μ. The next example shows that this is an incomplete guess.

Example 4.5. When Does a t Interval Cover the Population Mean? Again, in this example we consider random samples of size $n = 5$ from NORM(200, 10). The points $(\bar{Y}, S)$ for which $\bar{Y} \pm t^* S/\sqrt{n}$ covers μ satisfy the inequality

$$|T| = \left| \frac{\bar{Y} - \mu}{S/\sqrt{n}} \right| < t^*.$$

We use R code based on the program in Example 4.4 to make a scatterplot in Figure 4.5 of $m = 10\,000$ points to represent the joint distribution of $\bar{Y}$ and S. We find vectors `samp.mn` and `samp.sd` as before. In the code below, we first plot all points with light-colored dots and then plot in black the points corresponding to confidence intervals that cover μ.

```
plot(samp.mn, samp.sd, pch=".", col="darkgrey")
t.crit = qt(0.975, n-1)
t = sqrt(n)*(samp.mn - mu)/samp.sd
cover = (1:m)[abs(t) < t.crit]
points(samp.mn[cover], samp.sd[cover], pch=".")
xx = seq(min(samp.mn),max(samp.mn),length=1000)
ss = abs(sqrt(n)*(xx-mu)/t.crit)
lines(xx, ss, lwd=2)   # boundary lines
```

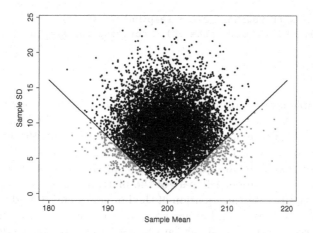

Figure 4.5. Scatterplot of the joint distribution of $\bar{Y}$ and S, showing the points (black) for which a t confidence interval covers the population mean. Especially for small sample sizes (here $n = 5$), the value of the sample standard deviation greatly influences whether the interval covers the mean. See Example 4.5.

Figure 4.5 shows that when a t confidence interval covers μ it is because of a combination of a sample mean sufficiently near μ and a sufficiently large sample standard deviation. Especially for sample sizes as small as $n = 5$, the effect of S is impressive. If S is too small, the confidence interval may be too short to cover μ even if $\bar{Y}$ is very near μ. However, if S is very large, then the interval may cover μ even if $\bar{Y}$ lies quite far from μ—and the interval may be so long that it is of little practical use. As the sample size n increases, the angle between the boundary lines becomes more acute. ◇

Because $\bar{Y}$ and S are independent only for normal data, it seems worthwhile to investigate the patterns of association between these two statistics when data are not normal. If $\bar{Y}$ and S are correlated, maybe the correlation is strong enough to see in a simulation. If they are not correlated, then maybe their scatterplot shows some obvious patterns of association. The next two examples briefly explore these approaches.

Example 4.6. Exponential Data. For exponentially distributed data, it is reasonable to anticipate positive association of the sample mean and standard deviation. Exponential data cannot be negative. Hence, as the sample mean increases, so does the opportunity for the standard deviation to increase. The following R code, very similar to that of Example 4.4, verifies this hunch and illustrates the pattern of positive association between $\bar{Y}$ and S.

```
# set.seed(12)
m = 100000;  n = 5;  lam = 2
DTA = matrix(rexp(m*n, lam), nrow=m)
samp.mn = rowMeans(DTA)
```

```
samp.sd = sqrt(rowSums((DTA - samp.mn)^2)/(n-1))
plot(samp.mn, samp.sd, pch=".");   cor(samp.mn, samp.sd)

> cor(samp.mn, samp.sd)
[1] 0.7744946
```

The sample correlation $r \approx 0.77$ clearly indicates the population correlation $\rho > 0$. Because $\bar{Y}$ and S are correlated, they cannot be independent.

Again in this example we print a plot that shows only 10 000 points (see Figure 4.10, p112), and you should plot all 100 000 for yourself. From this plot it appears there may be an upper boundary for the data cloud because there are no points in the upper-left part of the plotting region. We leave it as an exercise for you to find the equation of the boundary and to find probabilities showing directly that $\bar{Y}$ and S are not independent (see Problem 4.20). To avoid confusion, we stress that the data $Y_1, Y_2, \ldots, Y_5$ are independent because they are sampled at random; it is the statistics $\bar{Y}$ and S computed from the data that are not independent. $\diamond$

Example 4.7. Beta Data. In this example, we look at data simulated from BETA(0.1, 0.1), a symmetrical distribution on the interval $(0, 1)$, in which much of the probability is concentrated near the endpoints of that interval. Because of the symmetry, we might guess $\bar{Y}$ and S are uncorrelated. So we wonder what pattern of association we might detect in the scatterplot.

```
# set.seed(12)
m = 100000;  n = 5;   alpha = beta = 0.1
DTA = matrix(rbeta(m*n, alpha, beta), nrow=m)
samp.mn = rowMeans(DTA)
samp.sd = sqrt(rowSums((DTA - samp.mn)^2)/(n-1))
cor(samp.mn, samp.sd)
plot(samp.mn, samp.sd, pch=".")

> cor(samp.mn, samp.sd)
[1] 0.004622875
```

Here, as we anticipated, the sample correlation is consistent with $\rho = 0$. However, zero correlation does not imply independence, as is obvious from Figure 4.6 (again with only 10 000 of the 100 000 points). See Problem 4.22 for further exploration of the unusual shape of this bivariate distribution. $\diamond$

The examples in this section (and the associated problems) illustrate the extraordinarily important role of multivariate graphical methods in assessing whether variables in a multivariate distribution are independent or associated and the patterns of association that might appear when variables are not independent. Looking at sample correlations and univariate plots does not always tell the whole story.

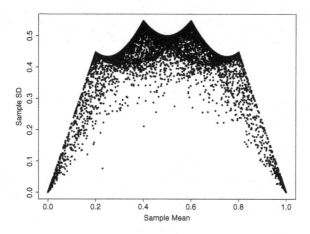

Figure 4.6. Joint distribution of $\bar{Y}$ and S for data from BETA$(0.1, 0.1)$, a symmetrical distribution on $(0, 1)$ that concentrates probability near 0 and 1. See Example 4.7.

4.4 Nonparametric Bootstrap Distributions

In making estimates of the parameters μ and σ for random observations from a normal distribution, we have noted that no information is lost if we summarize the data by using only the sample mean and sample standard deviation. That is the basis for the t confidence interval discussed in the previous section. If the distribution of the population is "close" to normal, then methods based on the t distribution are often still useful. However, to the degree that the population is not normal—especially if it is markedly skewed—these methods may give misleading results.

Of course, one might not know whether the population is nearly normal. In practice, one usually has to decide, based on evidence in the sample, whether t methods might be misleading. Traditionally, when there is strong evidence against normality in a sample, it has been common to use various methods based on ranks instead of t methods. Using ranks instead of actual data values causes some information to be lost. Traditionally, procedures that do not assume the data are normally distributed (or have any other specific shape) are called **nonparametric**.

In 1978, Bradley Efron proposed a computationally intensive type of interval estimation called the **bootstrap**, which uses all of the information in the sample to make an interval estimate. A nonparametric bootstrap does not require making the assumption that the population has a normal distribution. The purpose of this section is to give a very brief introduction to the computation and use of nonparametric bootstrap confidence intervals.

Example 4.8. Lead Poisoning. An important industry in a small town is the manufacture of car batteries, which contain lead. Especially for children, lead

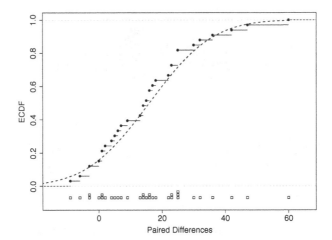

Figure 4.7. Strip chart and empirical cumulative distribution function (with best-fitting normal cumulative distribution function, dotted curve) of the paired differences for the lead poisoning data of Example 4.8.

is a serious neurotoxin. So workers who handle lead at the local battery factory are cautioned to shower, shampoo, and change clothes before leaving work in order to prevent carrying lead dust home and contaminating their children.

In a study to determine whether workers do carry lead dust home to bad effect, 33 of their children are selected as subjects. Children in this group are considered "at risk" for lead contamination. A blood test for each child determines the level of lead (in μg/dl) in his or her blood.

Because there are other possible environmental sources of lead contamination in the community, a control group of children is selected. For each at-risk child, another child, matched for age and neighborhood of residence and with no parental connection to the battery factory, is found and tested similarly. Thus we have 33 matched pairs of children. In each pair, one is at risk and the other serves as a control.

Preparatory to analysis in R, we enter the blood lead levels into two vectors, `Risk` and `Ctrl`, and find the differences for each matched pair. A simple and direct way to do this is as follows:

```
Risk = c(38, 23, 41, 18, 37, 36, 23, 62, 31, 34, 24,
         14, 21, 17, 16, 20, 15, 10, 45, 39, 22, 35,
         49, 48, 44, 35, 43, 39, 34, 13, 73, 25, 27)
Ctrl = c(16, 18, 18, 24, 19, 11, 10, 15, 16, 18, 18,
         13, 19, 10, 16, 16, 24, 13,  9, 14, 21, 19,
          7, 18, 19, 12, 11, 22, 25, 16, 13, 11, 13)
Pair.Diff = Risk - Ctrl
```

Because these are paired data, most of our analysis will be based on `Pair.Diff`, the differences in lead levels between the children with fathers working at the battery factory and their corresponding controls. However, to begin, we note that eight of the 33 children in the at-risk group have levels of lead above 40 μg/dl, a level that calls for medical treatment. (Moreover, according to the usual guidelines, the two children with levels above 60 μg/dl should receive emergency treatment.) In contrast, the highest level in the control group is 25. In R, sum(Risk >= 40) returns 8 and max(Ctrl) returns 25.

In this study, the at-risk subjects come close to being the entire population, so conclusions from the data cannot be rigorously extended beyond the situation at this one factory. But for purposes of illustration, we treat the paired data as if they were chosen at random from a theoretical population, and we seek a 95% confidence interval for the population mean μ of paired differences, as a way of assessing the amount of contamination a hypothetical randomly chosen child may acquire by having a father who works at the battery factory. Assuming for the moment that this population is nearly normal, we find the familiar t interval $\bar{d} \pm t^* S_d/\sqrt{n}$, where $\bar{d} = 15.97$ and $S_d = 15.86$ are the mean and standard deviation, respectively, of the sample of $n = 33$ differences, and $t^* = 2.04$ cuts off 2.5% from the upper tail of the distribution T(32). The resulting confidence interval is $(10.3, 21.6)$. The R code t.test(Pair.Diff) can be used for this computation.

A normal probability plot of the paired differences seems to fit a curve better than a straight line, suggesting that the population may not be exactly normal (see Problem 4.23 and Figure 4.11, p113). However, this relatively small dataset passes some standard tests for normality. Figure 4.7 shows a strip chart of the paired differences and also their empirical cumulative distribution function (ECDF) along with the cumulative distribution function of NORM$(\bar{d}, S_d)$. The R code for this figure is shown below.

```
plot(ecdf(Pair.Diff), pch=19, ylim=c(-.1,1))
xx = seq(-9,60, by = .01)
lines(xx, pnorm(xx, mean(Pair.Diff), sd(Pair.Diff)), lty="dashed")
stripchart(Pair.Diff, meth="stack", add=T, at=-.07, offset=1/2)
```

For any sample of size $n = 33$ that does not overwhelmingly fail a normality test, a t interval would probably give useful results because t intervals are robust against nonnormality. Nevertheless, we use these paired differences to illustrate a nonparametric bootstrap confidence interval.

Such a confidence interval is based on the fact that the ECDF contains all of the information we have about the population from which the data were sampled. Moreover, for a sufficiently large sample size, the ECDF is a reasonable approximation to the population cumulative distribution function. The idea behind the bootstrap is to treat the data as a substitute population and to take a large number B of resamples of size n with replacement from this substitute population.

For each resample, we compute the mean $\bar{d}_j^*$ for $j = 1, \ldots, B$. By taking many resamples, we can get a good notion of the variability of sample means selected from a population somewhat similar to the population of interest.

A simple nonparametric 95% confidence interval (L^*, U^*) for μ is found by cutting off 2.5% from each tail of the distribution consisting of B resampled means $\bar{d}_j^*$.

The following R code implements this procedure for the sample Pair.Diff provided above. Notice that here we use B, instead of the usual m, to indicate the number of iterations. Each of the B rows of the matrix RDTA is a resample.

```
# set.seed(1237)
n = length(Pair.Diff)                      # number of data pairs
d.bar = mean(Pair.Diff)                    # observed mean of diff's
B = 10000                                  # number of resamples
re.x = sample(Pair.Diff, B*n, repl=T)
RDTA = matrix(re.x, nrow=B)                # B x n matrix of resamples
re.mean = rowMeans(RDTA)                   # vector of B 'd-bar-star's
hist(re.mean, prob=T)                      # hist. of bootstrap dist.
bci = quantile(re.mean, c(.025, .975))     # simple bootstrap CI
alt.bci = 2*d.bar - bci[2:1]               # bootstrap percentile CI
bci; alt.bci

> bci; alt.bci
      2.5%     97.5%
  10.78788  21.45530
     97.5%      2.5%
  10.48409  21.15152
```

Figure 4.8 shows the histogram of the bootstrap distribution of resampled means $\bar{d}_j^*$ and indicates the corresponding **simple bootstrap confidence interval** $(10.8, 21.5)$. Because of the simulation, this interval is slightly different on each run of the program.

> Note: The more sophisticated alternate bootstrap confidence interval provided at the end of the program above is especially preferred if the histogram of resampled values $\bar{d}_j^*$ shows noticeable skewness. If we knew the .025 and .975 quantiles L and U, respectively, of the distribution of $\bar{d} - \mu$, then a 95% confidence interval $(\bar{d} - U, \bar{d} - L)$ for μ would arise from $P\{L \leq \bar{d} - \mu \leq U\} = P\{\bar{d} - U \leq \mu \leq \bar{d} - L\} = 0.95$. To get substitute values $\hat{L}$ of L and $\hat{U}$ of U, we view the distribution of $\bar{d}_j^* - \bar{d}$ as a substitute for the distribution of $\bar{d} - \mu$. Subtracting $\bar{d}$ from the bootstrap distribution of $\bar{d}_j^*$, we obtain $\hat{L} = L^* - \bar{d}$ and thus $\bar{d} - \hat{L} = \bar{d} - (L^* - \bar{d}) = 2\bar{d} - L^*$ as the upper bootstrap confidence limit and, similarly, $\bar{d} - \hat{U} = 2\bar{d} - U^*$ as the lower limit. This is often called the **bootstrap percentile method**.

By any reasonable method, it is clear that the children in the at-risk group have higher levels of lead than their peers in the control group. On average, the additional amount of lead associated with having a father who works at the battery factory is likely between a quarter and a half of the amount that is considered serious enough to warrant medical treatment. (The data are from [Mor82], where additional evidence is given to suggest that the highest levels

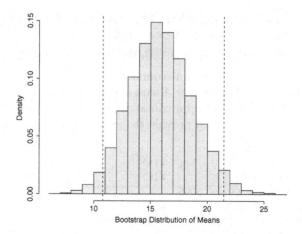

Figure 4.8. Histogram of the bootstrap distribution of means of paired differences in Example 4.8. The vertical dashed lines indicate the resulting nonparametric bootstrap confidence interval $(10.8, 21.5)$.

of contamination are found among children whose fathers have the highest on-job exposure to lead and the poorest adherence to end-of-day hygiene rules. These data have also been discussed in [Rosn93] and [Tru02].) $\diamondsuit$

Nonparametric bootstrap methods are useful in a vast variety of more complex situations where specific distributional assumptions are not warranted. Often they are the most effective way to use all of the information in the data when making inferences about a population. However, it is important to understand that the resampling involved in bootstrapping cannot create new information, nor make the original sample more representative of the population from which it was drawn.

4.5 Problems

Problems for Section 4.1 (Exponential-Based Models)

4.1   Let $N_t \sim \mathsf{POIS}(\lambda t)$ and $X \sim \mathsf{EXP}(\lambda)$, where $\lambda = 3$. Use R to evaluate $P\{N_1 = 0\}$, $P\{N_2 > 0\}$, $P\{X > 1\}$, and $P\{X \le 1/2\}$. Repeat for $\lambda = 1/3$.

Note: Equation (4.1) says that the first and third probabilities are equal.

4.2   Analytic results for $X \sim \mathsf{EXP}(\lambda)$. (Similar to problems in Chapter 2.)

a) Find η such that $P\{X \le \eta\} = 1/2$. Thus η is the median of X.
b) Show that $\mathrm{E}(X) = \int_0^\infty t f_X(t)\,dt = \int_0^\infty \lambda t e^{-\lambda t}\,dt = 1/\lambda$.
c) Similarly, show that $\mathrm{V}(X) = \mathrm{E}(X^2) - [\mathrm{E}(X)]^2 = 1/\lambda^2$, and $\mathrm{SD}(X) = 1/\lambda$.

4.3   Explain each step in equation (4.2). For $X \sim \text{EXP}(\lambda)$ and $r, s, t > 0$, why is $P\{X > r + t | X > r\} = P\{X > s + t | X > s\}$?

Hints: $P(A|B) = P(A \cap B)/P(B)$. If $A \subset B$, then what is $P(A \cap B)$?

4.4   In the R code below, each line commented with a letter (a)–(h) returns an approximate result related to the discussion at the beginning of Section 4.1. For each, say what method of approximation is used, explain why the result may not be exactly correct, and provide the exact value being approximated.

```
lam = .5;   i = 0:100
sum(dpois(i, lam))                      #(a)
sum(i*dpois(i, lam))                    #(b)
g = seq(0,1000,by=.001)-.0005
sum(dexp(g, lam))/1000                  #(c)
sum(g*dexp(g, lam))/1000                #(d)
x = rexp(1000000, lam)
mean(x)                                 #(e)
sd(x)                                   #(f)
mean(x > .1)                            #(g)
y = x[x > .05];   length(y)
mean(y > .15)                           #(h)
```

Hints: Several methods from Chapter 3 are used. Integrals over $(0, \infty)$ are approximated. For what values of s and t is the no-memory property illustrated?

4.5   Four statements in the R code below yield output. Which ones? Which two statements give the same results? Why? Explain what the other two statements compute. Make the obvious modifications for maximums, try to predict the results, and verify.

```
x1 = c(1, 2, 3, 4, 5, 0, 2);   x2 = c(5, 4, 3, 2, 1, 3, 7)
min(x1, x2);   pmin(x1, x2)
MAT = cbind(x1, x2);   apply(MAT, 1, min);   apply(MAT, 2, min)
```

4.6   *More simulations related to Example 4.1: Bank Scenarios 1 and 2.*

a) In Scenario 1, simulate the probability that you will be served by the female teller using `mean(x2==v)`. Explain the code. [Exact value is 5/9.]

b) In Scenario 1, simulate the expected waiting time using `mean(v)`, and compare it with the exact value, 20/9.

c) Now suppose there is only one teller with service rate $\lambda = 1/5$. You are next in line to be served. Approximate by simulation the probability it will take you more than 5 minutes to *finish* being served. This is the same as one of the probabilities mentioned under Scenario 2. Which one? What is the exact value of the probability you approximated? Discuss.

4.7   *Some analytic results for Example 4.1: Bank Scenario 3.*

a) Argue that $F_W(t) = (1 - e^{-t/5})(1 - e^{-t/4})$, for $t > 0$, is the cumulative distribution function of W.

b) Use the result of part (a) to verify the exact value $P\{W > 5\} = 0.5490$ given in Scenario 3.
c) Modify the program of Scenario 1 to approximate $E(W)$.
d) Use the result of part (a) to find the density function $f_W(t)$ of W, and hence find the exact value of $E(W) = \int_0^\infty t f_W(t)\, dt$.

4.8   Modify the R code in Example 4.2 to explore a parallel system of four CPUs, each with failure rate $\lambda = 1/5$. The components are more reliable here, but fewer of them are connected in parallel. Compare the ECDF of this system with the one in Example 4.2. Is one system clearly better than the other? (Defend your answer.) In each case, what is the probability of surviving for more than 12 years?

4.9   *Some analytic solutions for Example 4.2: Parallel Systems.*

a) A parallel system has n independent components, each with lifetime distributed as $\mathsf{EXP}(\lambda)$. Show that the cumulative distribution of the lifetime of this system is $F_W(t) = (1 - e^{-\lambda t})^n$. For $n = 5$ and $\lambda = 1/4$, use this result to show that $P\{W > 5\} = 0.8151$ as indicated in the example. Also evaluate $P\{W \le 15.51\}$.
b) How accurately does the ECDF in Example 4.2 approximate the cumulative distribution function F_W in part (a)? Use the same `plot` statement as in the example, but with parameters `lwd=3` and `col="green"`, so that the ECDF is a wide green line. Then overlay the plot of F_W with

    tt = seq(0, 30, by=.01); cdf = (1-exp(-lam*tt))^5; lines(tt, cdf)

and comment.
c) Generalize the result for F_W in part (a) so that the lifetime of the ith component is distributed as $\mathsf{EXP}(\lambda_i)$, where the λ_i need not be equal.
d) One could find $E(W)$ by taking the derivative of F_W in part (a) to get the density function f_W and then evaluating $\int_0^\infty t f_W(t)\, dt$, but this is a messy task. However, in the case where all components have the same failure rate λ, we can find $E(W)$ using the following argument, which is based on the no-memory property of exponential distributions.

   Start with the expected wait for the first component to fail. That is, the expected value of the minimum of n components. The distribution is $\mathsf{EXP}(n\lambda)$ with mean $1/\lambda n$. Then start afresh with the remaining $n - 1$ components, and conclude that the mean additional time until the second failure is $1/\lambda(n - 1)$. Continue in this fashion to show that the R code `sum(1/(lam*(n:1)))` gives the expected lifetime of the system. For a five-component system with $\lambda = 1/4$, as in Example 4.2, show that this result gives $E(W) = 9.1333$.

Note: The argument in (d) depends on symmetry, so it doesn't work in the case where components have different failure rates, as in part (c).

4.10   When a firm receives an invitation to bid on a contract, a bid cannot be made until it has been reviewed by four divisions: Engineering, Personnel,

Legal, and Accounting. These divisions start work at the same time, but they work independently and in different ways. Times from receipt of the offer to completion of review by the four divisions are as follows. Engineering: exponential with mean 3 weeks; Personnel: normal with mean 4 weeks and standard deviation 1 week; Legal: either 2 or 4 weeks, each with probability 1/2; Accounting: uniform on the interval 1 to 5 weeks.

a) What is the mean length of time W before all four divisions finish their reviews? Bids not submitted within 6 weeks are often rejected. What is the probability that it takes more than 6 weeks for all these reviews to be finished? **Write a program** to answer these questions by simulation. Include a histogram that approximates the distribution of the time before all divisions finish. Below is suggested R code for making a matrix DTA; proceed from there, using the code of Example 4.2 as a guide.

```
Eng = rexp(m, 1/3)
Per = rnorm(m, 4, 1)
Leg = 2*rbinom(m, 1, .5) + 2
Acc = runif(m, 1, 5)
DTA = cbind(Eng, Per, Leg, Acc)
```

b) Which division is most often the last to complete its review? If that division could decrease its mean review time by 1 week, by simply subtracting 1 from the values in part (a), what would be the improvement in the 6-week probability value?

c) How do the answers in part (a) change if the uniformly distributed time for Accounting starts precisely when Engineering is finished? Use the original distributions given in part (a).

Hints and answers: (a) Rounded results from one run with $m = 10\,000$; give more accurate answers: 5.1, 0.15. (b) The code **mean(w==Eng)** gives the proportion of the time Engineering is last to finish. Greatest proportion is 0.46. (c) Very little. Why? (Ignore the tiny chance that a normal random variable might be negative.)

4.11   Explain the similarities and differences among the five matrices produced by the R code below. What determines the dimensions of a matrix made from a vector with the **matrix** function? What determines the order in which elements of the vector are inserted into the matrix? What happens when the number of elements of the matrix exceeds the number of elements of the vector? Focus particular attention on MAT3, which illustrates a method we use in Problem 4.12.

```
a1 = 3;   a2 = 1:5;   a3 = 1:30
MAT1 = matrix(a1, nrow=6, ncol=5); MAT1
MAT2 = matrix(a2, nrow=6, ncol=5); MAT2
MAT3 = matrix(a2, nrow=6, ncol=5, byrow=T); MAT3
MAT4 = matrix(a3, 6); MAT4
MAT5 = matrix(a3, 6, byrow=T); MAT5
```

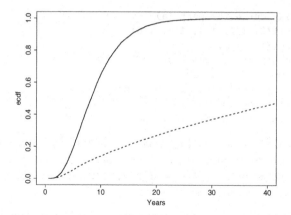

Figure 4.9. ECDFs of simulated lifetime distributions of the homogeneous parallel system of Example 4.2 (solid curve) and one much more reliable heterogeneous parallel system from reallocating foil as suggested in Problem 4.12(b).

4.12   In Example 4.2, each of the five component CPUs in the parallel system has failure rate $\lambda = 1/4$ because it is covered by a thickness of lead foil that cuts deadly radiation by half. That is, without the foil, the failure rate would be $\lambda = 1/2$. Because the foil is heavy, we can't afford to increase the total amount of foil used. Here we explore how the lifetime distribution of the system would be affected if we used the same amount of foil differently.

a) Take the foil from one of the CPUs (the rate goes to 1/2) and use it to double-shield another CPU (rate goes to 1/8). Thus the failure rates for the five CPUs are given in a 5-vector `lam` as shown in the simulation program below. Compare the mean and median lifetimes, probability of survival longer than 10 years, and ECDF curve of this heterogeneous system with similar results for the homogeneous system of Example 4.2. Notice that in order for each column of the matrix to have the same rate down all rows, it is necessary to fill the matrix by rows using the argument (`byrow=T`). Thus the vector of five rates "recycles" to provide the correct rate for each element in the matrix. (See Problem 4.11 for an illustrative exercise.)

```
# Curve for original example
m = 100000
n = 5;  lam.e = 1/4
x = rexp(m*n, lam.e)
DTA = matrix(x, nrow=m)
w.e = apply(DTA, 1, max)
mean(w.e);  quantile(w.e, .5);  mean(w.e > 10)
ecdf = (1:m)/m;  w.e.sort = sort(w.e)
plot(w.e.sort, ecdf, type="l", xlim=c(0,40), xlab="Years")
```

```
# Overlay curve for part (a)
lam = c(1/2, 1/4, 1/4, 1/4, 1/8)
x = rexp(m*n, lam)
DTA = matrix(x, nrow=m, byrow=T)
w.a = apply(DTA, 1, max)
mean(w.a);  quantile(w.a, .5);  mean(w.a > 10)
w.a.sort = sort(w.a)
lines(w.a.sort, ecdf, lwd=2, col="darkblue", lty="dashed")
```

b) Denote the pattern of shielding in part (a) as 01112. Experiment with other patterns with digits summing to 5, such as 00122, 00023, and so on. The pattern 00023 would have `lam = c(1/2, 1/2, 1/2, 1/8, 1/16)`. Which of your patterns seems best? Discuss.

Notes: The ECDF of one very promising reallocation of foil in part (b) is shown in Figure 4.9. Parallel redundancy is helpful, but "it's hard to beat" components with lower failure rates. In addition to the kind of radiation against which the lead foil protects, other hazards may cause CPUs to fail. Also, because of geometric issues, the amount of foil actually required for, say, triple shielding may be noticeably more than three times the amount for single shielding. Because your answer to part (b) does not take such factors into account, it might not be optimal in practice.

Problems for Section 4.2 (Range of Normal Data)

4.13   Repeat Example 4.3 for $n = 5$. For what value K is $R_{unb} = KR$ an unbiased estimate of σ. Assuming that $\sigma = 10$, what is the average length of a 95% confidence interval for σ based on the sample range?

4.14   Modify the code of Example 4.3 to try "round-numbered" values of n such as $n = 30, 50, 100, 200$, and 500. Roughly speaking, for what sample sizes are the constants $K = 1/4, 1/5$, and $1/6$ appropriate to make $R_{unb} = KR$ an unbiased estimator of σ? (Depending on your patience and your computer, you may want to use only $m = 10\,000$ iterations for larger values of n.)

4.15   This problem involves exploration of the sample standard deviation S as an estimate of σ. Use $n = 10$.

a) Modify the program of Example 4.3 to simulate the distribution of S. Use `x.sd = apply(DTA, 1, sd)`. Although $E(S^2) = \sigma^2$, equality (that is, unbiasedness) does not survive the nonlinear operation of taking the square root. What value a makes $S_{unb} = aS$ an unbiased estimator of σ?
b) Verify the value of $E(L_S)$ given in Example 4.3. To find the confidence limits of a 95% confidence interval for S, use `qchisq(c(.025,.975), 9)` and then use $E(S)$ in evaluating $E(L_S)$. Explain each step.
c) Statistical theory says that $V(S_{unb})$ in part (a) has the smallest possible variance among unbiased estimators of σ. Use simulation to show that $V(R_{unb}) \geq V(S_{unb})$.

Notes: $E(S_{10}) = 9.727$. For $n \geq 2$, $E(S_n) = \sigma\sqrt{\frac{2}{n-1}}\Gamma(\frac{n}{2})/\Gamma(\frac{n-1}{2})$. The Γ-function can be evaluated in R with gamma(). As n increases, the bias of S_n in estimating σ disappears. By Stirling's approximation of the Γ-function, $\lim_{n\to\infty} E(S_n) = \sigma$.

4.16   For a sample of size 2, show that the sample range is precisely a multiple of the sample standard deviation. [Hint: In the definition of S^2, express $\bar{X}$ as $(X_1 + X_2)/2$.] Consequently, for $n = 2$, the unbiased estimators of σ based on S and R are identical.

4.17   *(Intermediate)* The shape of a distribution dictates the "best" estimators for its parameters. Suppose we have a random sample of size n from a population with the uniform distribution $\mathsf{UNIF}(\mu - \sqrt{3}\sigma, \mu + \sqrt{3}\sigma)$, which has mean μ and standard deviation σ. Let R_{unb} and S_{unb} be unbiased multiples of the sample range R and sample standard deviation S, respectively, for estimating σ. (Here the distribution of S^2 is not related to a chi-squared distribution.) Use simulation methods in each of the parts below, taking $n = 10$, $\mu = 100$, and $\sigma = 10$, and using the R code of Example 4.3 as a pattern.

a) Find the unbiasing constants necessary to define R_{unb} and S_{unb}. These estimators are, of course, not necessarily the same as for normal data.
b) Show that $V(R_{\text{unb}}) < V(S_{\text{unb}})$. For data from such a uniform distribution, one can prove that R_{unb} is the unbiased estimator with minimum variance.
c) Find the quantiles of R_{unb} and S_{unb} necessary to make 95% confidence intervals for σ. Specify the endpoints of both intervals in terms of σ. Which confidence interval, the one based on R or the one based on S, has the shorter expected length?

Problems for Section 4.3 (Sample Mean and Standard Deviation)

4.18   Let $Y_1, Y_2, \ldots, Y_9$ be a random sample from $\mathsf{NORM}(200, 10)$.

a) Modify the R code of Example 4.4 to make a plot similar to Figure 4.4 based on $m = 100\,000$ and using small dots (plot parameter pch=". "). From the plot, try to estimate $E(\bar{Y})$ (balance point), $SD(\bar{Y})$ (most observations lie within two standard deviations of the mean), $E(S)$, and ρ. Then write and use code to simulate these values. Compare your estimates from looking at the plot and your simulated values with the exact values.
b) Based on the simulation in part (a), compare $P\{\bar{Y} \leq a\}P\{S \leq b\}$ with $P\{\bar{Y} \leq a, S \leq b\}$. Do this for at least three choices of a and b from among the values $a = 197, 200, 202$ and $b = 7, 10, 11$. Use normal and chi-squared distributions to find exact values for the probabilities you simulate. Comment.
c) In a plot similar to Figure 4.5, show the points for which the usual 95% confidence interval for σ covers the population value $\sigma = 10$. How does this differ from the display of points for which the t confidence interval for μ covers the true population value?

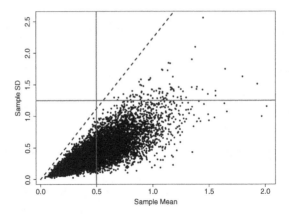

Figure 4.10. Joint distribution of $\bar{Y}$ and S for exponential data illustrated with 10 000 simulated points. Each dot is based on a sample of size $n = 5$ from $\mathsf{EXP}(2)$. No points can fall above the dashed line. See Example 4.6 and Problem 4.20.

4.19   Repeat the simulation of Example 4.5 twice, once with $n = 15$ random observations from $\mathsf{NORM}(200, 10)$ and again with $n = 50$. Comment on the effect of sample size.

4.20   More on Example 4.6 and Figure 4.10.

a) Show that there is an upper linear bound on the points in Figure 4.10. This boundary is valid for any sample in which negative values are impossible. Suggested steps: Start with $(n-1)S^2 = \sum_i Y_i^2 - n\bar{Y}^2$. For your data, say why $\sum_i Y_i^2 \leq (\sum_i Y_i)^2$. Conclude that $\bar{Y} \geq S/\sqrt{n}$.

b) Use `plot` to make a scatterplot similar to the one in Figure 4.10 but with $m = 100\,000$ points, and then use `lines` to superimpose your line from part (a) on the same graph.

c) For Example 4.6, show (by any method) that $P\{\bar{Y} \leq 0.5\}$ and $P\{S > 1.25\}$ are both positive but that $P\{\bar{Y} \leq 0.5,\, S > 1.25\} = 0$. Comment.

4.21   Figure 4.6 (p101) has prominent "horns." We first noticed such horns on $(\bar{Y}, S)$ plots when working with uniformly distributed data, for which the horns are not so distinct. With code similar to that of Example 4.7 but simulated samples of size $n = 5$ from $\mathsf{UNIF}(0, 1) = \mathsf{BETA}(1, 1)$, make several plots of S against $\bar{Y}$ with $m = 10\,000$ points. On most plots, you should see a few "straggling" points running outward near the top of the plot. The question is whether they are real or just an artifact of simulation. (That is, are they "signal or noise"?) A clue is that the stragglers are often in the same places on each plot. Next try $m = 20\,000$, $50\,000$, and $100\,000$. For what value of m does it first become obvious to you that the horns are real?

4.22   If observations $Y_1, Y_2, \ldots, Y_5$ are a random sample from $\mathsf{BETA}(\alpha, \beta)$, which takes values only in $(0, 1)$, then the data fall inside the 5-dimensional

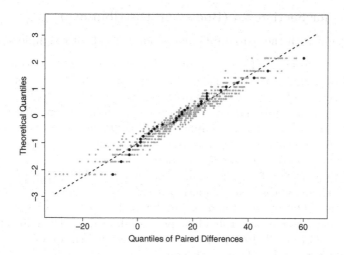

Figure 4.11. Normal probability plot (heavy black dots) of the 33 differences in Example 4.8. Under a transformation of the vertical axis, the cumulative distribution curve in Figure 4.7 (p102) becomes a straight line here. The lighter dots, from 20 simulated normal samples of size 33, suggest how far truly normal data may typically fall from the line. See Problem 4.23.

unit hypercube, which has $2^5 = 32$ vertices. Especially if we have parameters $\alpha, \beta < 1/2$, a large proportion of data points will fall near the vertices, edges, and faces of the hypercube. The "horns" in the plots of Example 4.7 (and Problem 4.21) are images of these vertices under the transformation from the 5-dimensional data space to the 2-dimensional space of $(\bar{Y}, S)$.

a) Use the code of Example 4.7 to make a plot similar to Figure 4.6 (p101), but with $m = 100\,000$ small dots. There are six horns in this plot, four at the top and two at the bottom. Find their exact $(\bar{Y}, S)$-coordinates. (You should also be able discern images of some edges of the hypercube.)

b) The horn at the lower left in the figure of part (a) is the image of one vertex of the hypercube, $(0, 0, 0, 0, 0)$. The horn at the lower right is the image of $(1, 1, 1, 1, 1)$. They account for two of the 32 vertices. Each of the remaining horns is the image of multiple vertices. For each horn, say how many vertices get mapped onto it, its "multiplicity."

c) Now make a plot with $n = 10$ and $m = 100\,000$. In addition to the two horns at the bottom, how many do you see along the top? Explain why the topmost horn has multiplicity $\binom{10}{5} = 252$.

Problems for Section 4.4 (Bootstrap Distributions)

4.23   Begin with the paired differences, `Pair.Diff`, of Example 4.8.

a) Use `mean(Pair.Diff)` to compute $\bar{d}$, `sd(Pair.Diff)` to compute S_d, `length(Pair.Diff)` to verify $n = 33$, and `qt(.975, 32)` to find t^*. Thus verify that the 95% t confidence interval is $(10.3, 21.6)$, providing two-place accuracy. Compare your interval with results from `t.test(Pair.Diff)`.

b) Modify the R code of the example to make a 99% nonparametric bootstrap confidence interval for the population mean difference μ. Compare with the 99% t confidence interval (see note).

c) Use `qqnorm(Pair.Diff)` to make a normal probability plot of the differences. Then use the `lines` function to overlay the line $y = (d - \bar{d})/S_d$ on your plot. The result should be similar to Figure 4.11, except that this figure also has normal probability plots (lighter dots) from 20 samples of size $n = 33$ from $\mathsf{NORM}(\bar{d}, S_d)$ to give a rough indication of the deviation of truly normal data from a straight line.

Notes: (b) Use parameter `conf.level=.99` in `t.test`. Approximate t interval: $(8.4, 23.5)$. (c) Although the normal probability plot of `Pair.Diff` seems to fit a curve better than a straight line, evidence against normality is not strong. For example, the Shapiro-Wilk test fails to reject normality: `shapiro.test(Pair.Diff)` returns a p-value of 0.22.

4.24   *Student heights.* In a study of the heights of young men, 41 students at a boarding school were used as subjects. Each student's height was measured (in millimeters) in the morning and in the evening, see [MR58]. Every student was taller in the morning. Other studies have found a similar decrease in height during the day; a likely explanation is shrinkage along the spine from compression of the cartilage between vertebrae. The 41 differences between morning and evening heights are displayed in the R code below.

```
dh = c(8.50,  9.75,  9.75,  6.00,  4.00, 10.75,  9.25, 13.25, 10.50,
      12.00, 11.25, 14.50, 12.75,  9.25, 11.00, 11.00,  8.75,  5.75,
       9.25, 11.50, 11.75,  7.75,  7.25, 10.75,  7.00,  8.00, 13.75,
       5.50,  8.25,  8.75, 10.25, 12.50,  4.50, 10.75,  6.75, 13.25,
      14.75,  9.00,  6.25, 11.75,  6.25)
```

a) Make a normal probability plot of these differences with `qqnorm(dh)` and comment on whether the data appear to be normal. We wish to have an interval estimate of the mean shrinkage in height μ in the population from which these 41 students might be viewed as a random sample. Compare the 95% t confidence interval with the 95% nonparametric bootstrap confidence interval, obtained as in Example 4.8 (but using `dh` instead of `Pair.Diff`).

b) Assuming the data are normal, we illustrate how to find a **parametric bootstrap confidence interval**. First estimate the parameters: the population mean μ by $\bar{d}$ and the population standard deviation σ by S_d. Then take B resamples of size $n = 41$ from the distribution $\mathsf{NORM}(\bar{d}, S_d)$, and find the mean of each resample. Finally, find confidence intervals as in the nonparametric case. Here is the R code.

```
B = 10000;  n = length(dh)

# Parameter estimates
dh.bar = mean(dh);  sd.dh = sd(dh)

# Resampling
re.x = rnorm(B*n, dh.bar, sd.dh)
RDTA = matrix(re.x, nrow=B)

# Results
re.mean = rowMeans(RDTA)
hist(re.mean)
bci = quantile(re.mean, c(.025, .975));  bci
2*dh.bar - bci[2:1]
```

Notes: (a) Nearly normal data, so this illustrates how closely the bootstrap procedure agrees with the t procedure when we know the latter is appropriate. The t interval is $(8.7, 10.5)$; in your answer, provide two decimal places. (b) This is a "toy" example because $T = n^{1/2}(\bar{d} - \mu)/S_d \sim \mathsf{T}(n-1)$ and $(n-1)S_d^2/\sigma^2 \sim \mathsf{CHISQ}(n-1)$ provide useful confidence intervals for μ and σ without the need to do a parametric bootstrap. (See [Rao89] and [Tru02] for traditional analyses and data, and see Problem 4.27 for another example of the parametric bootstrap.)

4.25   *Exponential data.* Consider $n = 50$ observations generated below from an exponential population with mean $\mu = 10$. (Be sure to use the seed shown.)

```
set.seed(1);  x = round(rexp(50, 1/10), 2);  x

> x
 [1]  7.55 11.82  1.46  1.40  4.36 28.95 12.30  5.40  9.57  1.47
[11] 13.91  7.62 12.38 44.24 10.55 10.35 18.76  6.55  3.37  5.88
[21] 23.65  6.42  2.94  5.66  1.06  0.59  5.79 39.59 11.73  9.97
[31] 14.35  0.37  3.24 13.20  2.04 10.23  3.02  7.25  7.52  2.35
[41] 10.80 10.28 12.92 12.53  5.55  3.01 12.93  9.95  5.14 20.08
```

a) For exponential data $X_1, \ldots, X_n$ with mean μ (rate $\lambda = 1/\mu$), it can be shown that $\bar{X}/\mu \sim \mathsf{GAMMA}(n, n)$. Use R to find L and U with $P\{L \le \bar{X}/\mu \le U\} = 0.95$ and hence find an exact formula for a 95% confidence interval for μ based on data known to come from an exponential distribution. Compute this interval for the data given above.

b) As an illustration, even though we know the data are not normal, find the t confidence interval for μ.

c) Set a fresh seed. Then replace `Pair.Diff` by x in the code of Example 4.8 to find a 95% nonparametric bootstrap confidence interval for μ.

d) Does a normal probability plot clearly show the data are not normal? The Shapiro-Wilk test is a popular test of normality. A small p-value indicates nonnormal data. In R, run `shapiro.test(x)` and comment on the result.

Answers: (a) $(7.6, 13.3)$, (b) $(7.3, 12.4)$. (c) On one run: $(7.3, 12.3)$.

4.26 *Coverage probability of a nonparametric bootstrap confidence interval.* Suppose we have a sample of size $n = 50$ from a normal population. We wonder whether an alleged 95% nonparametric bootstrap confidence interval really has nearly 95% coverage probability. Without loss of generality, we consider $m = 1000$ such random samples from $\mathsf{NORM}(0, 1)$, find the bootstrap confidence interval (based on $B = 1000$) for each, and determine whether it covers the true mean 0. (In this case, the interval covers the true mean if its endpoints have opposite sign, in which case the product of the endpoints is negative.) The fraction of the $m = 1000$ nonparametric bootstrap confidence intervals that cover 0 is an estimate of the coverage probability.

A suitable R program is given below. We chose relatively small values of m and B and simple bootstrap confidence intervals. The program has a computationally intensive loop and so it runs rather slowly with larger numbers of iterations. Do not expect a high-precision answer because with $m = 1000$ the final step alone has a margin of error of about 1.4%. Report results from three runs. Increase the values of m and B for improved accuracy, if you have some patience or a fast computer.

```
m = 1000;  cover = numeric(m);  B = 1000;  n = 50
for (i in 1:m)
  {
  x = rnorm(n)                       # simulate a sample
  re.x = sample(x, B*n, repl=T     # resample from it)
  RDTA = matrix(re.x, nrow=B)
  re.mean = rowMeans(RDTA)
  cover[i] = prod(quantile(re.mean, c(.025,.975)))
                                     # does bootstrap CI cover?
  }
mean(cover < 0)
```

4.27 *Mark-recapture estimate of population size.* To estimate the number τ of fish of a certain kind that live in a lake, we first take a random sample of c such fish, mark them with red tags, and return them to the lake. Later, after the marked fish have had time to disperse randomly throughout the lake but before there have been births, deaths, immigration, or emigration, we take a second random sample of n fish and note the number X of them that have red tags.

a) Argue that a reasonable estimate of τ is $\hat{\tau} = \lfloor cn/X \rfloor$, where $\lfloor\ \rfloor$ indicates the "largest included integer" or "floor" function. If $c = 900$, $n = 1100$, and $X = 95$, then evaluate $\hat{\tau}$.

b) For known values of τ, r, and n, explain why $P\{X = x\} = \binom{c}{x}\binom{\tau-c}{n-x}/\binom{\tau}{n}$, for $x = 0,\ldots,n$, where $\binom{a}{b}$ is defined as 0 if integer $b \geq 0$ exceeds a. We say that X has a hypergeometric distribution. Suppose $\tau = 10\,000$ total fish, $c = 900$ tagged fish, $\tau - c = 9100$ untagged fish, and $n = 1100$ fish in the second sample. Then, in R, use dhyper(95, 900, 9100, 1100) to evaluate $P\{X = 95\}$.

c) Now, with $c = 900$ and $n = 1100$, suppose we observe $X = 95$. For what value of τ is $P\{X = 95\}$ maximized? This value is the **maximum likelihood estimate** of τ. Explain why the following code evaluates this estimate. Compare your result with the value of $\hat{\tau}$ in part (a).

```
tau = 7000:15000;   like = dhyper(95, 900, tau-900, 1100)
mle = tau[like==max(like)];   mle
plot(tau, like, type="l");   abline(v=mle, lty="dashed")
```

d) The R code below makes a **parametric bootstrap** confidence interval for τ. For c, n, and X as in parts (a) and (c), we have the estimate $\hat{\tau} = 10\,421$ of the parameter τ. We resample $B = 10\,000$ values of X based on the known values of c and n and this estimate $\hat{\tau}$. From each resampled X, we reestimate τ. This gives a bootstrap distribution consisting of B estimates of τ, from which we obtain a confidence interval.

```
# Data
c = 900;   n = 1100;   x = 95

# Estimated population size
tau.hat = floor(c*n/x)

# Resample using estimate
B = 10000
re.tau = floor(c*n/rhyper(B, c, tau.hat-c, n))

# Histogram and bootstrap confidence intervals
hist(re.tau)
bci = quantile(re.tau, c(.025,.975));   bci   # simple bootstrap
2*tau.hat - bci[2:1]                           # percentile method
```

Notes: (a) How does each of the following fractions express the proportion of marked fish in the lake: c/τ and X/n? (d) Roughly $(8150, 12\,000)$ from the bootstrap percentile method, which we prefer here because of the skewness; report your seed and exact result. This is a "toy" example of the parametric bootstrap because standard methods of finding a confidence interval for τ require less computation and are often satisfactory. There is much literature on mark-recapture (also called *capture-recapture*) methods. For one elementary discussion, see [Fel57].

5

Screening Tests

Screening tests, also called **diagnostic tests**, are used in a wide variety of fields—public health, surgery, forensics, industrial quality management, satellite mapping, communications, banking, and many more. In the simplest case, a population is divided into two parts. A particular member of the population is infected with or free of a disease, a patient is about to reject a transplanted organ or not, a suspect is deceptive or telling the truth, a manufactured component is defective or fit for use, a remote mountainside is covered with snow or not, a transmitted bit of information is a 0 or a 1, or a prospective borrower will default on the loan or not.

If direct and sure classification is impossible, takes too long, or is too expensive, then a relatively convenient, quick, or inexpensive screening test may be used in an attempt to make the proper classification. Of course, one hopes that the screening test will usually give the correct indication. Our interest is mainly in a probability analysis of correct and incorrect test results.

Some of the terminology associated with screening tests varies with the field of application. For focus and consistency, we carry a single example on detecting a virus through much of this chapter, but other applications of screening tests appear in some of the examples and problems. In Chapters 6 and 9, our first examples of Gibbs sampling are based on screening tests.

5.1 Prevalence, Sensitivity, and Specificity

Suppose that international public health officials want to determine the prevalence of a particular virus in donated blood at several sites throughout the world. Also suppose that a relatively inexpensive **ELISA test** is available to screen units of blood for this virus. Accordingly, the study will be based on the results of ELISA tests performed on randomly chosen units of blood donated at each place to be surveyed. (ELISA is short for *enzyme-linked immunosorbent assay*. A specific ELISA test detects antibodies to a particular virus—such as HIV or one of several types of hepatitis.)

E.A. Suess and B.E. Trumbo, *Introduction to Probability Simulation and Gibbs Sampling with R*, Use R!, DOI 10.1007/978-0-387-68765-0_5, © Springer Science+Business Media, LLC 2010

It is convenient to define two random variables, D and T, corresponding to a randomly chosen unit of blood. Each of these random variables is binary, taking only the values 0 and 1:

$$\pi = P\{D = 1\} = P(\text{unit infected with virus}) = \textbf{prevalence},$$
$$1 - \pi = P\{D = 0\} = P(\text{unit not infected}),$$
$$\tau = P\{T = 1\} = P(\text{unit tests positive for virus}),$$
$$1 - \tau = P\{T = 0\} = P(\text{unit tests negative}).$$

The Greek letters used above are *pi* (π) and *tau* (τ). In this book, π does not ordinarily stand for the ratio of the diameter of a circle to its circumference; we will note the occasions when it does.

The proportion τ of the units in the sample for which the ELISA test indicates presence of the virus is not the same as the proportion π actually infected with the virus. The ELISA test is useful, but not perfect.

During its development, this ELISA test was performed on a large number of blood samples known to have come from subjects infected with the virus. Suppose that about 99% of these infected samples showed a positive result. That is to say, the ELISA test correctly detects the virus in 99% of infected units of blood; the remaining 1% are false-negative results. In terms of random variables and probabilities, we say that the **sensitivity** (denoted by the Greek letter *eta*) of the test is

$$\eta = P\{T = 1 | D = 1\} = P(\text{positive test} \mid \text{unit has virus})$$
$$= P\{T = 1, \ D = 1\}/P\{D = 1\}$$
$$= P(\text{true positive})/P(\text{infected}) = 0.99 = 99\%.$$

Here, and often throughout this chapter, we express probabilities and proportions as percentages.

In contrast, consider a group of units of blood known from more accurate and costly procedures to be free of the virus $\{D = 0\}$. When administered to such units of blood, the ELISA test was found to give negative results $\{T = 0\}$ for about 97% of them. That is, for some reason, the test gave false-positive results for 3% of the uninfected units of blood. We say that the **specificity** of the test (denoted by *theta*) is

$$\theta = P\{T = 0 | D = 0\} = P(\text{negative test} \mid \text{no virus})$$
$$= P\{T = 0, \ D = 0\}/P\{D = 0\}$$
$$= P(\text{true negative})/P(\text{not infected}) = 97\%.$$

The particular numerical values of η and θ that we have given above, and continue to use throughout this chapter, are reasonable but hypothetical values. The actual sensitivity and specificity of an ELISA test procedure for any particular virus would depend on whether tests are done once or several times on each unit of blood and whether borderline results are declared as "positive" or "negative." When the immediate purpose is to protect the blood supply from contamination with the virus, such borderline decisions would be made in favor of increasing the sensitivity. The consequence would be to decrease the specificity. (See [Gas87] for a discussion of sensitivity and specificity of diagnostic tests in a variety of applications.)

Example 5.1. The trade-off between sensitivity and specificity must be agreed upon before a screening test is used for a particular purpose. The nature of this trade-off becomes especially easy to quantify when the test criterion is a measurement X subject to random variation and the distribution of X is known for both the infected and the uninfected populations. That is, the conditional distributions of $X|D$ are known.

Specifically, suppose that $X|\{D = 0\} \sim \text{NORM}(50, 10)$ and $X|\{D = 1\} \sim \text{NORM}(70, 15)$. Let x be the cutoff value such that we declare a positive test result $\{T = 1\}$ for values of X above x and a negative test result $\{T = 0\}$ for values below x. Then it is easy to see how the sensitivity and specificity are affected by changes in the cutoff value:

$$\eta(x) = 1 - F_{X|D=1}(x) \quad \text{and} \quad \theta(x) = F_{X|D=0}(x). \tag{5.1}$$

One way to display the relationship between sensitivity and specificity is an **ROC plot** (for *Receiver-Operator Characteristic plot*, terminology that reflects the early use of these plots in the field of electronic communications). It is customary to plot $\eta(x)$ against $1 - \theta(x)$ as x varies, so that both axes involve conditional probabilities of testing positive. Below, we show the R code used to print the numerical results shown below and make Figure 5.1.

```
x = seq(40,80,1)
eta = 1 - pnorm(x, 70, 15)
theta = pnorm(x, 50, 10)
cbind(x, eta, theta)[x >= 54 & x <= 63, ]
plot(1-theta, eta, xlim=c(0,1), ylim=c(0,1), pch=19)
lines(c(0,1),c(1,0))

>   cbind(x, eta, theta)[x >= 54 & x <= 63, ]
           x       eta      theta
 [1,]     54 0.8569388 0.6554217
 [2,]     55 0.8413447 0.6914625
 [3,]     56 0.8246761 0.7257469
 [4,]     57 0.8069377 0.7580363
 [5,]     58 0.7881446 0.7881446
 [6,]     59 0.7683224 0.8159399
 [7,]     60 0.7475075 0.8413447
 [8,]     61 0.7257469 0.8643339
 [9,]     62 0.7030986 0.8849303
[10,]     63 0.6796308 0.9031995
```

If false-positive and false-negative results are about equally regrettable, then desirable cutoff values x correspond to points on the ROC curve that are near the line from the upper-left corner $(0, 1)$ to the lower-right corner $(1, 0)$, where $\eta = \theta$. There is one such point in the figure, and from the printout we see that it corresponds to a cutoff value $x = 58$, giving $\eta = \theta = 78.8\%$.

In general, useful tests correspond to points above the principal diagonal from $(0, 0)$ to $(1, 1)$ in such a diagram. A point on this line would correspond to a test that gives the same probability of a positive result whether $D = 0$ or $D = 1$. In this sense, all tests in the family considered in this example are useful. (For more on choosing the cutoff values, see Problem 5.5.)

The curve shown in Figure 5.1 is an idealized one. In a typical application, we would not know the exact conditional distributions, and the distributions would not necessarily be symmetrical, much less normal. Instead, we might have test data on some individuals (currently or eventually) known to have $D = 0$ and some known to have $D = 1$, enabling us to use each group to find a conditional ECDF. Then the two ECDFs could be used to plot points on an ROC plot that, with luck, would suggest the location of the ROC curve (see Problem 5.6). ◊

In summary, the two binary random variables D and T divide the entire population of units of blood into four disjoint parts.

- True positive $\{D = 1,\ T = 1\}$: Unit infected and correctly classified.
- True negative $\{D = 0,\ T = 0\}$: Unit uninfected and correctly classified.
- False positive $\{D = 0,\ T = 1\}$: Unit uninfected and incorrectly classified. This amounts to a "false alarm." As a result, a unit of blood that might have been used is destroyed. If a person is being screened for a disease, a false-positive result may lead to needless worry, further testing, and so on.
- False negative $\{D = 1,\ T = 0\}$: Unit infected and incorrectly classified. Here an infected unit of blood is introduced into the blood supply with the likely result of infecting its recipient. In the case of a test on a person, an opportunity may be lost to cure or manage a disease before it becomes more serious.

For a practical discussion of sensitivity and specificity in several applications of screening tests, see [Gas87].

5.2 An Attempt to Estimate Prevalence

Consider again our example of a screening test with sensitivity $\eta = 99\%$ and specificity $\theta = 97\%$. At one of the sites under study, suppose that we estimate $\tau = P\{T = 1\}$ as t, the proportion of positive tests in a sample. We have seen in Section 5.1 that t itself is not an appropriate estimate of the prevalence $\pi = P\{D = 1\}$. But can we use t indirectly to find an estimate p of π?

One proposed method of estimating prevalence is to use the following equation, which establishes a connection between τ and π:

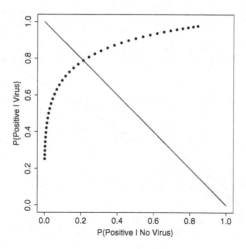

Figure 5.1. ROC plot for Example 5.1. $P(\text{Positive} \mid \text{Virus}) = \eta$ is plotted against $P(\text{Positive} \mid \text{NoVirus}) = 1 - \theta$ for various possible screening tests. Each test (dot) is defined by the cutoff value x above which the test result is declared "positive." The point on the diagonal line has $\eta = \theta$ (sensitivity equal to specificity).

$$\tau = P\{T = 1\} = P\{D = 1, T = 1\} + P\{D = 0, T = 1\}$$
$$= P\{D = 1\}P\{T = 1 \mid D = 1\} + P\{D = 0\}P\{T = 1 \mid D = 0\}$$
$$= \pi\eta + (1 - \pi)(1 - \theta). \tag{5.2}$$

Here we have partitioned all positive tests into true positives and false positives, applied the law of total probability, and twice used the multiplication rule $P(E \cap F) = P(E)P(F \mid E)$. Solving (5.2) for π, we obtain

$$\pi = (\tau + \theta - 1)/(\eta + \theta - 1). \tag{5.3}$$

Then, replacing τ by t, we have an estimator p of π:

$$p = (t + \theta - 1)/(\eta + \theta - 1). \tag{5.4}$$

For example, suppose that we have a random sample of $n = 1000$ units of blood from a particular site and that 49 of them test positive. We use the notation $A = \#\{T = 1\} = 49$. Then $t = A/n = 0.049 = 4.9\%$ and

$$p = (0.049 + 0.97 - 1)/(0.99 + 0.97 - 1)$$
$$= (4.9\% - 3\%)/(99\% - 3\%) = 0.0198 = 1.98\%.$$

Based on the normal approximation to the binomial distribution, the traditional 95% confidence interval for τ is $t \pm 1.96\sqrt{t(1 - t)/n}$. Here we get $4.9\% \pm 1.33\%$ or $(3.57\%, 6.23\%)$. Substituting the endpoints of this interval into equation (5.3), we obtain the corresponding 95% confidence interval

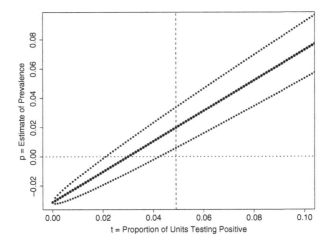

Figure 5.2. Estimates of prevalence π. For proportions t of units testing positive, the plot shows corresponding estimates p of prevalence—based on 1000 units tested with sensitivity 99% and specificity 97%. Curved bands show traditional 95% confidence intervals for π; in particular, $(0.59\%, 3.36\%)$ when $A = 49$ units test positive (vertical line). Unfortunately, this procedure can produce absurd negative estimates.

for π, $(0.59\%, 3.36\%)$. Similarly, the more accurate Agresti-Coull 95% confidence interval for τ gives the interval $(0.75\%, 3.58\%)$ for π. (See Figure 5.2 and Problem 5.7.)

Example 5.2. Absurd Estimates of Prevalence. Unfortunately, the method just discussed sometimes gives absurd estimates p of π. Here again, let $\eta = 99\%$ and $\theta = 97\%$. If we have a sample of $n = 250$ units of blood and six of them test positive, then $t = 2.4\%$ and $p = -0.62\%$. The difficulty here is that we would expect $100\% - \theta = 3\%$ of the tests to be positive even if the prevalence is 0, but sampling variation has given us a value of t less than 3%. In different circumstances, this method may give estimates of π that exceed 1.

Moreover, unreasonable estimates of π outside the interval $(0, 1)$ are not rare. Suppose this ELISA test is used in a population where, unknown to us, the prevalence of the virus is $\pi = 2\%$ and so, by equation (5.2), $\tau = 4.92\%$. We estimate π by administering the test to $n = 250$ units of blood, of which it happens that A test positive. What then is the probability that we get an absurd negative estimate of π? In this situation, $A \sim \mathsf{BINOM}(250, 0.0492)$ and $t = A/n$ estimates τ. Looking at equation (5.3), we see that the estimate p is negative when $t < 1 - \theta = 0.03$ or $A < 250(0.03) = 7.5$. Thus the answer to our question is that $P\{p < 0\} = P\{A \le 7\} = 0.0721$, which can be verified in R with `pbinom(7, 250, .0492)`.

A somewhat deeper question is how often the lower confidence limit of a 95% confidence interval for π will be negative. The following simulation with R provides an approximate answer.

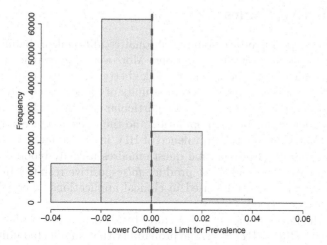

Figure 5.3. Absurd estimates of prevalence. In Example 5.2, negative point estimates of prevalence occur in about 7% of the $m = 100\,000$ simulated samples of size $n = 250$ from a population with prevalence $\pi = 2\%$. The histogram shows that the lower confidence limits for prevalence were negative about 75% of the time.

```
# set.seed(1212)
m = 100000;  n = 250;  sens = .99;  spec = .97;  prev = .02
tpos = prev*sens + (1-prev)*(1-spec)
a = rbinom(m, n, tpos);  t = a/n;  lcl.t = t - 1.96*sqrt(t*(1-t)/n)
p = (t + spec - 1)/(sens + spec - 1)
lcl.p = (lcl.t + spec - 1)/(sens + spec - 1)
hist(lcl.p, nclass=5)
mean(p < 0);  mean(lcl.p < 0)

> mean(p < 0);  mean(lcl.p < 0)
[1] 0.07241
[1] 0.74762
```

We see from the printout and the histogram in Figure 5.3 that the lower confidence limit for π is negative about 75% of the time. Also, notice that $P\{p < 0\}$ is simulated as approximately 0.07, which is in substantial agreement with the exact answer derived above. (For an example of an absurd estimate of prevalence based on real data, see [PG00], Chapter 6.) ◇

In many applications of screening tests, the estimation of prevalence is crucial—for example, in epidemiology to estimate the extent to which a disease has spread, in communications to estimate the percentage of bits transmitted in error, in banking to estimate the proportion of loans that will not be repaid, and so on. We return to this topic in Chapter 9, where we explore more satisfactory estimates of prevalence.

5.3 Predictive Values

In this section, we introduce some additional conditional probabilities, that are of importance in practical situations. Moreover, they provide a point of view that will eventually permit us to make better estimates of prevalence.

To continue our example of screening units of blood, we consider a hypothetical prevalence value of $\pi = 2\%$ at a particular site. In real life, prevalences range widely depending on the population and the virus of interest. For example, in the United States the prevalence of HIV in the donated blood supply is now essentially 0. (Pre-donation questionnaires used by blood banks even tend to eliminate donors likely to produce *false*-positive results.) In contrast, screening tests are sometimes used in clinical applications where the prevalence of a disease exceeds 50%.

During the development of a screening test, it is natural to focus on the sensitivity $\eta = P\{T = 1 | D = 1\}$, the probability that an infected unit of blood will test positive. After we have administered the test and have a positive result, it is natural to ask what proportion of units with positive results are actually infected with the virus. This can be viewed as the inverse conditional probability $P\{D = 1 | T = 1\}$, called the **predictive value of a positive test** (sometimes abbreviated to *predictive value positive*, or *PVP*). This is a property of the site as well as a property of the particular screening test used. Equation (5.2) gives

$$\tau = \pi\eta + (1 - \pi)(1 - \theta)$$
$$= (0.02)(0.99) + (0.98)(0.03) = 0.0492 = 4.92\%.$$

From this we can compute the predictive value of a positive test (denoted by *gamma*),

$$\gamma = P(D = 1 | T = 1) = \frac{P(D = 1, T = 1)}{P(T = 1)} = \frac{\pi\eta}{\pi\eta + (1 - \pi)(1 - \theta)}$$
$$= \frac{\pi\eta}{\tau} = \frac{0.0198}{0.0492} = 0.4024 = 40.24\%. \tag{5.5}$$

In this population, considerably less than half of the units that test positive are actually infected, so most of the units that are destroyed because they test positive are not really infected. Fortunately, in this instance, relatively few of the units test positive and so the overall consequence of these misclassifications is not great.

Similarly, the **predictive value of a negative test** (denoted by *delta* and sometimes abbreviated as *PVN*) is

$$\delta = P(D = 0 | T = 0) = (1 - \pi)\theta / (1 - \tau)$$
$$= 0.9506 / (1 - 0.0492) = 0.9998 = 99.98\%. \tag{5.6}$$

Thus, almost all of the units that test negative are uninfected. Of course, we can hope that the predictive values of both positive and negative tests are

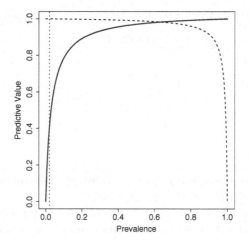

Figure 5.4. Predictive values. Conditional probabilities PVP γ (solid curve) and PVN δ (dashed) of a test depend on prevalence π and also on its sensitivity $\eta = 99\%$ and specificity $\theta = 97\%$. The vertical dotted line illustrates equations (5.5) and (5.6), where $\pi = 2\%$.

high (near 1), but even when one or both of these values is low, the screening test may still be useful. (See Figure 5.4 and Problems 5.10 and 5.11.)

One way to get direct information about predictive values is to perform a **gold standard** procedure on some of the units of blood. In concept, a gold standard provides essentially a 100% accurate determination as to whether or not the virus is present in a unit, but at a cost of administration that prevents its use on every unit of blood. (If such a gold standard were inexpensive, why bother with imperfect ELISA tests for screening?)

> Procedures called Western blot tests are regarded as a gold standard for some viruses. They use a different technology than ELISA tests and are considerably more accurate, and expensive, to use than ELISA tests. However, in practice, no such procedure is absolutely perfect. Both the Western blot test and the ELISA test actually detect antibodies to a specific virus. In most circumstances, the presence of antibodies corresponds to presence of the virus itself. Exceptions might be units of blood from people who have taken vaccines (having antibodies, but no virus) or who have been very recently infected (virus, but no antibodies yet). Blood banks use pre-donation questionnaires to try to avoid accepting such units of blood.

Unless the prevalence of the virus is extremely high at a particular location, the actual number of units with ELISA-positive tests found there may be small enough that we could check them all against the gold standard. Without knowing π, we could then estimate γ for this site as the proportion of ELISA-

Table 5.1. Counts used in the program of Example 5.3.

TEST	BATCH Bad	Good	Total
Fail	n - n.p - n.g + n.gp	n.g - n.gp	n - n.p
Pass	n.p - n.gp	n.gp	n.p
Total	n - n.g	n.g	n

positive units proved by subsequent gold-standard procedures actually to have the virus; that is, $\#(T = 1, D = 1)/\#(T = 1)$.

Although we would not ordinarily be able to apply the gold standard to all units of blood that tested ELISA-negative, we might be able to check some of them against the gold standard to get an estimate of δ (if only to verify that δ really is very nearly 1, as in our example).

Example 5.3. Batches of a drug are synthesized using a process that gives somewhat variable results. The potency of a randomly chosen batch is $S \sim \text{NORM}(110, 5)$. A batch is called good (G)—that is, it meets design specifications—if its potency exceeds 100. Otherwise, it is called bad (B).

However, it is difficult to assay potency. When a batch with potency s is assayed according to a method accepted for use in production, the value observed is a random variable $X \sim \text{NORM}(s, 1)$. By agreement with the appropriate regulatory agency, a batch passes inspection (P) and can be packaged for sale if $X > 101$. Otherwise, the batch fails inspection (F) and must be destroyed.

Of all the batches produced, we wish to know the proportion $\pi = P(B)$ of bad batches and the proportion $\tau = P(F)$ that must be destroyed. Each of these probabilities can be found as an area under a normal curve.

- Clearly, $\pi = P\{S \le 100\} = \Phi(-2) = 0.02275$ from tables of the standard normal distribution, or in R as `pnorm(100, 110, 5)` or `pnorm(-2)`.
- To find τ, we express X as the sum $X = S + E$ of two independent random variables, $S \sim \text{NORM}(110, 5)$ and $E \sim \text{NORM}(0, 1)$. Then $X \sim N(110, \sqrt{1^2 + 5^2})$ and $\tau = P\{X \le 101\} = \Phi(-9/\sqrt{26}) = 0.03878$.

We see that more batches are being destroyed (almost 4%) than are actually bad (about 2.3%). Clearly, some good batches fail the test. This is one of the consequences of using a less than perfect testing procedure.

In economic terms, it is natural for the manufacturer to be concerned about the specificity θ, the probability that a batch will pass inspection given that it is good, and the predictive value of a negative test γ, the proportion of discarded batches that is actually bad. However, an analytical evaluation of these two conditional probabilities is a little more difficult than for π and τ above. (See Problem 5.14.)

Below is R code to simulate θ and δ (and, for an easy bit of verification, also π). Notice that the step x = rnorm(n, s, sd.x) has two vectors of length n; the ith entry in s is used as the mean when the ith entry in x is generated. We use pp for our π because R reserves pi for the usual constant. Table 5.1 displays counts used in the program.

```
# set.seed(1237)
n = 500000
mu.s = 110;   sd.s = 5;   cut.s = 100
sd.x = 1;   cut.x = 101
s = rnorm(n, mu.s, sd.s);   x = rnorm(n, s, sd.x)
n.g  = length(s[s > cut.s])              # number Good
n.p  = length(x[x > cut.x])              # number Pass
n.gp = length(x[s > cut.s & x > cut.x])  # number Good & Pass
n.bf = n - n.p - n.g + n.gp              # number Bad & Fail
pp   = (n - n.g)/n                       # prevalence pi
theta = n.gp/n.g
gamma = n.bf/(n - n.p)
pp;   theta;   gamma


> pp;   theta;   gamma
[1] 0.023292
[1] 0.982742
[1] 0.572313
```

The simulated value of π is correct to three decimal places: all but 2.3% of the batches are good. Also, more than 98% of good batches pass inspection, so relatively few batches are destroyed. Of the batches that are destroyed, only about 57% are actually bad. $\diamondsuit$

By now we have accumulated a number of probabilities and conditional probabilities. In order to avoid confusion, it is necessary to interpret each probability by focusing on the population or subpopulation to which it refers.

- For example, $P\{T = 1, D = 1\}$, the probability of a true positive outcome, refers to the *entire population*. It is the proportion of all units of blood that are infected *and* also test positive.
- The sensitivity $\eta = P\{T = 1 | D = 1\}$ refers to the *subpopulation of infected units*. It is the rate of (true) positive results within this infected subpopulation. (To what subpopulation does θ refer?)
- In contrast, $\gamma = P\{D = 1 | T = 1\}$, the predictive value of a positive test, refers to the *subpopulation of units that test positive*. It is the rate of infection within this subpopulation. (To what subpopulation does δ refer?)

In Chapter 9, we see that if reliable estimates of the conditional probabilities γ and δ are available, they can provide the basis for an improved estimate of π. In fact, this possibility provides a simple illustration of the important estimation technique known as the Gibbs sampler.

5.4 Bayes' Theorem for Events

In Section 5.1, we defined sensitivity $\eta = P\{T = 1|D = 1\}$ and specificity $\theta = P\{T = 0|D = 0\}$ using the conditional distributions of T given D. Then, in Section 5.3 we inverted the conditioning to obtain predictive values $\gamma = P\{D = 1|T = 1\}$ and $\delta = P\{D = 0|T = 0\}$, which involve the conditional distributions of D given T. This process is a special case of an important general principle, which we now explore a little more deeply.

Consider sets $A_0, A_1, \ldots, A_K$ that form a **partition** of the population into subpopulations. That is, the A_i are disjoint and together they exhaust the population. Suppose we know the probabilities $P(A_i)$ of all of these subpopulations. Also, for some event E of interest, suppose we know the probability $P(E|A_i)$ relative to each subpopulation. Then **Bayes' Theorem** shows how to find any one of the conditional probabilities $P(A_j|E)$ as follows:

$$P(A_j|E) = \frac{P(A_j)P(E|A_j)}{\sum_{i=0}^{K} P(A_i)P(E|A_i)}, \quad \text{for } j = 0, \ldots, K. \tag{5.7}$$

Example 5.4 below indicates the method of proving this result. (Many elementary probability textbooks give formal proofs, for example [Pit93]. For a more general version of Bayes' Theorem, see Problem 5.19.)

When we computed the predictive value of a positive test in equation (5.5), we used a special case of equation (5.7). There the partition consisted of the subpopulations $A_0 = \{D = 0\}$ and $A_1 = \{D = 1\}$, with $P(A_0) = 1 - \pi$ and $P(A_1) = \pi$. The event of interest was that the unit tested positive $E = \{T = 1\}$, and we knew $P(E|A_0) = 1 - \theta$ and $P(E|A_1) = \eta$. We evaluated the reverse conditional probability $\gamma = P\{D = 1|T = 1\} = P(A_1|E)$.

An important point of view associated with Bayes' Theorem has to do with changing one's assessment of a probability based on data. Prior to seeing any data, we judge the probability of the jth subpopulation to be $P(A_j)$. This is called a **prior probability**. After seeing some relevant data for a randomly chosen member of the population, we may reassess the probability that it belongs to the jth subpopulation. The conditional probability based on data is called a **posterior probability**. Bayes' Theorem gives the posterior probability $P(A_j|E)$ of belonging to the jth subpopulation given the data E.

In terms of our example of testing units of blood, the prior probability of infection (belonging to subpopulation A_1) is the prevalence π. After seeing that a unit tests positive (data), our revised probability that the unit is infected is γ, the predictive value of a positive test.

Example 5.4. At a manufacturing company, any one of four injection-molding machines can be used to produce a particular kind of plastic part. Thus, the population of all parts is partitioned into four subpopulations depending on the machine. Machines A and B are each used to make 40% of the parts, high-precision Machine C is used to make 15% of the parts, and older Machine D is used to make the remaining 5% of them. The respective error rates for

Machines A–D are 1%, 1%, 0.5%, and 3%. For example, 1% of the parts made by Machine A are bad; $P(E|A) = 1\%$.

Suppose one of these parts is chosen at random from the company warehouse. The probability that it is bad is given by the *denominator* of Bayes' Theorem, where the contributions of all four machines to the overall error rate are summed:

$$\begin{aligned}
P(E) &= P(E \cap A) + P(E \cap B) + P(E \cap C) + P(E \cap D)\\
&= P(A)P(E|A) + P(B)P(E|B) + P(C)P(E|C) + P(D)P(E|D)\\
&= 0.40(0.01) + 0.40(0.01) + 0.15(0.005) + 0.05(0.03)\\
&= 0.01025 = 1.025\%.
\end{aligned}$$

Thus, of all the parts of this type that the company makes, a little more than 1% are bad. Of all bad parts, the proportion made on Machine D is given by Bayes' Theorem as

$$\begin{aligned}
P(D|E) &= P(D \cap E)/P(E) = P(D)P(E|D)/P(E)\\
&= 0.05(0.03)/0.01025 = 0.0015/0.01025 = 14.63\%.
\end{aligned}$$

Even though Machine D makes only 5% of the parts (prior probability), it is responsible for almost 15% of all the bad parts produced (posterior probability, given bad). Perhaps Machine D should be retired. (See Problem 5.16 for a variation of this example.) ◇

In Chapter 8, we show applications of a more general version of Bayes' Theorem in estimation. Chapters 6 and 7 introduce basic ideas of Markov chains. Screening tests, Markov chains, and a Bayesian approach to estimation are important elements of Chapters 9 and 10.

5.5 Problems

Problems for Section 5.1 (Prevalence, Sensitivity, and Specificity)

5.1 In a newspaper trivia column, L. M. Boyd [Boy99] ponders why lie detector results are not admissible in court. His answer is that "lie detector tests pass 10 percent of the liars and fail 20 percent of the truth-tellers." If you use these percentages and take $\{D = 1\}$ to mean being deceitful and $\{T = 1\}$ to mean failing the test, what are the numerical values of the sensitivity and specificity for such a lie detector test? (Continued in Problem 5.12.)

5.2 In a discussion of security issues, Charles C. Mann [Man02] considers the use of face-recognition software to identify terrorists at an airport terminal:

[One of the largest companies marketing face-recognition technology] contends that...[its] software has a success rate of 99.32 percent—that is, when the software matches a passenger's face with a face on a list of terrorists, it is mistaken only 0.68 percent of the time. Assume for the moment that this claim is credible; assume, too, that good pictures of suspected terrorists are readily available. About 25 million passengers used Boston's Logan Airport in 2001. Had face-recognition software been used on 25 million faces, it would have wrongly picked out just 0.68 percent of them–but that would have been enough...to flag as many as 170,000 innocent people as terrorists. With almost 500 false alarms a day, the face-recognition system would quickly become something to ignore.

Interpret the quantities η, θ, and π of Section 5.1 in terms of this situation. As far as possible, say approximately what numerical values of these quantities Mann seems to assume.

5.3   Consider a bogus test for a virus that always gives positive results, regardless of whether the virus is present or not. What is its sensitivity? What is its specificity? In describing the usefulness of a screening test, why might it be misleading to say how "accurate" it is by stating its sensitivity but not its specificity?

5.4   Suppose that a medical screening test for a particular disease yields a continuum of numerical values. On this scale, the usual practice is to take values less than 50 as a negative indication for having the disease $\{T = 0\}$ and take values greater than 56 as positive indications $\{T = 1\}$. The borderline values between 50 and 56 are usually also read as positive, and this practice is reflected in the published sensitivity and specificity values of the test. If the borderline values were read as negative, would the sensitivity increase or decrease? Explain your answer briefly.

5.5   Many criteria are possible for choosing the "best" (η, θ)-pair from an ROC plot. In Example 5.1, we mentioned the pair with $\eta = \theta$. Many references vaguely suggest picking a pair "close to" the upper-left corner of the plot. Two ways to quantify this are to pick the pair on the curve that maximizes the **Youden index** $\eta + \theta$ or the pair that maximizes $\eta^2 + \theta^2$.

a) As shown below, modify the line of the program in Example 5.1 that prints numerical results. Use the expanded output to find the (η, θ)-pair that satisfies each of these maximization criteria.

```
cbind(x, eta, theta, eta + theta, eta^2 + theta^2)
```

b) Provide a specific geometrical interpretation of each of the three criteria: $\eta = \theta$, maximize $\eta + \theta$, and maximize $\eta^2 + \theta^2$. (Consider lines, circles, and tangents.)

Hint and notes: (b) When the ROC curve is only roughly estimated from data, it may make little practical difference which criterion is used. Also, if false-positive results are much more (or less) consequential errors than false-negative ones, then criteria different from any of these may be appropriate.

Table 5.2. Sensitivities and specificities for values of CREAT and B2M.

	CREAT			B2M	
Value	Sensitivity	Specificity	Value	Sensitivity	Specificity
1.2	.939	.123	1.2	.909	.067
1.3	.939	.203	1.3	.909	.074
1.4	.909	.281	1.4	.909	.084
1.5	.818	.380	1.5	.909	.123
1.6	.758	.461	1.6	.879	.149
1.7	.727	.535	1.7	.879	.172
1.8	.636	.649	1.8	.879	.215
1.9	.636	.711	1.9	.879	.236
2.0	.545	.766	2.0	.818	.288
2.1	.485	.773	2.1	.818	.359
2.2	.485	.803	2.2	.818	.400
2.3	.394	.811	2.3	.788	.429
2.4	.394	.843	2.4	.788	.474
2.5	.364	.870	2.5	.697	.512
2.6	.333	.891	2.6	.636	.539
2.7	.333	.894	2.7	.606	.596
2.8	.333	.896	2.8	.576	.639
2.9	.303	.909	2.9	.576	.676

5.6 *Empirical ROC.* DeLong et al. [DVB85] investigate blood levels of createnine (CREAT) in mg% and β_2 microglobulin (B2M) in mg/l as indicators of imminent rejection $\{D = 1\}$ in kidney transplant patients. Based on data from 55 patients, of whom 33 suffered episodes of rejection, DeLong and her colleagues obtained the data in Table 5.2.

For example, as a screening test for imminent rejection, we might take a createnine level above 2.5 to be a positive test result. Then we would estimate its sensitivity as $\eta(2.5) = 24/33 = 0.727$ because 24 patients with such creatinine levels had a rejection episode soon after. Similarly, we estimate its specificity as $\theta(2.5) \approx 0.535$ because that is the proportion of occasions on which no rejection episode closely followed such a creatinine level. If we want a test that "sounds the alarm" more often, we can use a level smaller than 2.5. Then we will "predict" more rejection episodes, but we will also have more false alarms.

Use these data to make approximate ROC curves for both CREAT and B2M. Put both sets of points on the same plot, using different symbols (or colors) for each, and try to draw a smooth curve through each set of points (imitating Figure 5.1). Compare your curves to determine whether it is worthwhile to use a test based on the more expensive B2M determinations. Would you use CREAT or B2M? If false positives and false negatives were equally serious, what cutoff value would you use? What if false negatives are somewhat more serious? Defend your choices.

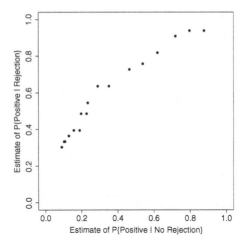

Figure 5.5. Approximating an ROC curve. An ROC curve can be approximated by drawing a smooth curve through these points, estimated from the createnine data of Problem 5.6. Compare this with Figure 5.1 (p123).

Notes: Data can be coded as follows. Use `plot` for the first set of points (as shown in Figure 5.5), then `points` to overlay the second.

```
cre.sens = c(.939, .939, .909, .818, .758, .727, .636, .636, .545,
             .485, .485, .394, .394, .364, .333, .333, .333, .303)
cre.spec = c(.123, .203, .281, .380, .461, .535, .649, .711, .766,
             .773, .803, .811, .843, .870, .891, .894, .896, .909)
b2m.sens = c(.909, .909, .909, .909, .879, .879, .879, .879, .818,
             .818, .818, .788, .788, .697, .636, .606, .576, .576)
b2m.spec = c(.067, .074, .084, .123, .149, .172, .215, .236, .288,
             .359, .400, .429, .474, .512, .539, .596, .639, .676)
```

In practice, a combination of the two determinations, including their day-to-day changes, may provide better predictions than either determination alone. See [DVB85] for an exploration of this possibility and also for a general discussion (with further references) of a number of issues in diagnostic testing. The CREAT data also appear in [PG00] along with the corresponding approximate ROC curve.

Problems for Section 5.2 (Estimates of Prevalence)

5.7   *Confidence intervals for prevalence.*

a) Compute the 95% confidence interval for τ given in Section 5.2. Show how the corresponding 95% confidence interval for π is obtained from this confidence interval.

b) The traditional approximate confidence interval for a binomial probability used in Section 5.2 can seriously overstate the confidence level. Especially for samples of small or moderate size, the approximate confidence interval suggested by Agresti and Coull [AC98] is more accurate. Their procedure is to "add two successes and two failures" when estimating the probability of success. Here, this amounts to using $t' = (A+2)/n'$, where $n' = n+4$, as the point estimate of τ and then computing the confidence interval $t' \pm 1.96\sqrt{t'(1-t')/n'}$. Find the 95% Agresti-Coull confidence interval for τ and from it the confidence interval for π given in Section 5.2. (See Chapter 1 for a discussion of the Agresti-Coull adjustment.)

5.8   Suppose that a screening test for a particular parasite in humans has sensitivity 80% and specificity 70%.

a) In a sample of 100 from a population, we obtain 45 positive tests. Estimate the prevalence.
b) In a sample of 70 from a different population, we obtain 62 positive tests. Estimate the prevalence. How do you explain this result?

5.9   Consider the ELISA test of Example 5.2, and suppose that the prevalence of infection is $\pi = 1\%$ of the units of blood in a certain population.

a) What proportion of units of blood from this population test positive?
b) Suppose that $n = 250$ units of blood are tested and that A of them yield positive results. What values of $t = A/n$ and of the integer A yield a negative estimate of prevalence?
c) Use the results of part (b) to find the proportion of random samples of size 250 from this population that yields negative estimates of prevalence.

Problems for Section 5.3 (Predictive Values)

5.10   Write a program to make a figure similar to Figure 5.4 (p127). What are the exact values of PVP γ and PVN δ when $\pi = 0.05$?

5.11   Suppose that a screening test for a particular disease is to be given to all available members of a population. The goal is to detect the disease early enough that a cure is still possible. This test is relatively cheap, convenient, and safe. It has sensitivity 98% and specificity 96%. Suppose that the prevalence of the disease in this population is 0.5%.

a) What proportion of those who test positive will actually have the disease? Even though this value may seem quite low, notice that it is much greater than 0.5%.
b) All of those who test positive will be subjected to more expensive, less convenient (possibly even somewhat risky) diagnostic procedures to determine whether or not they actually have the disease. What percentage of the population will be subjected to these procedures?

c) The entire population can be viewed as having been split into four groups: true and false positive, true and false negative. What proportion of the entire population falls into each of these four categories? Suppose you could change the sensitivity of the test to 99% with a consequent change in specificity to 94%. What factors of economics, patient risk, and preservation of life would be involved in deciding whether to make this change?

Note: (b) This is a small fraction of the population. It would have been prohibitively expensive (and depending on risks, possibly even unethical) to perform the definitive diagnostic procedures on the entire population. But the screening test permits focus on a small subpopulation of people who are relatively likely to have the disease and in which it may be feasible to perform the definitive diagnostic procedures.

5.12   Recall the lie detector test of Problem 5.1. In the population of interest, suppose 5% of the people are liars.

a) What is the probability that a randomly chosen member of the population will fail the test?
b) What proportion of those who fail the test are really liars? What proportion of those who fail the test are really truth-tellers?
c) What proportion of those who pass the test is really telling the truth?
d) Following the notation of this chapter, express the probabilities and proportions in parts (a), (b), and (c) in terms of the appropriate Greek letters.

5.13   In Example 5.3, a regulatory agency may be concerned with the values of η and γ. Interpret these two conditional probabilities in terms of testing a batch for potency. Extend the program in this example to obtain approximate numerical values for η and γ.

Note: For verification, the method of Problem 5.14 provides values accurate to at least four decimal places.

5.14   The results of Example 5.3 can be obtained without simulation through a combination of analytic and computational methods.

a) Express the conditional probabilities η, θ, γ, and δ in terms of π, τ, and $P(G \cap P) = P\{S > 100, X > 101\}$.
b) Denote the density function of $\mathrm{NORM}(\mu, \sigma)$ by $\varphi(\,\cdot\,, \mu, \sigma)$ and its CDF by $\Phi(\,\cdot\,, \mu, \sigma)$. Manipulate a double integral to show that

$$P(G \cap P) = \int_{100}^{\infty} \varphi(s, 110, 5)[1 - \Phi(101, s, 1)]\, ds.$$

c) **Write a program** in R to evaluate $P(G \cap P)$ by Riemann approximation and to compute the four conditional probabilities of part (a). We suggest including (and explaining) the following lines of code.

```
mu.s = 110;   sd.s = 5;   cut.s = 100   # as in the Example
s = seq(cut.s, mu.s + 5 * sd.s, .001)
```

```
int.len = mu.s + 5 * sd.s - cut.s
integrand = dnorm(s, mu.s, sd.s) * (1 - pnorm(cut.x, s, sd.x))
pr.gp = int.len * mean(integrand);  pr.gp
```

Compare your results for θ and δ with the simulated values shown in Example 5.3. (If you did Problem 5.13, then also compare your results for η and γ with the values simulated there.)

5.15   In Example 5.3, change the rule for "passing inspection" as follows. Each batch is assayed twice; if either of the two assays gives a result above 101, then the batch passes.

a) Change the program of the example to simulate the new situation; some useful R code is suggested below. What is the effect of this change on τ, θ, and γ?

```
x1 = rnorm(n,s,sd.x);   x2 = rnorm(n,s,sd.x);   x = pmax(x1, x2)
```

b) If you did Problem 5.13, then also compare the numerical values of η and γ before and after the change in the inspection protocol.

Problems for Section 5.4 (Bayes' Theorem)

5.16   In Example 5.4, suppose that Machine D is removed from service and that Machine C is used to make 20% of the parts (without a change in its error rate). What is the overall error rate now? If a defective part is selected at random, what is the probability that it was made by Machine A?

5.17   There are three urns, identical in outward appearance. Two of them each contain 3 red balls and 1 white ball. One of them contains 1 red ball and 3 white balls. One of the three urns is selected at random.

a) Neither you nor John has looked into the urn. On an intuitive "hunch," John is willing to make you an even-money bet that the urn selected has one red ball. (You each put up $1 and then look into the urn. He gets both dollars if the urn has exactly one red ball, otherwise you do.) Would you take the bet? Explain briefly.

b) Consider the same situation as in (a), except that one ball has been chosen at random from the urn selected, and that ball is white. The result of this draw has provided both of you with some additional information. Would you take the bet in this situation? Explain briefly.

5.18   According to his or her use of an illegal drug, each employee in a large company belongs to exactly one of three categories: frequent user, occasional user, or abstainer (never uses the drug at all). Suppose that the percentages of employees in these categories are 2%, 8%, and 90%, respectively. Further suppose that a urine test for this drug is positive 98% of the time for frequent users, 50% of the time for occasional users, and 5% of the time for abstainers.

a) If employees are selected at random from this company and given this drug test, what percentage of them will test positive?
b) Of those employees who test positive, what percentage are abstainers?
c) Suppose that employees are selected at random for testing and that those who test positive are severely disciplined or dismissed. How might an employee union or civil rights organization argue against the fairness of drug testing in these circumstances?
d) Can you envision different circumstances under which it might be appropriate to use such a test in the workplace? Explain.

Comment: (d) Consider, as one example, a railroad that tests only train operators who have just crashed a train into the rear of another train.

5.19   *A general form of Bayes' Theorem.*

a) Using f with appropriate subscripts to denote joint, marginal, and conditional density functions, we can state a form of Bayes' Theorem that applies to distributions generally, not just to probabilities of events,

$$f_{S|X}(s|x) = \frac{f_{X,S}(x,s)}{f_X(x)} = \frac{f_{X,S}(x,s)}{\int f_{X,S}(x,s)\,ds} = \frac{f_S(s)f_{X|S}(x|s)}{\int f_S(s)f_{X|S}(x|s)\,ds},$$

where the integrals are taken over the real line. Give reasons for each step in this equation. (Compare this result with equation (5.7).)
b) In Example 5.3, $S \sim$ NORM(110, 5) is the potency of a randomly chosen batch of drug, and it is assayed as $X|\{S = s\} \sim$ NORM(s, 1). The expression for the posterior density in part (a) allows us to find the probability that a batch is good given that it assayed at 100.5, thus barely failing inspection. Explain why this is not the same as $1 - \delta$.
c) We seek $P\{S > 100|X = 100.5\}$. Recalling the distribution of X from Example 5.3 and using the notation of Problem 5.14, show that this probability can be evaluated as follows:

$$\int_{100}^{\infty} f_{S|X}(s|100.5)\,ds = \frac{\int_{100}^{\infty} \varphi(s,110,5)\varphi(100.5,s,1)\,ds}{\varphi(100.5,110,5.099)}.$$

d) The R code below implements Riemann approximation of the probability in part (c). Run the program and provide a numerical answer (roughly 0.8) to three decimal places. In Chapter 8, we will see how to find the exact conditional distribution $S|\{X = 100.5\}$, which is normal. For now, carefully explain why this R code can be used to find the required probability.

```
s = seq(100, 130, 0.001)
numer = 30 * mean(dnorm(s, 110, 5) * dnorm(100.5, s, 1))
denom = dnorm(100.5, 110, 5.099)
numer/denom
```

6

Markov Chains with Two States

A **stochastic process** is a collection of random variables, usually considered to be indexed by time. In this book we consider sequences of random variables $X_1, X_2, \ldots$, viewing the subscripts $1, 2, \ldots$ as successive **steps** in time. The values assumed by the random variables X_n are called **states** of the process, and the set of all its states is called the **state space**. Sometimes it is convenient to think of the process as describing the movement of a particle over time. If $X_1 = i$ and $X_2 = j$, then we say that the process (or particle) has made a **transition** from state i at step 1 to state j at step 2. Often we are interested in the behavior of the process over the long run—after many transitions.

The definition of a stochastic process is very broad. There is no restriction on the variety of probability distributions that the various X_n may follow the complexity of the associations among these distributions. In applications, we seek models that strike an appropriate compromise between mathematical simplicity and practical realism. Here are a few examples of stochastic processes we have already considered.

- Example 3.5 (p58) shows an especially simple process of independent tosses of a fair coin with $P(\text{Heads}) = \pi$. Here the state space is $S = \{0, 1\}$, the (marginal) distribution at each step n is given by $X_n \sim \text{BINOM}(1, \pi)$, and the steps are *independent* of one another. We saw that, over the long run, the proportion of Heads (the proportion of the time the process is "in state 1") is approximately equal to π.

- In Example 3.6 (p61), we simulated a somewhat more complex process in which the probability of rain on one day *depends* on the weather of the previous day. Here again, the state space is $S = \{0, 1\}$, where $0 = \text{Sunny}$ and $1 = \text{Rainy}$. A longer simulation run than the one we did there would show the long-run proportion of rainy days to be $1/3$. (Problem 6.2 explores joint distributions of two dependent random variables, each taking only values 0 and 1.)

E.A. Suess and B.E. Trumbo, *Introduction to Probability Simulation and Gibbs Sampling with R*, 139
Use R!, DOI 10.1007/978-0-387-68765-0_6, © Springer Science+Business Media, LLC 2010

- Problem 3.20 (p80) provides examples of two dependent weather processes that converge more quickly. (If you have not already worked this problem as you read Chapter 3, now may be a good time to do so.)

The purpose of this chapter is to introduce Markov chains, which we use for computational purposes throughout the rest of this book. A Markov chain is a particular kind of stochastic process that allows for the possibility of a "limited" dependence among the random variables X_n. (Such processes were first considered by the Russian mathematician A. A. Markov in 1907. *Markov* is usually pronounced and sometimes spelled as *Markoff*.)

As we see in the next section, all the examples itemized above are Markov chains with two states. In this chapter, we look at the structure, applications, and long-run behavior of 2-state Markov chains.

6.1 The Markov Property

A 2-state Markov chain is a sequence of random variables X_n, $n = 1, 2, \ldots,$ that take only two values (which we call 0 and 1). The random variables X_n are not necessarily independent, but any dependence is of a restricted nature. In particular,

$$
\begin{aligned}
p_{01} &= P\{X_n = 1 | X_{n-1} = 0, X_{n-2} = i_{n-2}, \ldots, X_1 = i_1\} \\
&= P\{X_n = 1 | X_{n-1} = 0\} = \alpha,
\end{aligned}
\tag{6.1}
$$

where $0 \leq \alpha \leq 1$. That is, the probability α of making a transition from state 0 at step $n - 1$ to state 1 at step n does not depend on the states (0 or 1) that were occupied before step $n - 1$. Of course, it follows that

$$
p_{00} = P\{X_n = 0 | X_{n-1} = 0\} = 1 - \alpha.
$$

Similarly, we define

$$
\begin{aligned}
p_{10} &= P\{X_n = 0 | X_{n-1} = 1, X_{n-2} = i_{n-2}, \ldots, X_1 = i_1\} \\
&= P\{X_n = 0 | X_{n-1} = 1\} = \beta,
\end{aligned}
\tag{6.2}
$$

where $0 \leq \beta \leq 1$, so that

$$
p_{11} = P\{X_n = 1 | X_{n-1} = 1\} = 1 - \beta.
$$

A consequence of equations (6.1) and (6.2) is that only the state at the most recently observed step may be relevant in predicting what will happen next. This one-step-at-most kind of dependence, known as the **Markov property**, characterizes Markov chains.

Here we consider only **homogeneous** Markov chains, for which the probability of any particular transition from one step to the next remains constant over time. For example,

$$p_{01} = P\{X_2 = 1 | X_1 = 0\} = P\{X_3 = 1 | X_2 = 0\} = \cdots$$
$$= P\{X_n = 1 | X_{n-1} = 0\} = P\{X_{n+1} = 1 | X_n = 0\} = \alpha,$$

for any n. Similarly, β is constant over time. To keep the language simple, we will usually say *Markov chain* instead of *homogeneous Markov chain*.

Sometimes Markov dependence provides a good fit to real data and sometimes it does not.

- Experience with actual data has shown that Markov dependence works well for modeling some kinds of weather patterns. The dependence of today's weather on yesterday's weather may be very slight or even nonexistent in some climates. In others, the dependence is stronger but of a kind well-suited to Markov modeling. (Cox and Miller [CM65] report winter rainfall data from Tel Aviv that fits a strongly dependent Markov chain.)
- A Markov chain would be better than a strictly independent process for modeling whether a worker holds a blue-collar (0) or white-collar (1) job at the end of each month, but the fit to actual data would not be precise. Human memory extends farther than a month back in time. So in predicting the next job choices of accountants (1) who have just been laid off, a Markov chain could not distinguish between an accountant who has previously held only white-collar jobs and one who made a good living as an electrician (0) a few years ago.
  Such a Markov model was introduced in a monograph by Blumen et al. [BKM55] as an approximate method to study industrial labor mobility. They called it a **mover-stayer model**, and its modifications have been used in a wide variety of fields, including finance, criminology, geography, and epidemiology. See [Goo61] and [Fry84].

In many important applications, all four transition probabilities p_{00}, p_{01}, p_{10}, and p_{11} are positive. That is, $0 < \alpha, \beta < 1$. However, our definition of a 2-state Markov chain allows for situations where certain movements from one step to the next are prohibited or required.

First, we consider two strictly **deterministic cases**. If $\alpha = \beta = 0$, then no movement among states is possible. In this "never-move" chain, the value of X_1 determines the values of all other $X_n = X_1$. In contrast, if $\alpha = \beta = 1$, the process cycles between 0 and 1 on alternate steps. In such a "flip-flop" chain, the value $X_1 = 0$ would determine the values of all other X_n, with $X_n = 0$ for all odd values of n and $X_n = 1$ for all even values of n.

Second, two **absorbing cases** are also included in our definition of a Markov chain. If $\alpha = 0$ and $\beta > 0$, then the chain will eventually move to state 0, where it will stay ("be absorbed") for all subsequent steps n. Suppose that $X_1 = 1$. Then, if $\beta = 1$, absorption into 0 occurs precisely at step 2. Otherwise, the length of the run of 1s before absorption into 0 is governed by a geometric distribution with mean $1/\beta$. Similarly, if $\beta = 0$ and $\alpha > 0$, then 1 is the absorbing state. (See Problem 6.3.)

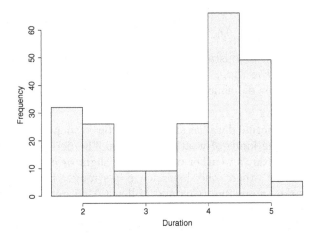

Figure 6.1. Histogram of the durations of 222 eruptions of Old Faithful geyser. In this bimodal distribution, we arbitrarily choose 3 minutes as the boundary between "Short" and "Long" eruptions.

Finally, among these restrictive models, we explore an example in which $\alpha = 1$ and $0 < \beta < 1$. (The case where $\beta = 1$ and $0 < \alpha < 1$ is similar.)

Example 6.1. Modeling Eruptions of Old Faithful Geyser. According to data collected on 16 days during the summers of 1978 and 1979, eruptions of Old Faithful geyser in Yellowstone National Park can be classified as either Short (0) or Long (1). We take the Long eruptions to be those lasting 3 minutes or more. The histogram in Figure 6.1, based on 222 eruptions, illustrates the rationale for this somewhat arbitrary choice. (Data were collected by park rangers and reported by Weisberg in [Wei85].)

Looking at the plot of 205 *adjacent* pairs of eruptions in Figure 6.2, we conclude that the behavior of successive eruptions can be modeled as a Markov chain with the estimated values $\alpha = 1$ and $\beta = 0.44$. That is, in the available data, Short eruptions are always followed by Long ones, and about 44% of Long eruptions are followed by Short ones. (Although 222 eruptions were observed, only 205 could be paired with an immediately following one.)

The probability rules governing the long-run behavior of this Markov chain are relatively simple. Once a Long eruption occurs, there is some tendency for the next eruption to be Long also: 56% of the time. The eventual transition to a Short eruption is governed by a geometric distribution with probability 0.44. So the average length of a run of Long eruptions will be $1/0.44 = 2.2727$. Then a single Short eruption will be followed immediately by a Long one and the cycle will repeat. On average, there will be 2.2727 Long eruptions out of the 3.2727 eruptions in a complete cycle. Thus, over the long run, the proportion of Long eruptions will be $2.2727/3.2727 = 69.44\%$, which agrees well with the data: 69.8% Long eruptions among the 222 observed.

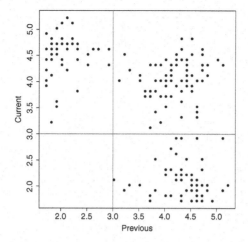

Figure 6.2. Dependence between durations of adjacent eruptions of Old Faithful geyser. Each of 205 eruptions is plotted against the one just before. No Short eruption (less than 3 minutes) was immediately followed by another Short eruption.

The following R script simulates $m = 2000$ steps of this Markov chain. From the results, we approximate both the long-run proportion of Long eruptions and the average cycle length. For simplicity, the number a of cycles is taken to be the number of $0 \rightarrow 1$ transitions and the average cycle length to be m/a. This may be very slightly inaccurate because the last cycle is truncated at m and so may be incomplete. Because $\alpha = 1$, we could replace rbinom(1, 1, alpha) by 1, but for consistency with our other simulations of Markov chains, we use code that works for $0 \leq \alpha \leq 1$.

```
# Preliminaries
# set.seed(1237)
m = 2000;  n = 1:m;  x = numeric(m);  x[1] = 0
alpha = 1;  beta = 0.44

# Simulation
for (i in 2:m)
{
   if (x[i-1]==0)  x[i] = rbinom(1, 1, alpha)
   else            x[i] = rbinom(1, 1, 1 - beta)
}
y = cumsum(x)/n                 # Running fractions of Long eruptions

# Results
y[m]         # Fraction of Long eruptions among m.  Same as: mean(x)
a = sum(x[1:(m-1)]==0 & x[2:m]==1);  a  # No. of cycles
m/a                              # Average cycle length
plot(x[1:20], type="b", xlab="Step", ylab="State")
```

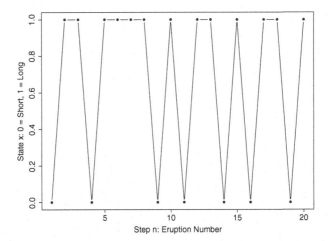

Figure 6.3. Simulated Old Faithful process. The alternation between Short and Long eruptions is shown for the first 20 eruptions. Each return to Short (0) starts a new cycle of this 2-state chain.

```
> y[m]          # Fraction of Long eruptions among m.  Same as: mean(x)
[1] 0.697
> a = sum(x[1:(m-1)]==0 & x[2:m]==1);  a  # No. of cycles
[1] 605
> m/a                               # Average cycle length
[1] 3.305785
```

The proportion 0.70 of the $m = 2000$ simulated steps that result in Long eruptions and the average 3.3 of the cycle lengths (run of Long eruptions followed by one Short eruption) are in good agreement with the theoretical values derived above. Figure 6.3 shows the first few cycles in our simulation.

Moreover, Figure 6.4 plots the cumulative proportion of Long eruptions at each simulated step, showing that this Markov chain stabilizes quickly to its limiting distribution as the process continues to cycle between Long and Short states. It is made with the following additional line of code,

```
plot(y, type="l", ylim=c(0,1), xlab="Step", ylab="Proportion Long")
```

If we know the current state of this chain, then we have some strong information about the next state. Specifically, a Short eruption is always followed by a Long one, and about 44% of Long eruptions are followed by Short ones. However, this one-step Markov dependence dissipates rather quickly, so that knowledge of the current step does not tell us much about what will happen several steps later. By looking at autocorrelations of the process, the additional code below illustrates one way to quantify this "wearing away" of dependence over time.

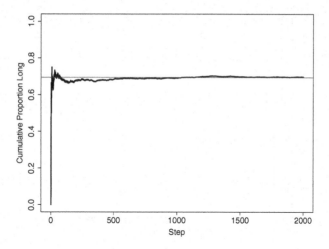

Figure 6.4. Simulated Old Faithful process. For each simulated step the proportion of long eruptions to date is plotted. This proportion converges to about 70%.

```
acf(x)              # Autocorrelation plot
acf(x, plot=F)      # Printed output

> acf(x, plot=F)     # Printed output
  Autocorrelations of series 'x', by lag

   0       1       2       3       4       5       6       7       8
1.000  -0.434   0.163  -0.072   0.038  -0.062   0.051  -0.043   0.028
  ...
```

As always, the autocorrelation of order 0 is 1. The estimated autocorrelation of order 1 is about −0.43, and autocorrelations of larger orders tend to alternate in sign and trend quickly to values that are not significantly different from 0 (see Figure 6.5). This autocorrelation structure provides additional evidence that the simulated values of X_n have a limiting distribution that does not depend on the initial state.◇

Is a Markov chain a realistic model for the eruptions of Old Faithful? The answer is that it is about as good as we can do with the data at hand. First, our Markov chain is clearly better than an independent process with 70% Long eruptions and 30% Short ones. Under this independent model, 30% of Short eruptions would be followed by another Short one, and that never happens in our data. Second, it might be possible to construct a model that makes more detailed use of past information and is somewhat more accurate than our Markov chain. But with the limited amount of data available, it would be difficult to verify whether this is so.

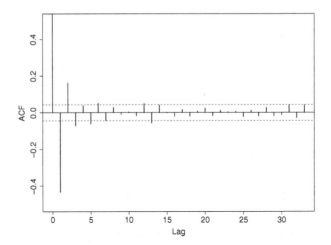

Figure 6.5. Simulated Old Faithful process. A plot of the autocorrelation function. As the lag increases, the correlation shrinks to nonsignificant levels.

6.2 Transition Matrices

We have adopted the notation $p_{ij} = P\{X_{n+1} = j | X_n = i\}$ for the probability of a 1-step transition from state i to state j, for $i, j = 0, 1$, and for any $n = 1, 2, \ldots$. It is useful to arrange these probabilities in a matrix,

$$\mathbf{P} = \begin{bmatrix} p_{00} & p_{01} \\ p_{10} & p_{11} \end{bmatrix} = \begin{bmatrix} 1 - \alpha & \alpha \\ \beta & 1 - \beta \end{bmatrix},$$

where the ith row gives the conditional distribution of $X_{n+1} | X_n = i$. Thus each row of a transition matrix sums to 1.

The 2-step transition probability $p_{ij}(2) = P\{X_{n+2} = j | X_n = i\}$ is found by taking into account the two possible states of X_{n+1}. For example, let $n = 1$, $i = 0$, and $j = 1$. The sequence of states for stages 1, 2, and 3 must be either $0 \to 0 \to 1$ or $0 \to 1 \to 1$. Thus

$$p_{01}(2) = P\{X_3 = 1 | X_1 = 0\} = p_{00}p_{01} + p_{01}p_{11} = \alpha(2 - \alpha - \beta), \quad (6.3)$$

where the first equality is by definition, the second is intuitively reasonable (with rigorous details of its proof relegated to Problem 6.7), and the third follows from simple algebra.

More generally, we have the **Chapman-Kolmogorov equations**

$$p_{ij}(2) = \sum_{k \in S} p_{ik}p_{kj}, \quad (6.4)$$

for $i, j \in S$, where S is the state space of the Markov chain. This means that the 2-step transition matrix $\mathbf{P}^2$ of elements $p_{ij}(2)$ can be found as the square of the 1-step transition matrix $\mathbf{P}$ by using ordinary matrix multiplication. After some algebra (see Problem 6.8), we have

$$\mathbf{P}^2 = \frac{1}{\alpha + \beta} \begin{bmatrix} \beta & \alpha \\ \beta & \alpha \end{bmatrix} + \frac{(1 - \alpha - \beta)^2}{\alpha + \beta} \begin{bmatrix} \alpha & -\alpha \\ -\beta & \beta \end{bmatrix}, \tag{6.5}$$

for $\alpha + \beta > 0$. (As usual, the coefficient of a matrix multiplies each of its elements, and the sum of two matrices is found by adding corresponding elements.) Thus, for example, the upper-right element of $\mathbf{P}^2$ is

$$p_{01}(2) = \frac{\alpha}{\alpha + \beta} + \frac{(1 - \alpha - \beta)^2(-\alpha)}{\alpha + \beta} = \alpha(2 - \alpha - \beta),$$

which agrees with equation (6.3).

Similarly, the r-step transition matrix for a Markov chain is found by taking the rth power of $\mathbf{P}$. It can be shown by mathematical induction (see Problem 6.9) that, for $\alpha + \beta > 0$,

$$\mathbf{P}^r = \frac{1}{\alpha + \beta} \begin{bmatrix} \beta & \alpha \\ \beta & \alpha \end{bmatrix} + \frac{(1 - \alpha - \beta)^r}{\alpha + \beta} \begin{bmatrix} \alpha & -\alpha \\ -\beta & \beta \end{bmatrix}. \tag{6.6}$$

Example 6.2. Old Faithful Chain (continued). Recall that in Example 6.1 the transition matrix $\mathbf{P}$ is specified by $p_{01} = \alpha = 1$ and $p_{10} = \beta = 0.44$. Given that the first eruption was Short (0), let us find the conditional probability that the third eruption will be Long (1). Evaluating $p_{01}(2)$ according to equation (6.3), we have $p_{01}(2) = (0)(1) + (1)(0.56) = 0.56$, where one of the two 2-step paths turns out to have 0 probability.

To find the probability that $X_5 = 1$ subject to the same condition, we can substitute the numerical values of r, α, and β into the upper-right element in the matrix equation (6.6) to obtain

$$p_{01}(4) = \frac{\alpha + (1 - \alpha - \beta)^r(-\alpha)}{\alpha + \beta} = \frac{1 - (-0.44)^4}{1.44} = 0.6684. \tag{6.7}$$

In R, the computation of $\mathbf{P}^r$, for $r = 2$, 4, 8, and 16, can be done as shown below. Because R labels elements of matrices starting at 1 rather than 0, you will find $p_{01}(r)$ by looking at row [1,] and column [,2] in the appropriate section of output.

```
P = matrix(c(    0,   1,
               .44, .56), nrow=2, ncol=2, byrow=T)

P
P2  = P %*% P;    P2
P4  = P2 %*% P2;  P4
P8  = P4 %*% P4;  P8
P16 = P8 %*% P8;  P16
```

```
>  P
       [,1] [,2]
[1,]  0.00 1.00
[2,]  0.44 0.56
>  P2   = P %*% P;   P2
          [,1]    [,2]
[1,]  0.4400 0.5600
[2,]  0.2464 0.7536
>  P4   = P2 %*% P2; P4
            [,1]      [,2]
[1,]  0.3315840 0.668416
[2,]  0.2941030 0.705897
>  P8   = P4 %*% P4; P8
           [,1]       [,2]
[1,]  0.3065311 0.6934689
[2,]  0.3051263 0.6948737
>  P16 = P8 %*% P8; P16
           [,1]       [,2]
[1,]  0.3055569 0.6944431
[2,]  0.3055550 0.6944450
```

This computation confirms the calculation $p_{01}(4) = 0.6684$ we made above from equation (6.6). Also notice that $p_{11}(4) \approx 0.71$ and $p_{01}(4) \approx 0.69$ are not far apart. This means that, 4 steps into the future (at step 5), the probability of a Long eruption is roughly 0.7 regardless of whether the eruption at step 1 is Short or Long. After 16 transitions, the probability of a Long eruption is 0.6944—essentially independent of step 1. ◇

This tendency for dependence to "wear off" over time is a property of many chains used in practical probability modeling. In the next section, we explore when and how rapidly Markov dependence dissipates.

6.3 Limiting Behavior of a 2-State Chain

Equation (6.6) expresses $\mathbf{P}^r$ as the sum of two matrices. The first term does not depend on r, and the second term depends on r only as an exponent of $\Delta = 1 - \alpha - \beta$. Thus, for $|\Delta| = |1 - \alpha - \beta| < 1$, the second term vanishes as r increases. In symbols,

$$
\boldsymbol{\Lambda} = \lim_{r \to \infty} \mathbf{P}^r = \frac{1}{\alpha + \beta} \begin{bmatrix} \beta & \alpha \\ \beta & \alpha \end{bmatrix} + \frac{\lim_{r \to \infty} \Delta^r}{\alpha + \beta} \begin{bmatrix} \alpha & -\alpha \\ -\beta & \beta \end{bmatrix}
$$

$$
= \frac{1}{\alpha + \beta} \begin{bmatrix} \beta & \alpha \\ \beta & \alpha \end{bmatrix} = \begin{bmatrix} \boldsymbol{\lambda} \\ \boldsymbol{\lambda} \end{bmatrix}, \tag{6.8}
$$

where the limiting matrix consists of two identical row vectors, each expressing the long-run distribution, $\boldsymbol{\lambda} = [\lambda_0, \lambda_1] = [\frac{\beta}{\alpha+\beta}, \frac{\alpha}{\alpha+\beta}]$.

The condition $|\Delta| < 1$ excludes the deterministic cases where $\alpha + \beta = 0$ (never move) and $\alpha + \beta = 2$ (flip-flop). Notice that the rate of convergence is **geometric**; that is, the discrepancy from the limiting value decreases as a power of r. Thus, convergence is very rapid unless $|\Delta|$ is very near 1. This accounts for the near equality in the two rows of $\mathbf{P}^{16}$ seen in Example 6.2. There $\Delta = -0.44$, and so the chain essentially converges after very few transitions.

For a 2-state Markov chain, we conclude that it is easy to determine from the values of α and β whether the chain has a long-run probability distribution. If so, it is easy to compute the limiting probabilities λ_1 and $\lambda_0 = 1 - \lambda_1$ and also to see how quickly Markov dependence dissipates.

However, knowing the probability structure does not reveal exactly how a chain will behave in practice in a particular instance. If we approximate the long-run probability λ_1 by simulating $\bar{X}_m$, the convergence can be relatively slow, requiring m to be in the thousands in order to obtain a useful approximation. The examples below illustrate that, for a chain with transition matrix $\mathbf{P}$, the rate of convergence of $\mathbf{P}^r$ is not the same thing as the rate of convergence of a simulation of the chain.

Example 6.3. An Independent Chain. Suppose that we take independent samples from a population in which the proportion of individuals infected with a certain disease is $\alpha = 0.02$. If the nth individual sampled is infected, then $X_n = 1$; if not, then $X_n = 0$. This is a special case of a Markov chain with $\alpha = 1 - \beta = 0.02$. For this chain, there is no dependence to wear off, and $\mathbf{P} = \mathbf{P}^2 = \mathbf{P}^3 = \cdots = \mathbf{\Lambda} = \begin{bmatrix} 0.98 & 0.02 \\ 0.98 & 0.02 \end{bmatrix}$.

Using a result about binomial confidence intervals from Chapter 1, we see that it takes a simulation of length $m = (1.96/E)^2(0.02)(0.98) \approx 750$ to be 95% sure that $\bar{X}_m$ is within $E = 0.01$ of its limit $\lambda_1 = 0.02$. To be reasonably sure of two-place accuracy ($E = 0.005$), we would require $m \approx 3000$. ◇

Example 6.4. An Almost-Absorbing Chain. In Section 6.4, we use a Markov chain with $\alpha = 0.0123$ and $\beta = 0.6016$. For this chain, $\lim_{r\to\infty} p_{01}(r) = \lim_{r\to\infty} p_{11}(r) = \lambda_1 = \alpha/(\alpha + \beta) = 0.0123/0.6139 = 0.0200$. The convergence of $\mathbf{P}^r$ as $r \to \infty$ is rapid, with $\mathbf{P}^{16} \approx \begin{bmatrix} 0.9800 & 0.0200 \\ 0.9800 & 0.0200 \end{bmatrix}$.

However, because $\alpha = p_{01}$ is very small, once this chain enters state 0 it is unlikely to leave at any one step. Because $\alpha > 0$, state 0 is not an absorbing state: The chain *eventually* moves to state 1—after a run of 0s of average length $1/\alpha = 81.30$. The average length of a cycle of this 2-state Markov chain (a run of 0s followed by a run of 1s) is $1/\alpha + 1/\beta = 82.96$.

To get a good approximation of λ_1 by simulation, we need to go through many cycles. So this chain requires a larger run length m than does the independent chain (with its cycle length of 51.02) to achieve similar accuracy. Specifically, five successive runs of a program similar to that of Example 6.1 gave $\bar{X}_{3000} = 0.0147, 0.0210, 0.0203, 0.0263,$ and 0.0223. (In the notation of that program, $\bar{X}_{3000}$ is y[3000].) Clearly, $m = 3000$ does not consistently give two-place accuracy in simulating a chain with such a long cycle length. However, the first-order autocorrelations of the X_n in these runs were mostly between 0.3 and 0.5 and, in each run, higher-order autocorrelations trended towards 0. So longer runs should give satisfactory accuracy. (In contrast, Problem 6.10 shows a chain for which the powers of $\mathbf{P}$ converge more slowly than the simulated values $\bar{X}_n$.) $\diamondsuit$

6.4 A Simple Gibbs Sampler

A Gibbs sampler is a simulated Markov chain $X_1, X_2, \ldots$ that is constructed to have a desired long-run distribution. By observing enough simulated values of X_n after the simulation has stabilized, we hope that the distribution of sampled X_n will approximate its long-run distribution. In practice, Gibbs samplers are often used to approximate distributions that would be difficult to derive analytically. Here we illustrate a very simple Gibbs sampler in a screening test problem where we can find the correct answer by other means.

Consider a screening test. As in Chapter 5, the random variable $D = 1$ or 0 according to whether a tested sample is infected or not and the random variable $T = 1$ or 0 according to whether the sample tests positive or not. Suppose the conditional distributions $T|D$ and $D|T$ for a particular population are known in terms of the following four numbers:

$$\text{Sensitivity} = \eta = P\{T = 1|D = 1\},$$
$$\text{Specificity} = \theta = P\{T = 0|D = 0\},$$
$$\text{PV Positive} = \gamma = P\{D = 1|T = 1\},$$
$$\text{PV Negative} = \delta = P\{D = 0|T = 0\}.$$

In Chapter 5, we saw that, for a particular population, the two predictive values γ and δ can be computed easily from the prevalence $\pi = P\{D = 1\}$ of the infection in that population knowing the sensitivity η and specificity θ of the screening test. So it seems that we should be able to compute the distribution $\boldsymbol{\lambda} = (\lambda_0, \lambda_1) = (1 - \pi, \pi)$ of D from the four quantities given. But going from η, θ, γ, and δ to π is not quite as straightforward as going from π, η, and θ to γ and δ. Later in this section, we show one way to compute π analytically, but now we show how to approximate π with a Gibbs sampler.

The idea is to construct a Markov chain $D_1, D_2 \ldots$. We begin the simulation with an arbitrarily chosen value of $D_1 = 0$ or 1. Depending on the value of D_{n-1}, we obtain D_n in two "half-steps." In the first of these half-steps, we use the conditional distributions $T|D$ to simulate a value of T_{n-1}, and in the second we use the conditional distributions of $D|T$ to simulate a value of D_n. Specifically,

$$T_{n-1} \sim \mathsf{BINOM}(1, \eta) \text{ if } D_{n-1} = 1 \text{ or}$$
$$T_{n-1} \sim \mathsf{BINOM}(1, 1 - \theta) \text{ if } D_{n-1} = 0;$$
$$D_n \sim \mathsf{BINOM}(1, \gamma) \text{ if } T_{n-1} = 1 \text{ or}$$
$$D_n \sim \mathsf{BINOM}(1, 1 - \delta) \text{ if } T_{n-1} = 0.$$

Clearly, the D_n form a Markov chain. Once D_{n-1} is known, this procedure specifies all the distributional information for obtaining D_n. It is not necessary to know the value of D_{n-2} or earlier values.

We can use graphical methods to see if this chain converges to a long-run distribution. Assume that the chain does stabilize satisfactorily after an appropriate **burn-in period**. Then it seems that $\lim_{n \to \infty} P\{D_n = 1\} = \lambda_1$ can be approximated as the proportion of cases after burn-in for which $D_n = 1$.

Example 6.5. To illustrate this Gibbs sampler, we use the values $\eta = 0.99$, $\theta = 0.97$, $\gamma = 0.4024$, and $\delta = 0.9998$ of Chapter 5. Because the latter two values were computed from the former two for a population with $\pi = 0.02$, we expect the result $\lambda = (0.98, 0.02)$. The desired chain is simulated by the R script shown below.

```
# set.seed(1234)
m = 80000
eta = .99; theta = .97            # T|D
gamma = .4024;  delta = .9998     # D|T
d = numeric(m);  d[1] = 0         # vector of D's; start at 0
t = numeric (m)                   # vector of T's

for (n in 2:m)
{
   if (d[n-1]==1)  t[n-1] = rbinom(1, 1, eta)
      else         t[n-1] = rbinom(1, 1, 1 - theta)

   if (t[n-1]==1)  d[n] = rbinom(1, 1, gamma)
      else         d[n] = rbinom(1, 1, 1 - delta)
}

runprop = cumsum(d)/1:m           # running proportion infected
mean(d[(m/2+1):m])                # prevalence after burn-in

par(mfrow=c(1,2))
   plot(runprop, type="l", ylim=c(0,.05),
      xlab="Step", ylab="Running Proportion Infected")
   acf(d, ylim=c(-.1,.4), xlim=c(1,10))
par(mfrow=c(1,1))
acf(d, plot=F)

> mean(d[(m/2 + 1):m])            # prevalence after burn-in
[1] 0.01915
> acf(d, plot=F)
Autocorrelations of series 'd', by lag
```

0	1	2	3	4	5	6	7	8
1.000	0.395	0.156	0.055	0.017	0.006	-0.002	-0.009	-0.004

. . .

The left panel of Figure 6.6 shows a plot of running proportions of prevalence, which seem to stabilize nicely about halfway through the $m = 80\,000$ iterations programmed. We consider the first $40\,000$ iterations to constitute the burn-in period. The average of the last $40\,000$ simulated values of D is 0.019, which is consistent with the anticipated value $P\{D = 1\} = \lambda_1 = 0.02$.

Ten additional runs starting with $D_1 = 0$ gave results between 0.0184 and 0.0218, averaging 0.0200. Ten runs starting with $D_1 = 1$ gave results between 0.0180 and 0.0222, averaging 0.0199. This confirms that the starting value has a negligible effect on the result.

Moreover, estimated first-order autocorrelations of the simulated D_n from these additional runs averaged about 0.39, and, for each run, autocorrelations of higher orders converge quickly to 0. That the autocorrelations are mainly positive shows poorer mixing than in Example 6.2, but the quick convergence of higher-order autocorrelations to 0 provides further evidence that the Markov dependence wears off quickly. Autocorrelations for the particular simulation shown above are shown in the right panel of Figure 6.6.

An intuitive argument may help you to understand how this Gibbs sampler works. It is futile to try to simulate $\pi = P\{D = 1\} = 0.02$ from the conditional distributions $D|T$ alone. The probability $\gamma = P\{D = 1|T = 1\} = 0.4024$ is much too big and the probability $1 - \delta = P\{D = 1|T = 0\} = 0.0002$ much too small. However, as the simulation runs through the steps of our Markov chain, the conditional distributions of $T|D$ come into play to ensure that—in the long run—the probabilities γ and $1 - \delta$ are each used the appropriate proportion of the time, so we get 0.02 as a long-run weighted average of 0.4024 and 0.0002. ◇

Now we show explicitly that the random variables D_n of the Gibbs sampler are a 2-state Markov chain with the desired limiting distribution. We use the conditional distributions $T|D$ and $D|T$ to find the transition matrix of this chain. For $i, j, k \in S = \{0, 1\}$ and $n = 1, 2, \ldots$,

$$P\{D_n = j|D_{n-1} = i\} = \sum_{k \in S} P\{T_{n-1} = k, D_n = j|D_{n-1} = i\}$$

$$= \sum_{k \in S} P\{T_{n-1} = k|D_{n-1} = i\} P\{D_n = j|T_{n-1} = k\}$$

$$= \sum_{k in S} q_{ik} r_{kj},$$

where the second equation uses $P(A \cap B|C) = P(A|C)P(B|A \cap C)$ together with the Markov property, and the last equation defines q_{ik} and r_{kj}. The Markov property holds for this "half-step" because the earlier condition $D_n = i$ is irrelevant given the later information that $T_n = k$. When the symbols η, θ, γ, and δ are substituted as appropriate, the equation above can be written in matrix form as

$$\mathbf{P} = \mathbf{QR} = \begin{bmatrix} \theta & 1 - \theta \\ 1 - \eta & \eta \end{bmatrix} \begin{bmatrix} \delta & 1 - \delta \\ 1 - \gamma & \gamma \end{bmatrix}.$$

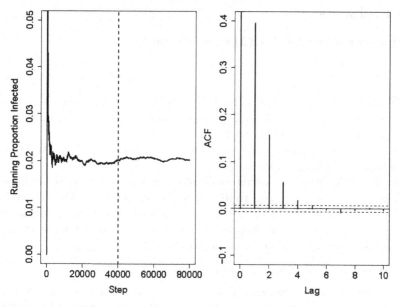

Figure 6.6. Diagnostic graphics for the simple Gibbs sampler of Example 6.5. The left panel shows relatively stable convergence of the simulation after a burn-in of 40 000 iterations. The right panel shows high-order autocorrelations tending to 0.

Example 6.6. Returning to the specific example of Chapter 5, we have the transition matrix

$$\mathbf{P} = \begin{bmatrix} 0.97 & 0.03 \\ 0.01 & 0.00 \end{bmatrix} \begin{bmatrix} 0.9998 & 0.0002 \\ 0.5976 & 0.4024 \end{bmatrix} = \begin{bmatrix} 0.9877 & 0.0123 \\ 0.6016 & 0.3984 \end{bmatrix}$$

of the "almost absorbing" 2-state chain with $\alpha = 0.0123$, $\beta = 0.6016$, and $\boldsymbol{\lambda} = (0.98, 0.02)$ that we discussed in Example 6.4. If the characteristics of our screening test had produced a chain in which the values 0 and 1 "mix" better (that is, tend to have shorter runs), then our Gibbs sampler would have converged more rapidly. (For a similar Gibbs sampler—with a less satisfactory clinical situation but more agreeable convergence properties—see Problem 6.13.) ◇

Especially in Bayesian estimation, Gibbs samplers are used to approximate distributions that would be very difficult or impossible to derive analytically. These more practical uses of Gibbs samplers rely upon Markov chains with larger state spaces, which we consider in the next chapter. Also, before we look at the most important applications Gibbs samplers, in Chapters 9 and 10, we introduce some of the basic ideas of Bayesian estimation in Chapter 8.

6.5 Problems

Problems for Section 6.1 (Markov Property)

6.1   In each part below, consider the three Markov chains specified as follows: (i) $\alpha = 0.3, \beta = 0.7$; (ii) $\alpha = 0.15, \beta = 0.35$; and (iii) $\alpha = 0.03, \beta = 0.07$.

a) Find $P\{X_2 = 1|X_1 = 1\}$ and $P\{X_n = 0|X_{n-1} = 0\}$, for $n \geq 2$.
b) Use the given values of α and β and means of geometric distributions to find the average cycle length for each chain.
c) For each chain, modify the program of Example 6.1 to find the long-run fraction of steps in state 1.
d) For each chain, make and interpret plots similar to Figures 6.3 (where the number of steps is chosen to illustrate the behavior clearly), 6.4, and 6.5.
e) In summary, what do these three chains have in common, and in what respects are they most remarkably different. Is any one of these chains an independent process, and how do you know?

Answers for one of the chains: (a) 0.85 and 0.65, (b) 9.52, (c) 0.3.

6.2   There is more information in the joint distribution of two random variables than can be discerned by looking only at their marginal distributions. Consider two random variables X_1 and X_2, each distributed as $\mathsf{BINOM}(1, \pi)$, where $0 < \pi < 1$.

a) In general, show that $0 \leq Q_{11} = P\{X_1 = 1, X_2 = 1\} \leq \pi$. In particular, evaluate Q_{11} in three cases: where X_1 and X_2 are independent, where $X_2 = X_1$, and where $X_2 = 1 - X_1$.
b) For each case in part (a), evaluate $Q_{00} = P\{X_1 = 0, X_2 = 0\}$.
c) If $P\{X_2 = 1|X_1 = 0\} = \alpha$ and $P\{X_2 = 0|X_1 = 1\} = \beta$, then express π, Q_{00}, and Q_{11} in terms of α and β.
d) In part (c), find the correlation $\rho = \mathrm{Cov}(X_1, X_2)\,/\,\mathrm{SD}(X_1)\mathrm{SD}(X_2)$, recalling that $\mathrm{Cov}(X_1, X_2) = \mathrm{E}(X_1 X_2) - \mathrm{E}(X_1)\mathrm{E}(X_2)$.

    Hints and partial answers: $P(A \cap B) \leq P(A)$. Make two-way tables of joint distributions, showing marginal totals. $\pi = \alpha/(\alpha + \beta)$. $\mathrm{E}(X_1 X_2) = P\{X_1 X_2 = 1\}$. $\rho = 1 - \alpha - \beta$. Independence (hence $\rho = 0$) requires $\alpha + \beta = 1$.

6.3   *Geometric distributions.* Consider a coin with $0 < P(\text{Heads}) = \pi < 1$. A geometric random variable X can be used to model the number of independent tosses required until the first Head is seen.

a) Show that the probability function $P\{X = x\} = p(x) = (1 - \pi)^{x-1}\pi$, for $x = 1, 2, \ldots$.
b) Show that the geometric series $p(x)$ sums to 1, so that one is sure to see a Head eventually.
c) Show that the moment generating function of X is $m(t) = \mathrm{E}(e^{tX}) = \pi e^t/[1 - (1 - \pi)e^t]$, and hence that $\mathrm{E}(X) = m'(0) = \frac{dm(t)}{dt}|_{t=0} = 1/\pi$. (You may assume that the limits involved in differentiation and summing an infinite series can be interchanged as required.)

Problems for Section 6.2 (Transition Matrices)

6.4   Suppose the weather for a day is either Dry (0) or Rainy (1) according to a homogeneous 2-state Markov chain with $\alpha = 0.1$ and $\beta = 0.5$. Today is Monday ($n = 1$) and the weather is Dry.

a) What is the probability that both tomorrow and Wednesday will be Dry?
b) What is the probability that it will be Dry on Wednesday?
c) Use equation (6.5) to find the probability that it will be Dry two weeks from Wednesday ($n = 17$).
d) Modify the R code of Example 6.2 to find the probability that it will be Dry two weeks from Wednesday.
e) Over the long run, what will be the proportion of Rainy days? Modify the R code of Example 6.1 to simulate the chain and find an approximate answer.
f) What is the average length of runs of Rainy days?
g) How do the answers above change if $\alpha = 0.15$ and $\beta = 0.75$?

6.5   Several processes $X_1, X_2, \ldots$ are described below. For each of them evaluate (i) $P\{X_3 = 0 | X_2 = 1\}$, (ii) $P\{X_{13} = 0 | X_{12} = 1, X_{11} = 0\}$, (iii) $P\{X_{13} = 0 | X_{12} = 1, X_{11} = 1\}$, and (iv) $P\{X_{13} = 0 | X_{11} = 0\}$. Also, (v) say whether the process is a 2-state homogeneous Markov chain. If not, show how it fails to satisfy the Markov property. If so, give its 1-stage transition matrix **P**.

a) Each X_n is determined by an independent toss of a coin, taking the value 0 if the coin shows Tails and 1 if it shows Heads, with $0 < P(\text{Heads}) = \pi < 1$.
b) The value of X_1 is determined by whether a toss of a fair coin is Tails (0) or Heads (1), and X_2 is determined similarly by a second independent toss of the coin. For $n > 2$, $X_n = X_1$ for odd n, and $X_n = X_2$ for even n.
c) The value of X_1 is 0 or 1, according to whether the roll of a fair die gives a Six (1) or some other value (0). For each step $n > 1$, if a roll of the die shows Six on the nth roll, then $X_n \neq X_{n-1}$; otherwise, $X_n = X_{n-1}$.
d) Start with $X_1 = 0$. For $n > 1$, a fair die is rolled. If the maximum value shown on the die at any of the steps $2, \ldots, n$ is smaller than 6, then $X_n = 0$; otherwise, $X_n = 1$.
e) At each step $n > 1$, a fair coin is tossed, and U_n takes the value -1 if the coin shows Tails and 1 if it shows Heads. Starting with $V_1 = 0$, the value of V_n for $n > 1$ is determined by

$$V_n = V_{n-1} + U_n \pmod 4.$$

The process V_n is sometimes called a "random walk" on the points 0, 1, 2, and 3, arranged around a circle (with 0 adjacent to 3). Finally, $X_n = 0$, if $V_n = 0$; otherwise $X_n = 1$.

Hints and partial answers: (a) Independence is consistent with the Markov property. (b) Steps 1 and 2 are independent. Show that the values at steps 1 and 2 determine the value at step 3 but the value at step 2 alone does not. (c) $\mathbf{P} = \frac{1}{6} \begin{bmatrix} 5 & 1 \\ 1 & 5 \end{bmatrix}$. (d) Markov chain. (e) (ii) > 0, (iii) $= 0$.

6.6   To monitor the flow of traffic exiting a busy freeway into an industrial area, the highway department has a TV camera aimed at traffic on a one-lane exit ramp. Each vehicle that passes in sequence can be classified as Light (for example, an automobile, van, or pickup truck) or Heavy (a heavy truck). Suppose data indicate that a Light vehicle is followed by another Light vehicle 70% of the time and that a Heavy vehicle is followed by a Heavy one 5% of the time.

a) What assumptions are necessary for the Heavy-Light process to be a homogenous 2-state Markov chain? Do these assumptions seem realistic? (One reason the process may not be *independent* is a traffic law that forbids Heavy trucks from following one another within a certain distance on the freeway. The resulting tendency towards some sort of "spacing" between Heavy trucks may carry over to exit ramps.)

b) If I see a Heavy vehicle in the monitor now, what is the probability that the second vehicle after it will also be Heavy? The fourth vehicle after it?

c) If I see a Light vehicle in the monitor now, what is the probability that the second vehicle after it will also be Light? The fourth vehicle after it?

d) In the long run, what proportion of the vehicles on this ramp do you suppose is Heavy?

e) How might an observer of this Markov process readily notice that it differs from a purely independent process with about 24% Heavy vehicles.

f) In practice, one would estimate the probability that a Heavy vehicle is followed by another Heavy one by taking data. If about 1/4 of the vehicles are Heavy, *about* how many Heavy vehicles (paired with the vehicles that follow immediately behind) would you need to observe in order to estimate this probability accurately enough to distinguish meaningfully between a purely independent process and a Markov process with dependence?

6.7   For a rigorous proof of equation (6.3), follow these steps:

a) Show that $p_{01}(2)$ is a fraction with numerator

$$P\{X_1 = 0, X_3 = 1\} = P\{X_1 = 0, X_2 = 0, X_3 = 1\}$$
$$+ P\{X_1 = 0, X_2 = 1, X_3 = 1\}.$$

b) Use the relationship $P(A \cap B \cap C) = P(A)P(B|A)P(C|A \cap B)$ and the Markov property to show that the first term in the numerator can be expressed as $p_0 p_{00} p_{01}$, where $p_0 = P\{X_1 = 0\}$.

c) Complete the proof.

6.8   To verify equation (6.5) do the matrix multiplication and algebra necessary to verify each of the four elements of $\mathbf{P}^2$.

6.9   Prove equation (6.6), by mathematical induction as follows:

Initial step: Verify that the equation is correct for $r = 1$. That is, let $r = 1$ in (6.6) and verify that the result is $\mathbf{P}$.

Induction step: Do the matrix multiplication $\mathbf{P} \cdot \mathbf{P}^r$, where $\mathbf{P}^r$ is given by the right-hand side of (6.6). Then simplify the result to show that the product $\mathbf{P}^{r+1}$ agrees with the right-hand side of (6.6) when r is replaced by $r + 1$.

Problems for Section 6.3 (Limiting Behavior)

6.10   Consider the 2-state Markov chain with $\alpha = \beta = 0.9999$. This is almost a "flip-flop" chain. Find λ_1, the cycle length, and $\mathbf{P}^{132}$. Simulate $\bar{X}_{100}$ and $\bar{X}_{101}$ several times. Also, look at the autocorrelations of X_n in several simulation runs. Comment on your findings.

Note: The autocorrelations for small lags have absolute values near 1 and they alternate in sign; for larger lags, the trend towards 0 is extremely slow.

6.11   A single strand of a DNA molecule is a sequence of nucleotides. There are four possible nucleotides in each position (step), one of which is cytosine (C). In a particular long strand, it has been observed that C appears in 34.1% of the positions. Also, in 36.8% of the cases where C appears in one position along the strand, it also appears in the next position.

a) What is the probability that a randomly chosen pair of adjacent nucleotides is CC (that has cytosine in both locations).
b) If a position along the strand is not C, then what is the probability that the next position is C?
c) If a position n along the strand is C, what is the probability that position $n + 2$ is also C? How about position $n + 4$?
d) Answer parts (a)–(c) if C appeared *independently* in any one position with probability 0.341.

Hint: Find the transition matrix of a chain consistent with the information given.

6.12   Consider a 2-state Markov chain with $\mathbf{P} = \left[\begin{smallmatrix} 1-\alpha & \alpha \\ \beta & 1-\beta \end{smallmatrix}\right]$. The elements of the row vector $\boldsymbol{\sigma} = (\sigma_1, \sigma_2)$ give a **steady-state** distribution of this chain if $\boldsymbol{\sigma}\mathbf{P} = \boldsymbol{\sigma}$ and $\sigma_1 + \sigma_2 = 1$.

a) If $\boldsymbol{\lambda}$ is the limiting distribution of this chain, show that $\boldsymbol{\lambda}$ is a steady-state distribution.
b) If $|1 - \alpha - \beta| < 1$, show that the solution of the vector equation $\boldsymbol{\lambda}\mathbf{P} = \boldsymbol{\lambda}$ is the long-run distribution $\boldsymbol{\lambda} = \left[\frac{\beta}{\alpha+\beta}, \frac{\alpha}{\alpha+\beta}\right]$.
c) It is possible for a chain that does not have a long-run distribution to have a steady-state distribution. What is the steady-state distribution of the "flip-flop" chain? What are the steady-state distributions of the "never move" chain?

Problems for Section 6.4 (Simple Gibbs Sampler)

6.13   Suppose a screening test for a particular disease has sensitivity $\eta = 0.8$ and specificity $\theta = 0.7$. Also suppose, for a particular population that is especially at risk for this disease, PV Positive $\gamma = 0.4$ and PV Negative $\delta = 0.9$.

a) Use the analytic method of Example 6.6 to compute π.
b) As in Example 6.5, use a Gibbs sampler to approximate the prevalence π. As seems appropriate, adjust the vertical scale of the plot, the run length m, and the burn-in period. Report any adjustments you made and the reasons for your choices. Make several runs of the modified simulation, and compare your results with the value obtained in part (a). Compare the first-order autocorrelation with that of the example.

Answer: $\pi \approx 0.22$; your answer to part (a) should show four places.

6.14   Mary and John carry out an iterative process involving two urns and two dice as follows:

(i) Mary has two urns: Urn 0 contains 2 black balls and 5 red balls; Urn 1 contains 6 black balls and 1 red ball. At step 1, Mary chooses one ball at random from Urn 1 (thus $X_1 = 1$). She reports its color to John and returns it to Urn 1.

(ii) John has two fair dice, one red and one black. The red die has three faces numbered 0 and three faces numbered 1; the black die has one face numbered 0 and five faces numbered 1. John rolls the die that corresponds to the color Mary reported to him. In turn, he reports the result X_2 to Mary. At step 2, Mary chooses the urn numbered X_2 (0 or 1).

(iii) This process is iterated to give values of $X_3, X_4, \ldots$.

a) Explain why the X-process is a Markov chain, and find its transition matrix.
b) Use an algebraic method to find the percentage of steps on which Mary samples from Urn 1.
c) Modify the program of Example 6.5 to approximate the result in part (b) by simulation.

Hint: Drawing from an urn and rolling a die are each "half" a step; account for both possible paths to each full-step transition: $\mathbf{P} = \frac{1}{7}\begin{bmatrix} 2 & 5 \\ 6 & 1 \end{bmatrix} \cdot \frac{1}{6}\begin{bmatrix} 1 & 5 \\ 3 & 3 \end{bmatrix}$.

7

Examples of Markov Chains with Larger State Spaces

In Chapter 6, we took advantage of the simplicity of 2-state chains to introduce fundamental ideas of Markov dependence and long-run behavior using only elementary mathematics. Markov chains taking more than two values are needed in many simulations of practical importance. These chains with larger state spaces can behave in very intricate ways, and a rigorous mathematical treatment of them is beyond the scope of this book. Our approach in this chapter is to provide examples that illustrate some of the important behaviors of more general Markov chains.

We give examples of Markov chains that have K-states, countably many states, and a continuum of states. Among our examples are chains that have served as useful models for important scientific problems. We focus mainly on their long-run behavior.

7.1 Properties of K-State Chains

First, we consider homogeneous Markov chains with a finite number K of states. Usually, we denote the state space as $S = \{1, 2, \ldots, K\}$. The most basic definitions and results are much the same as for a 2-state chain. The Markov property is

$$
\begin{aligned}
p_{ij} &= P\{X_n = j | X_{n-1} = i, X_{n-2} = i_{n-2}, \ldots, X_1 = i_1\} \\
&= P\{X_n = j | X_{n-1} = i\},
\end{aligned}
\tag{7.1}
$$

for $i, i_{n-2}, \ldots, i_1$ and $j \in S$, and for $n = 1, 2, \ldots$.

Here again, we consider only homogeneous chains. That is, the p_{ij} in (7.1) do not depend on n. Accordingly, the 1-step transition probabilities can be arranged in a $K \times K$ matrix $\mathbf{P}$, where the ith row specifies the conditional distribution of $X_{n+1} | X_n = i$.

E.A. Suess and B.E. Trumbo, *Introduction to Probability Simulation and Gibbs Sampling with R*, 159
Use R!, DOI 10.1007/978-0-387-68765-0_7, © Springer Science+Business Media, LLC 2010

Also, the Chapman-Kolmogorov equations for $p_{ij}(2)$ take a familiar form,

$$p_{ij}(2) = \sum_{k \in S} p_{ik} p_{kj}, \tag{7.2}$$

for $i, j \in S$. These equations imply that r-step transition probabilities $p_{ij}(r)$ are the elements of $\mathbf{P}^r$.

As in Chapter 6, we want to know whether a Markov chain has a long-run distribution. For a K-state chain, this means that $\lim_{r \to \infty} \mathbf{P}^r = \mathbf{\Lambda}$, where the limiting matrix $\mathbf{\Lambda}$ has K identical rows $\boldsymbol{\lambda} = [\lambda_1, \dots, \lambda_K]$. In terms of individual states, $\lim_{r \to \infty} P\{X_r = j | X_1 = i\} = \lambda_j$, for any $i, j \in S$.

When there are more than two states, the paths of movement from one state to another can be complex, and formulating a condition for convergence to a long-run distribution requires a more general approach than we used in Chapter 6. Specifically, a long-run distribution is guaranteed if we can find a number N for which $\mathbf{P}^N$ has all positive elements, written $\mathbf{P}^N > \mathbf{0}$. If this condition holds, then it follows directly from the Chapman-Kolmogorov equations that $\mathbf{P}^r > \mathbf{0}$ for all $r \geq N$.

We call such a chain **ergodic**, and we use this terminology for its transition matrix $\mathbf{P}$ also. (Some authors use the term *regular* instead of *ergodic*.) For any two of its states i and j, an ergodic chain can move from i to j and back to i in exactly $2N$ steps or any larger number of steps.

For 2-state chains, the never-move, flip-flop, and absorbing cases are non-ergodic. If a 2-state chain is ergodic, then $\mathbf{P}^2$ has all positive elements. Problem 7.1 provides some examples of ergodic and nonergodic 4-state chains.

Sketch of proof. We do not give a formal proof that $\lim_{n \to \infty} \mathbf{P}^r = \mathbf{\Lambda}$ for an ergodic chain, but it is not difficult to show a key step in the proof that the rows of $\mathbf{P}^r$ become more and more nearly identical as r increases. Assume that $\mathbf{P}^M > \mathbf{0}$ and $\mathbf{P}^N > \mathbf{0}$. Then consider the jth column of $\mathbf{P}^{M+N}$. The element

$$p_{ij}(M + N) = \sum_{k \in S} p_{ik}(M) p_{kj}(N)$$

is a weighted average of the elements of the jth column of $\mathbf{P}^N$, where the weights are the elements of the ith row of $\mathbf{P}^M$ and sum to 1. A weighted average, with all positive weights, of diverse quantities has a value strictly between the extremes of the quantities averaged. So if the elements of the jth column of $\mathbf{P}^N$ differ, then the elements of the jth column of $\mathbf{P}^{M+N}$ cannot differ by as much:

$$\min_{i' \in S} p_{i'j}(N) < p_{ij}(N + M) < \max_{i' \in S} p_{i'j}(N),$$

for any $i \in S$. A more technical version of this argument is required to show that all rows converge to the same vector $\boldsymbol{\lambda}$—and at a geometric rate. (See [Goo88] for a very clear exposition of the averaging lemma just described and its use in proving the ergodic theorem.)

7.2 Computing Long-Run Distributions

In practice, the long-run distribution λ of a K-state ergodic Markov chain can be derived or approximated in a variety of ways described briefly below. Most of our examples and problems use the first two of these methods.

Powers of the transition matrix. Compute $\mathbf{P}^r$ for large enough r that all rows are approximately the same—that is, equal to λ. This method works well for many K-state ergodic chains (as in Example 7.1 and Problem 7.2). However, even by today's standards, it may require excessive computation if a chain converges very slowly or if the number of states K is very large.

Simulation. For a 2-state chain, we have seen that λ_1 can be approximated as the proportion $\bar{X}_m$ of visits to state 1 in a large number m of simulated steps of a chain. For a K-state ergodic chain, the empirical distribution of X_n, for $n = 1, 2, \ldots, m$, approximates the limiting distribution. Often we make a histogram of the X_n to visualize the approximate shape of the long-run distribution. However, to use this method we must be sure that the chain is really ergodic (see Example 7.2).

Steady-state distribution. A distribution vector σ that satisfies the matrix equation $\sigma \mathbf{P} = \sigma$ is called a **steady-state** distribution of the chain because this distribution does not change from one step to the next. A K-state ergodic chain has a unique steady-state distribution that is the same as its long-run distribution. We illustrate this method for a particularly simple special case in Problem 7.5. Here again, it is important to know that the chain is ergodic; Example 7.2 shows a nonergodic chain with a more than one steady-state distribution.

Explicit algebraic solution. In Chapter 6, we showed that a 2-state chain with $|1 - \alpha - \beta| < 1$ has $\lim_{r \to \infty} \mathbf{P}^r = \Lambda$, where the two identical rows of Λ are $\lambda = [\lambda_0, \lambda_1] = (\alpha + \beta)^{-1}[\beta, \alpha]$. For an ergodic K-state chain, the long-run distribution vector λ can be found by the methods of linear algebra. Specifically, λ is the row eigenvector of $\mathbf{P}$ corresponding to the eigenvalue 1, multiplied by a constant so its entries sum to 1. (For an illustration, see Problem 7.10; for the theory, see [CM65].)

Means of geometric distributions. In Chapter 6, we show that the mean cycle length of a 2-state chain (with $\alpha, \beta > 0$) is the sum $1/\alpha + 1/\beta$ of the means of two geometric distributions, from which we can find the long-run distribution. This method may be intractable in a K-state chain ($K > 2$) if there are many possible paths leading from one state to another. (The chain in Problem 7.5 is simple enough for this method to work).

In the following example, we use the first two of these methods to find the long-run distribution of a 4-state ergodic chain.

Example 7.1. In most regions of the human genome, the dinucleotide cytosine-guanine (often denoted CpG for "C preceding G") along a DNA strand is susceptible to mutation to thymine-guanine (TpG). In certain regions, called

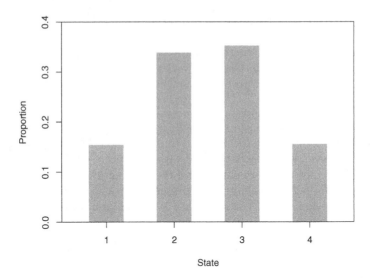

Figure 7.1. CpG islands process. The long-run distribution of this chain is approximated by the proportions of steps in which the four states are occupied in a simulation run with $m = 100\,000$. See Example 7.1 for a description of this chain.

CpG islands, this kind of mutation is naturally somewhat suppressed. In the mutation-prone regions, CpG dinucleotides are far less common than would be anticipated from the independent occurrence of C and G. In the CpG islands, one sees more C, G, and CpG than in the rest of the genome.

Experience has shown that the progression of nucleotides along a strand of DNA is well-modeled as a 4-state Markov chain. Above we have mentioned nucleotides C, G, and T, a fourth nucleotide occurring in DNA is adenine (A). We label the states as 1 = A, 2 = C, 3 = G, and 4 = T. The probabilities in the transition matrix below were estimated from 48 supposed CpG islands totaling about $60\,000$ nucleotides in length (see [DEKM98]):

$$\mathbf{P} = \begin{bmatrix} 0.180 & 0.274 & 0.426 & 0.120 \\ 0.171 & 0.368 & 0.274 & 0.188 \\ 0.161 & 0.339 & 0.375 & 0.125 \\ 0.079 & 0.355 & 0.384 & 0.182 \end{bmatrix}.$$

To three-place accuracy, each row of $\mathbf{P}^5$ is $\boldsymbol{\lambda} = (0.155, 0.341, 0.350, 0.154)$. Before doing this computation, we reduced each entry in the second row of the transition matrix by 0.00025 so that the second row adds exactly to 1; otherwise the computed powers of the matrix are not stable.

The following R program simulates $m = 100\,000$ steps of this chain. From Chapter 1, recall useful arguments of the **sample** function. Here these are the states of the chain from which we sample, the number of values generated (one on each pass through the loop), and **prob** to specify the respective probabilities

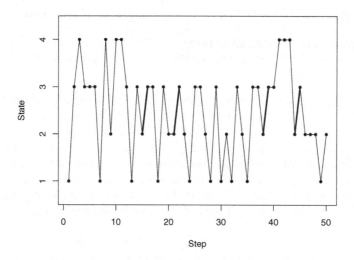

Figure 7.2. CpG islands process. The trace of the first 50 simulated steps shows that the process moves readily among its four states ($1 = $ A, $2 = $ C, $3 = $ G, and $4 = $ T). This suggests that the proportions in Figure 7.1 accurately approximate the long-run distribution. Four CpG transitions are emphasized with heavy line segments.

of states. The expression `as.factor(x)` treats elements of the vector `x` as categories to be tallied, and division by `m` converts counts to proportions.

```
#Preliminaries
# set.seed(1237)
m = 100000
x = numeric(m);   x[1] = 1

#Simulation
for (i in 2:m)
{
  if (x[i-1] == 1)
     x[i] = sample(1:4, 1, prob=c(.180,.274,.426,.120))
  if (x[i-1] == 2)
     x[i] = sample(1:4, 1, prob=c(.171,.368,.274,.188))
  if (x[i-1] == 3)
     x[i] = sample(1:4, 1, prob=c(.161,.339,.375,.125))
  if (x[i-1] == 4)
     x[i] = sample(1:4, 1, prob=c(.079,.355,.384,.182))
}

#Results
summary(as.factor(x))/m          # Table of proportions
mean(x[1:(m-1)]==2 & x[2:m]==3)  # Est. Proportion of CpG
hist(x, breaks=0:4+.5, prob=T, xlab="State", ylab="Proportion")
```

```
> summary(as.factor(x))/m            # Table of proportions
      1        2        3        4
0.15400 0.33826 0.35252 0.15522
> mean(x[1:(m-1)]==2 & x[2:m]==3)  # Est. Proportion of CpG
[1] 0.09348093
```

To two-place accuracy, the proportion of steps that the simulated process spends in each of the four states agrees well with the long-run distribution λ. A bar chart of the simulated results is shown in Figure 7.1. We started the chain at $X_1 = 1$. However, Figure 7.2, made in R with plot(x[1:50], type="o", pch=19), shows that the chain moves readily enough among states that the starting point makes no practical difference in a run of $m = 100\,000$.

In the sample of CpG islands used here, the proportion of CpGs among about $60\,000$ (overlapping) dinucleotides must have been about

$$P\{X_n = 2, X_{n+1} = 3\} = P\{X_n = 2\}P\{X_{n+1} = 3 | X_n = 2\}$$
$$\approx \lambda_2 p_{23} = (0.341)(0.274) = 0.093 = 9.3\%,$$

based on a chain at steady state. The proportion of CpGs in our simulated sample of $99\,999$ dinucleotides is $0.09348 \approx 9.3\%$. In contrast, an independent process with about 34% Cs and about 35% Gs would give about 12% CpGs. In Problem 7.2 we explore data from a part of the genome where CpGs are even less likely because of their tendency to mutate into TpGs. ◇

For population geneticists, the following example gives a clue how even harmful mutations can come to predominate in a small inbred population. For us, it also illustrates the difficulty in trying to draw conclusions about the behavior of a nonergodic chain via simulation.

Example 7.2. Brother-Sister Mating. Suppose that two individuals are mated (generation or step $n = 1$). From their offspring (step $n = 2$), two individuals are selected and mated. Then their offspring are mated (step $n = 3$), and so on. This scheme is called **brother-sister mating**.

If a gene has two alleles A and a, then three genotypes AA, Aa, and aa are possible. The pairs to be mated at each step will exhibit one of six genotype crosses, which we take to be the states of a Markov chain: $1 = AA \times AA$, $2 = AA \times Aa$, $3 = Aa \times Aa$, $4 = Aa \times aa$, $5 = AA \times aa$, and $6 = aa \times aa$.

The elements of $\mathbf{P}$ are found by considering the probabilities of the types of offspring that can result from each cross (state). For example, Cross 1 can yield only offspring of genotype AA, so that the only available cross of its offspring is again Cross 1. Similarly, Cross 5 can lead only to Cross 3 in the next generation because each offspring necessarily inherits A from one parent and a from the other. Two equally likely genotypes, AA and Aa, result from Cross 2, yielding the transition probabilities in the second row of $\mathbf{P}$ below. (You should verify the remaining transition probabilities; see Problem 7.3.)

$$\mathbf{P} = \frac{1}{16} \begin{bmatrix} 16 & 0 & 0 & 0 & 0 & 0 \\ 4 & 8 & 4 & 0 & 0 & 0 \\ 1 & 4 & 4 & 4 & 2 & 1 \\ 0 & 0 & 4 & 8 & 0 & 4 \\ 0 & 0 & 16 & 0 & 0 & 0 \\ 0 & 0 & 0 & 0 & 0 & 16 \end{bmatrix}.$$

This transition matrix is not ergodic because the "homozygotic" states 1 and 6 are absorbing states. That is, the chain will eventually move to one of these two states and stay there forever. The output of R computations below shows $\mathbf{P}^{16}$ (rounded to 3 places)

```
       [,1]   [,2]   [,3]   [,4]   [,5]   [,6]
[1,]  1.000  0.000  0.000  0.000  0.000  0.000
[2,]  0.735  0.009  0.011  0.009  0.002  0.235
[3,]  0.481  0.011  0.013  0.011  0.002  0.481
[4,]  0.235  0.009  0.011  0.009  0.002  0.735
[5,]  0.477  0.013  0.017  0.013  0.003  0.477
[6,]  0.000  0.000  0.000  0.000  0.000  1.000
```

and $\mathbf{P}^{128}$ (rounded to 15 places).

```
       [,1]        [,2]        [,3]        [,4]        [,5]  [,6]
[1,]  1.00   0.00e+000   0.00e+000   0.00e+000   0.00e+000  0.00
[2,]  0.75   4.33e-013   5.35e-013   4.33e-013   8.30e-014  0.25
[3,]  0.50   5.35e-013   6.62e-013   5.35e-013   1.02e-013  0.50
[4,]  0.25   4.33e-013   5.35e-013   4.33e-013   8.30e-014  0.75
[5,]  0.50   6.62e-013   8.18e-013   6.62e-013   1.26e-013  0.50
[6,]  0.00   0.00e+000   0.00e+000   0.00e+000   0.00e+000  1.00
```

We see that 0s persist in the first and last rows of $\mathbf{P}^{128}$ because escape from states 1 and 6 is impossible. Also, in rows 2 through 5 all entries except those in the first and last columns are very nearly 0. No matter what the starting state, absorption into either state 1 or 6 by generation 16 is very likely.

Often it is of interest to find the probability f_{ij} of absorption into state j starting from state i. Look at row 3 of $\mathbf{P}^{128}$. When the chain starts in state 3, the sibling-mating chain is equally likely to get absorbed into state 1 or 6: $p_{31}(128) \approx f_{31} = 0.5$ and $p_{36}(128) \approx f_{36} = 0.5$. But when the chain starts in state 4, the absorbing states have different probabilities: $f_{41} = 0.25$ and $f_{46} = 0.75$. It is possible to find analytic solutions for absorption probabilities and also for mean times until absorption. (See [Fel57], [CM65], or [Goo88].)

As for most absorbing chains, steady-state distributions are uninformative for brother-sister mating. Any vector of the form $\boldsymbol{\lambda} = (p, 0, 0, 0, 0, q)$, where $0 \le p, q \le 1$ and $p + q = 1$, satisfies the matrix equation $\boldsymbol{\lambda P} = \boldsymbol{\lambda}$.

We simulated this process with an R script similar to that of Example 7.1. We changed the Results section to plot the trace of each run and to give the numerical output below concerning the step before absorption, denoted sba. Specifically, we started one run at $X_1 = 3$, and absorption occurred with

$X_6 = 6$, following $X_5 = 4$. Figure 7.3 shows the trace of this run superimposed on another trace showing absorption at $X_9 = 1$. (Also see Problem 7.4.)

```
> plot(x, type="b", ylim=c(1,6))
> sba = length(x[(x > 1) & (x < 6)]);   sba
[1] 5
> x[sba]
[1] 4
```

Notes: Brother-sister mating has been used to produce homozygotic plant and animal populations of commercial and research interest. This mating scheme has a long history as a topic in genetics and probability. See [Fel57] for mathematical discussions beyond the scope of this book and historical references. ◇

In general, when a chain has one absorbing state, the main question of interest is how long we expect to wait before absorption. Also, when a chain has more than one absorbing state, we may want to know the probabilities of getting trapped in the various absorbing states. The answers to these questions typically depend on the starting state. In neither case will a single simulation (or even a few) give meaningful answers to these questions. However, algebraic methods are often available (see, for example, [CM65]).

A careless observer who believes such a chain to be ergodic may mistake absorption (or near absorption) in a simulation run for convergence to a long-run distribution—especially if absorption happens slowly through a long path of intercommunicating states. Furthermore, chains with many states may have "absorbing clusters" of states. So, after absorption, the process can continue to "mix" within a subset of S but not within the whole state space S. Making several runs with careful attention to diagnostic methods, especially graphical ones, can provide some protection against mistaking absorption for ergodicity.

7.3 Countably Infinite State Spaces

Many Markov chains of practical interest have infinitely many states. Such chains can exhibit behaviors not seen in chains with finite state spaces. In this section and the next, we look at a few examples.

Consider a chain with a countably infinite state space: for example, $S = \{0, 1, 2, \ldots\}$ or $S = \{\ldots, -2, -1, 0, 1, 2, \ldots\} = \{0, \pm 1, \pm 2, \ldots\}$. Here again, the Markov property is specified by equation (7.1). It may or may not be convenient to show transition probabilities $p_{ij}, i, j \in S$ in the form of a matrix with infinite dimensions, but the Chapman-Kolmogorov equations are the same as in (7.2), where the sum is now an infinite series.

An important difference between finite and infinite chains is illustrated by the example below. Even if an infinite chain can move freely among all its states in the sense that $p_{ij}(n) > 0$ for all n exceeding some N (which may depend on i and j), this does not guarantee the existence of a long-run distribution.

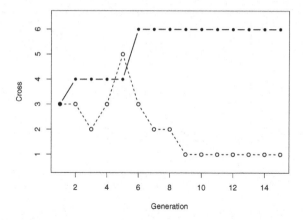

Figure 7.3. Brother-sister mating. Two simulations of the chain of Example 7.2 are shown, both starting at state $3 = Aa \times Aa$. One is absorbed in state $6 = aa \times aa$ at generation 6, the other in state $1 = AA \times AA$ at generation 9.

Example 7.3. For some purposes, the diffusion of molecules in a gas can be modeled approximately as a **random walk on the integers** of the line. Here we focus on a single molecule; it moves only in one dimension (along a line), and movements are all of unit size. Specifically, consider independent random displacements D_n that take the values $1, -1$, and 0 with probabilities α, β, and γ, respectively (with $\alpha + \beta + \gamma = 1$ and $\alpha, \beta, \gamma > 0$). Starting at $X_1 = 0$, the location of a molecule at stages $n = 2, 3, \ldots$ is $X_n = X_{n-1} + D_n$. Thus $p_{i,i+1} = \alpha$, $p_{i,i-1} = \beta$, and $p_{ii} = \gamma$, for $i - 1, i, i + 1 \in S = \{0, \pm 1, \pm 2, \ldots\}$.

What is the distribution of X_n as n increases? It is helpful to look at means and variances. Because $E(D_n) = \mu_D = (-1)\beta + (0)\gamma + (1)\alpha = \alpha - \beta$, $E(D_n^2) = \alpha + \beta$, and $V(D_n) = \sigma_D^2 = \alpha + \beta - (\alpha - \beta)^2 > 0$, we have

$$E(X_n) = E(D_2) + E(D_3) + \cdots + E(D_n) = (n-1)\mu_D,$$
$$V(X_n) = V(D_2) + V(D_3) + \cdots + V(D_n) = (n-1)\sigma_D^2.$$

Suppose $\alpha > \beta$ so that the process has a **drift** to the right (moves more readily to the right than to the left). Then, as n increases, the mean and the variance both become arbitrarily large. We say that the probability "escapes to infinity," and there is no long-run distribution. (In a symmetrical case where $\alpha = \beta > 0$, we have $E(X_n) \equiv 0$, but $V(X_n)$ still becomes infinite.)

The behavior of such a process X_n can be illustrated by simulation. The R script below simulates $m = 1000$ steps of a random walk with a strong positive drift: $\alpha = 1/2$, $\beta = 1/4$, and $\gamma = 1/4$. Because of the simple additive structure of the chain, we can use the R function cumsum to write vectorized code for this simulation.

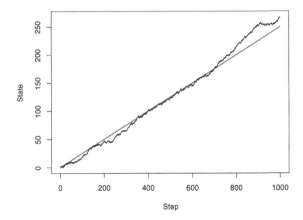

Figure 7.4. Random walk on the integers. The process of Example 7.3 has a drift towards larger-numbered states. The simulation run illustrated here stays unusually close to the mean function of the process, represented by the straight line.

```
# set.seed(1237)
m = 1000
d = c(0, sample(c(-1,0,1), m-1, replace=T, c(1/4,1/4,1/2)))
x = cumsum(d)
plot(x, pch=".", xlab="Step", ylab="State")
lines(c(0,m), c(0,m/4), type="l")
```

Figure 7.4 shows the drift towards higher-numbered states. The mean function is $\mu(n) = E(X_n) = (n - 1)/4$, so the theoretical slope of the random path is $1/4$. The standard deviation of X_n increases with n. For example, $SD(X_{100}) = \sqrt{99(3/4 - 1/16)} = 8.25$ is smaller than $SD(X_{1000}) = 26.21$.

Note: For such random walks, a curious technical distinction can be proved. Suppose the process has a tendency to drift either to the left or to the right ($\alpha \neq \beta$). Of course, the molecule may happen to return to its starting place 0. For example, the probability that it is in state 0 again at step 3 is $2\alpha\beta + \gamma^2$. However, because of the drift, the molecule can visit 0 only finitely many times. One says it is "absorbed" towards $\pm\infty$, depending on the direction of the drift. A paradox arises in the case where $\alpha = \beta$ precisely (a condition seldom achieved in practice). Then the molecule is "sure" to return to 0 infinitely often, even though the expected period between returns is infinite.

In contrast, a molecule of a more realistic 3-dimensional diffusion model can visit its starting place only a finite number of times even if movement is symmetrical—with probability 1/6 at each step of moving one unit left, right, up, down, forward, or backward. (For more precise formulations and proofs, see [Fel57]). $\diamondsuit$

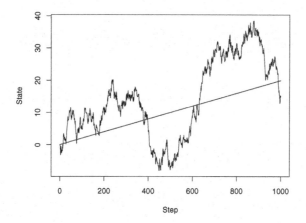

Figure 7.5. Random walk with normal displacements. This simulation of the process of Example 7.4 illustrates its slight positive drift and relatively large volatility.

For examples of *convergent* Markov chains with countably infinite state spaces, see Problems 7.11 and 7.12. Intuitively, the idea is that a Markov chain will have a long-run distribution if its extreme states are so unlikely that the particle cannot "slip away" to infinity in either direction.

7.4 Continuous State Spaces

Markov chains with **continuous state spaces** are especially important in computational methods such as Gibbs sampling. For a chain of continuous random variables, we need to take a different mathematical approach than we have taken for discrete random variables.

- For a chain of *discrete* random variables X_n, the transitional behavior is specified in terms of probabilities $p_{ij} = P\{X_n = j | X_{n-1} = i\}$. We have often displayed the p_{ij} as a matrix, of which the jth row gives the conditional distribution of $X_n | X_{n-1} = i$.
- For a chain of *continuous* random variables X_n, we cannot show the transitional behavior in terms of a matrix. Here our approach is to show how the observed value $x_{n-1} = s \in S$ of the process at step $n-1$ determines the conditional density function $f(t|s)$ of $X_n | X_{n-1} = s$. Then the 1-step probability of a transition from the point s to the interval $T = (t_L, t_U)$, with states $t_L < t_U$, is expressed as

$$P\{t_L < X_n < t_U | X_{n-1} = s\} = \int_{t_L}^{t_U} f(t|s)\, dt.$$

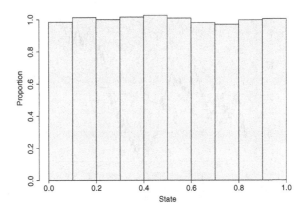

Figure 7.6. Random walk on a circle of unit circumference. This histogram summarizes the states occupied in 50 000 iterations of the simulation of Example 7.5. The long-run distribution of this Markov chain is UNIF$(0, 1)$.

Example 7.4. Consider a random walk with $S = (-\infty, \infty)$, the initial value $X_1 = 0$, and $X_n = X_{n-1} + D_n$, for $n = 2, 3, \ldots$, and where the independent displacements D_n are distributed as NORM$(0.02, 1)$. This process is a Markov chain because the distribution of X_n depends only on the outcome at step $n-1$. Specifically, the conditional distribution of $X_n | X_{n-1} = s$ is NORM$(s+0.02, 1)$.

This process has a slight positive drift because $E(D_n) = \mu_D = 0.02$. Both $E(X_n) = (n-1)\mu_D \to \infty$ and $V(X_n) = (n-1)\sigma_D^2 = n-1 \to \infty$ are evaluated as in Example 7.3. Thus there is no long-run distribution. Figure 7.5 shows a simulation of this process based on the R script below.

```
# set.seed(1237)
m = 1000
d = c(0, rnorm(m-1,0.02,1));   x = cumsum(d)
plot(x, type="l", xlab="Step", ylab="State")
lines(c(0,m), c(0, 0.02*m), type="l")
```

Such a process, with appropriate choices of $\mu_D > 0$ and $\sigma_D > 0$, might be used as a short-term model of a stock index for which one supposes there is a drift towards higher prices but with considerable random volatility from step to step. A step may be taken as any specified length of time: a trading day, a week, and so on. While the process will eventually increase without bound, there may be very large fluctuations in value along the way. ◊

If the continuous state space S of a Markov chain is an interval of finite length, then the distribution of X_n cannot "escape" to infinity in the long run. Our next example illustrates this.

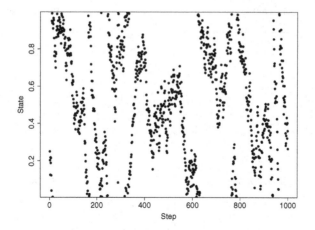

Figure 7.7. Random walk on a circle of unit circumference. This plot, based on the first 1000 iterations of the simulation of Example 7.5, is best imagined as a cylinder (with the top and bottom edges adjacent).

Example 7.5. Consider a random walk on the circumference (of length 1) of a circle. That is, the state space is $S = [0, 1)$. Then define $X_1 = 0$ and $X_n = X_{n-1} + D_n \pmod{1}$, for $n = 2, 3, \ldots$, where the D_n are independently distributed as $\mathsf{UNIF}(-0.1, 0.1)$. The following R script can be used to simulate this random walk.

```
# set.seed(1212)
m = 50000
d = c(0, runif(m-1, -.1, .1))
x = cumsum(d) %% 1
hist(x, breaks=10, prob=T, xlab="State", ylab="Proportion")
```

As suggested by the histogram of Figure 7.6, the long-run distribution of this chain is $\mathsf{UNIF}(0, 1)$. Figure 7.7, made with the additional code `plot(x[1:1000], pch=".", xlab="Step", ylab="State")`, shows the first 1000 steps of the simulation. It is best imagined as being bent around to form a horizontal cylinder. Notice that a few "paths" of outcomes run off the plot at the top but continue at the bottom. (Also see Problem 7.13.) ◊

A state space of finite length does not ensure *useful* long-run behavior. The following example shows a chain on the unit interval that mixes badly and converges slowly to a bimodal long-run distribution.

Example 7.6. The beta family of distributions $\mathsf{BETA}(\alpha, \beta)$ has density functions of the form $f(x) \propto x^{\alpha-1}(1-x)^{\beta-1}$, for $0 < x < 1$, and both parameters positive. The symbol $\propto$ (read *proportional to*) means that the constant necessary to make the density function integrate to 1 is not written. [This constant, not explicitly involved in our discussion, is $\Gamma(\alpha + \beta)/(\Gamma(\alpha)\Gamma(\beta))$.]

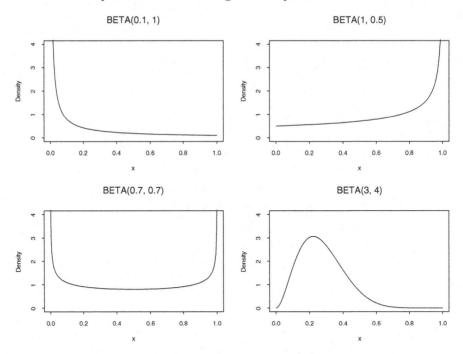

Figure 7.8. Four beta density functions, illustrating some of the many different shapes of the densities in the beta family.

Because beta random variables take values in $(0, 1)$, this family of distributions is often used to model random proportions. Figure 7.8 shows examples of four beta densities. Other choices of parameters give density functions with a wide variety of shapes.

Roughly speaking, the parameter α controls the shape of the beta density curve $y = f(x)$ near 0. If α is near 0, then the density is very high for x near 0; if $\alpha > 2$, then the density is low for x near 0. Similarly, the density near 1 is very high or low according as β is near 0 or larger than 2.

In this example, each of the random variables X_n, $n = 2, 3, \ldots$, of a Markov chain has a beta distribution where the parameters α and β are determined by the observed value of X_{n-1}. Although this is an artificial example, contrived to show a particular kind of bad long-run behavior, many useful Markov chains consist of sequences of beta random variables.

If we take $X_n \sim \text{BETA}(0.001 + 3X_{n-1}, 3.001 - 3X_{n-1})$, for $n = 2, 3, \ldots$, then a value of X_{n-1} near 0 will yield values of α and β in the distribution of X_n that tend to produce values of X_n near 0 and away from 1. Similarly, if the value of X_{n-1} is near 1, then the value of X_n will also tend to be near 1. That is, X_{n-1} and X_n will be highly correlated. The following R program simulates this process.

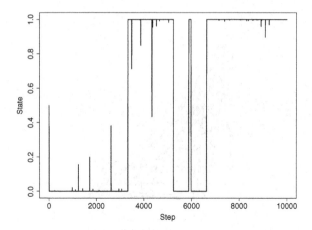

Figure 7.9. A Markov chain with poor mixing. An accurate assessment of its long-run distribution would require a very large number of iterations. See Example 7.6.

```
# set.seed(1212)
m = 10000
x = numeric(m)
x[1] = .5          # Arbitrary starting value in (0,1)
for (i in 2:m)
{
   x[i] = rbeta(1, .001+3*x[i-1], 3.001-3*x[i-1])
}
plot(x, type="l")
```

Figure 7.9, shows the plot of successive values of X_n. The limiting distribution of this process is a mixture of beta distributions, many with high densities near 0, many with high densities near 1, and few with high densities near the middle of $(0, 1)$. Not surprisingly, the histogram of the simulated distribution, made with the additional code hist(x) but omitted here, shows a bimodal distribution with tall bars near 0 and 1 and negligible numbers of visits elsewhere. In several runs of this simulation, none of the autocorrelations through lag 40 were ever smaller than 0.95. These results can be seen by using the additional code acf(x). In summary, this chain mixes poorly, tending to "stick" in the neighborhood of 0 or of 1. (Problem 7.14 shows two variants of this example with very different limiting behaviors.) ◇

An important use of Markov chains in computation is to find a chain that has a desired long-run distribution. The distribution may be of interest in its own right, and then we might find its mean, quantiles, and so on. Alternatively, sampling from the long-run distribution might be the basis of a further computation, such as a numerical integration; this is illustrated by the next example.

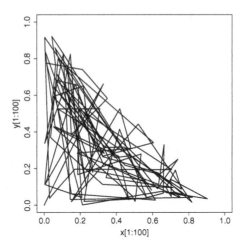

Figure 7.10. A triangular state space. The first 100 states visited in a simulation of the chain of Example 7.7 show excellent mixing. To three-place accuracy, the first four points visited are (0, 0), (0.328, 0.641), (0.008, 0.117), and (0.173, 0.661).

Example 7.7. Define a continuous Markov chain on the triangle $\triangle ABC$ with vertices at $A = (0,0)$, $B = (0,1)$, and $C = (1,0)$ as follows. Begin at the origin: $(X_1, Y_1) = (0,0)$. Then select $X_2 \sim \mathsf{UNIF}(0,1)$ along the base of the triangle and Y_2 at random along the vertical line segment from $(X_2, 0)$ to the hypotenuse as $Y_2 \sim \mathsf{UNIF}(0, 1 - X_2)$. Next select X_3 at random along the horizontal line segment within the triangle and passing through (X_2, Y_2) and Y_3 at random along the vertical line segment within the triangle and passing through (X_3, Y_2), and so on.

The following R script simulates this process and makes Figure 7.10.

```
# set.seed (1237)
m = 5000
x = y = numeric(m)
x[1] = y[1] = 0
for (i in 2:m)
{
  x[i] = runif(1, 0, 1-y[i-1])
  y[i] = runif(1, 0, 1-x[i])
}
plot(x[1:100], y[1:100], type="l", xlim=c(0,1), ylim=c(0,1))
```

Figure 7.10 shows the movement of this Markov chain within the triangle over the first 100 steps. Figure 7.11, made with the additional code `plot(x, y, pch=".", xlim=c(0,1), ylim=c(0,1))`, shows the 5000 simulated points, illustrating that the long-run distribution of this chain is uniform on the triangle. (Compare this with the Markov chain in Problem 7.16, which has a much more intricate long-run distribution. Also see Figure 7.16.)

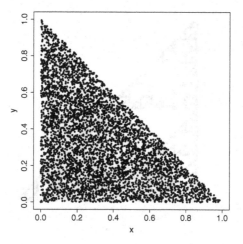

Figure 7.11. A triangular state space. Here we show 5000 simulated points of the chain of Example 7.7. The long-run distribution of this chain is uniform on the triangle.

In Chapter 3, we use Monte Carlo integration to approximate the volume

$$J = \iint_{\triangle ABC} \varphi(x; 0, 1)\varphi(y; 0, 1)\, dx\, dy \approx 0.0677$$

above the triangle and beneath the bivariate standard normal density function. There, the method was to generate points uniformly distributed in the unit square and reject the points outside the triangle (about half of them). Here, we have found a way to generate only points we will actually use. The following additional code does the Monte Carlo integration and estimates the standard error.

```
h = dnorm(x)*dnorm(y)
.5*mean(h);   .5*sd(h)/sqrt(m)

> .5*mean(h);   .5*sd(h)/sqrt(m)
[1] 0.06777085
[1] 0.00009937692
```

The result from this run, $0.0678 \pm 2(0.0001)$, is consistent with the exact value 0.06773. We see no way to vectorize the current script, so in R the "wasteful," but vectorizable, procedure of Chapter 3 runs a little faster—even though the number of steps there would have to be set at $m = 10\,000$ to yield approximately 5000 accepted points. ◇

   Although we have not even approached a thorough mathematical treatment of Markov chains that have continuous state spaces, we hope our examples have illustrated some issues of practical importance. In particular, we

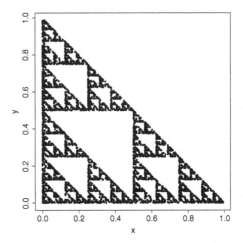

Figure 7.12. Sierpinski's triangle. The plot shows 5000 simulated points of the chain of Problem 7.16. The definition of the chain may seem similar to that of Example 7.7, but comparison with Figure 7.11 shows a surprisingly different long-run distribution.

hope we have provided memorable evidence that not all Markov chains converge and that some convergent ones behave strangely either because they mix badly and converge slowly or because they converge to unexpected limiting distributions.

When simulating Markov chains in practice, one must rely heavily on descriptive diagnostic methods to be confident that apparent limiting distributions are what they seem to be. The use of autocorrelations and graphs of simulated transition paths can help to assess mixing. Depending on the dimensionality of the state space, histograms or scatterplots can help to reveal shapes of distributions. Also, multiple runs can be used to check whether there are multiple "sticking points" or "modes" that are absorbing or nearly absorbing.

7.5 Uses of Markov Chains in Computation

Ergodic Markov chains are often used as an aid in simulating complex probability models. Such simulation methods are sometimes called **Markov chain Monte Carlo** (MCMC) methods. Typically, one specifies a chain that has the target model as its stationary distribution. In order to do this, it is necessary to use key information about the model in defining how the chain moves from one step to the next. Examples of MCMC in this section, based on bivariate normal distributions, provide relatively simple illustrations of two methods of specifying computationally useful Markov chains, the Metropolis algorithm and the Gibbs sampler.

Consider a possibly correlated bivariate normal distribution of random variables X and Y with zero means and unit standard deviations. The joint density function is

$$\phi(x,y) = K \, \exp\{-[2(1-\rho^2)]^{-1}\,[x^2 - 2\rho xy + y^2]\},$$

where $K = [2\pi(1-\rho^2)]^{-1/2}$ and $\rho = \text{Cor}(X,Y)$. Specifically, suppose two achievement tests on a particular field of knowledge are administered to subjects in some population. If the population mean of a test is μ and its standard deviation is σ, then a subject who has raw score U is said to have standard score $X = (U - \mu)/\sigma$.

One can show that the marginal distributions of X and Y are standard normal, and that the conditional distribution of $X|Y = y$ is NORM$(\rho y, \sqrt{1-\rho^2})$, and symmetrically for the conditional distribution of Y given X. Thus, in R the joint density function of X and Y can be expressed by either of the following lines of code.

```
dnorm(x)*dnorm(y, rho*x, sqrt(1-rho^2))
dnorm(y)*dnorm(x, rho*y, sqrt(1-rho^2))
```

In the computational examples below, we use standard scores X and Y corresponding to the two tests, respectively. Because the tests have a substantial amount of subject matter in common, we have $\rho = 0.8$. Suppose a subject for whom either $X \geq 1.25$ or $Y \geq 1.25$ receives a certificate. One question of interest is to evaluate the proportion of subjects taking both tests who receive certificates.

Because this bivariate model can be simulated in a more traditional way (as shown in Problem 7.18), our computational examples using Markov chains are "toy" ones. Nevertheless, they provide useful introductory illustrations of two important computational methods.

Example 7.8. Metropolis Algorithm. This algorithm is an acceptance-rejection procedure that can be considered a generalization of the ideas in Example 3.3 and Problem 3.11. An arbitrary starting point (x_1, y_1) within the support of the desired model is chosen as the state of the chain at step 1. Then the state (x_i, y_i), at each successive step i, is simulated in two stages:

- A symmetrical random jump function provides a proposed (or candidate) state (x_p, y_p) for step i.
- The proposal is either accepted or rejected. The acceptance criterion uses the density function of the target distribution. If the proposed state (x_p, y_p) is at a point of higher density than the previous state (x_{i-1}, y_{i-1}), then the proposal is accepted. If it is a point of lower density, then it is accepted only with a certain probability that involves the ratio of the two densities.

For our bivariate model of standard test scores, the following R code implements such a procedure.

```
set.seed(1234);   m = 40000;   rho = .8;   sgm = sqrt(1 - rho^2)
xc = yc = numeric(m)            # vectors of state components
xc[1] = -3;  yc[1] = 3          # arbitrary starting values
jl = 1;  jr = 1                 # l and r limits of proposed jumps
for (i in 2:m)
{
  xc[i] = xc[i-1];  yc[i] = yc[i-1]      # if jump rejected
  xp = runif(1, xc[i-1]-jl, xc[i-1]+jr)  # proposed x coord
  yp = runif(1, yc[i-1]-jl, yc[i-1]+jr)  # proposed y coord
  nmtr = dnorm(xp)*dnorm(yp, rho*xp, sgm)
  dntr = dnorm(xc[i-1])*dnorm(yc[i-1], rho*xc[i-1], sgm)
  r = nmtr/dntr                          # density ratio
  acc = (min(r, 1) > runif(1))           # jump if acc == T
  if (acc) {xc[i] = xp;  yc[i] = yp}
}
x = xc[(m/2+1):m];  y = yc[(m/2+1):m]    # states after burn-in
round(c(mean(x), mean(y), sd(x), sd(y), cor(x,y)), 4)
mean(diff(x)==0)              # proportion or proposals rejected
mean(pmax(x,y) >= 1.25)      # prop. of subj. getting certificates
par(mfrow = c(1,2), pty="s")
  plot(xc[1:100], yc[1:100], xlim=c(-4,4), ylim=c(-4,4), type="l")
  plot(x, y, xlim=c(-4,4), ylim=c(-4,4), pch=".")
par(mfrow = c(1,1), pty="m")

> round(c(mean(x), mean(y), sd(x), sd(y), cor(x,y)), 4)
[1] -0.0348 -0.0354  0.9966  0.9992  0.7994
> mean(diff(x)==0)                # proportion or proposals rejected
[1] 0.4316216
> mean(pmax(x,y) >= 1.25)    # prop. of subj. getting certificates
[1] 0.14725
```

Assuming that the initial $m/2$ steps (called the burn-in period) are enough for the simulation to stabilize at the stationary distribution of the chain, we use only the second half of the simulated states to get our approximate values. For validation, we note that the simulated means are near 0, the simulated standard deviations near 1, and the simulated correlation near 0.8, as required. About 15% of subjects receive certificates. (From the standard normal distribution, we know that about 10% will get a standard score above 1.25 on a particular one of the exams.)

The upper left-hand panel in Figure 7.13 shows the movement of the chain over the first 100 steps. Notice that when a proposed state is rejected (which happens about 43% of the time), then the state at the previous step is repeated. Thus some dots in the scatterplot represent more than one step in the simulated chain. There is a trade-off in selecting the random jump distribution. If it is too disperse, the proportion of acceptances can be small. If it is too concentrated, the chain cannot move freely among the values of the target distribution. ◇

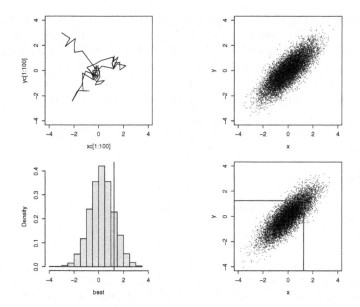

Figure 7.13. Bivariate normal distributions simulated by Metropolis (upper panels) and Gibbs methods. States (X, Y) after burn-in are sampled from a bivariate normal distribution of standard test scores with zero means, unit standard deviations, and correlation 0.8. In the bottom panels, the heavy lines separate cases where the best of two scores exceeds 1.25 from those where it does not. See Examples 7.8 and 7.9.

As in this example, the Metropolis algorithm requires a symmetrical jump distribution. Problem 7.19 shows the incorrect results from using an asymmetrical one and illustrates a more general algorithm, called the **Metropolis-Hastings algorithm**, that permits and corrects for an asymmetrical jump distribution. Problem 8.6 uses the Metropolis algorithm to find Bayesian posteriors in a conceptually simple and computationally problematic situation. For more rigorous developments and various applications of MCMC methods in general and these algorithms in particular, see [CG94], [GCSR04], [Lee04], [Has55], and [RC04].

Next we use a Gibbs sampler to simulate the target bivariate distribution. In a sense that we will not make explicit here, a Gibbs sampler is a special case of the Metropolis-Hastings algorithm in which all proposals are accepted.

Example 7.9. A Gibbs Sampler. Here we solve exactly the same problem as in Example 7.8, but with a Gibbs sampler. Again we make a Markov chain that has the target correlated bivariate normal distribution as its stationary distribution.

The simulation starts at an arbitrary state (x_1, y_1) at step 1. Then we use the conditional distribution $X|Y = y_1$ to simulate a value x_2 and the conditional distribution $Y|X = x_2$ to simulate a value y_2. Thus we have determined the state of the chain (x_2, y_2) at step 2. This process is iterated for $i = 3, 4, \ldots, m$, where m is suitably large, to obtain successive states of the chain. After a burn-in period in which the simulation stabilizes, the interplay between the two conditional distributions ensures that all states are appropriately representative of the target distribution. The R code below implements this simulation for our distribution of test scores.

```
set.seed(1235);  m = 20000;  rho = .8;  sgm = sqrt(1 - rho^2)
xc = yc = numeric(m)        # vectors of state components
xc[1] = -3;  yc[1] = 3      # arbitrary starting values

for (i in 2:m)
{
  xc[i] = rnorm(1, rho*yc[i-1], sgm)
  yc[i] = rnorm(1, rho*xc[i], sgm)
}

x = xc[(m/2+1):m];  y = yc[(m/2+1):m]   # states after burn-in
round(c(mean(x), mean(y), sd(x), sd(y), cor(x,y)), 4)
best = pmax(x,y);  mean(best >= 1.25)   # prop. getting certif.
summary(best)
par(mfrow = c(1,2), pty="s")
hist(best, prob=T, col="wheat", main="")
  abline(v = 1.25, lwd=2, col="red")
  plot(x, y, xlim=c(-4,4), ylim=c(-4,4), pch=".")
  lines(c(-5, 1.25, 1.25), c(1.25, 1.25, -5), lwd=2, col="red")
par(mfrow = c(1,1), pty="m")

> round(c(mean(x), mean(y), sd(x), sd(y), cor(x,y)), 4)
[1] 0.0083 0.0077 1.0046 1.0073 0.8044
> best = pmax(x,y);  mean(best >= 1.25)   # prop. getting certif.
[1] 0.1527

> summary(best)
   Min. 1st Qu.  Median    Mean 3rd Qu.    Max.
-3.6050 -0.3915  0.2626  0.2589  0.9087  3.3850
```

The numerical results and the scatterplot (lower right in Figure 7.13) are essentially the same as results from the Metropolis algorithm. Here we also show a histogram of the best-of-two test scores (lower left). In both of these graphs, lines separate test results that would earn a certificate from those that would not. ◇

Typically, as in the example just above, Gibbs samplers use Markov chains with continuous state spaces. A Markov chain is chosen because it has a long-run distribution of interest that is difficult to describe or simulate by more direct means. Following an introduction to Bayesian estimation in Chapter 8, some practical uses of Gibbs samplers to solve Bayesian problems are shown in Chapters 9 and 10.

7.6 Problems

Problems for Sections 7.1 and 7.2 (Finite State Spaces)

7.1 *Ergodic and nonergodic matrices.* In the transition matrices of the six 4-state Markov chains below, elements 0 are shown and * indicates a positive element. Identify the ergodic chains, giving the smallest value N for which $\mathbf{P}^N$ has all positive elements. For nonergodic chains, explain briefly what restriction on the movement among states prevents ergodicity.

$$\text{a)} \quad \mathbf{P} = \begin{bmatrix} * & * & 0 & 0 \\ 0 & * & * & 0 \\ 0 & 0 & * & * \\ * & 0 & 0 & * \end{bmatrix}, \quad \text{b)} \quad \mathbf{P} = \begin{bmatrix} * & * & * & 0 \\ * & * & * & 0 \\ * & * & * & 0 \\ 0 & 0 & 0 & * \end{bmatrix}, \quad \text{c)} \quad \mathbf{P} = \begin{bmatrix} * & * & 0 & 0 \\ * & * & 0 & 0 \\ * & 0 & * & * \\ 0 & 0 & * & 0 \end{bmatrix},$$

$$\text{d)} \quad \mathbf{P} = \begin{bmatrix} 0 & * & 0 & 0 \\ 0 & 0 & * & * \\ * & 0 & 0 & 0 \\ * & 0 & 0 & 0 \end{bmatrix}, \quad \text{e)} \quad \mathbf{P} = \begin{bmatrix} 0 & * & * & 0 \\ 0 & 0 & 0 & * \\ * & 0 & 0 & 0 \\ 0 & * & * & * \end{bmatrix}, \quad \text{f)} \quad \mathbf{P} = \begin{bmatrix} * & * & 0 & 0 \\ * & * & 0 & 0 \\ 0 & * & * & * \\ 0 & 0 & 0 & * \end{bmatrix}.$$

Answers: In each chain, let the state space be $S = \{1, 2, 3, 4\}$. (a) Ergodic, $N = 3$. (b) Class $\{1, 2, 3\}$ does not intercommunicate with $\{4\}$. (d) Nonergodic because of the period 3 cycle $\{1\} \rightarrow \{2\} \rightarrow \{3, 4\} \rightarrow \{1\}$; starting in $\{1\}$ at step 1 allows visits to $\{3, 4\}$ only at steps 3, 6, 9, ... (f) Starting in $\{3\}$ leads eventually to absorption in either $\{1, 2\}$ or $\{4\}$.

7.2 *Continuation of Example 7.1, CpG islands.* We now look at a Markov chain that models the part of the genome where mutation of CpGs to TpGs is not inhibited. In the transition matrix below, note particularly that the probability p_{23} is much smaller than in the matrix of Example 7.1 (p161).

$$\mathbf{P} = \begin{bmatrix} 0.300 & 0.205 & 0.285 & 0.210 \\ 0.322 & 0.298 & 0.078 & 0.302 \\ 0.248 & 0.246 & 0.298 & 0.208 \\ 0.177 & 0.239 & 0.292 & 0.292 \end{bmatrix}.$$

a) Find a sufficiently high power of $\mathbf{P}$ to determine the long-run distribution of this chain. Comment on how your result differs from the long-run distribution of the chain for CpG islands.

b) Modify the R program of Example 7.1 to simulate this chain, approximating its long-run distribution and the overall proportion of CpGs. How does this compare with the product $\lambda_2 p_{23}$? With the product $\lambda_2 \lambda_3$? Comment. How does it compare with the proportion of CpGs in the CpG-islands model?

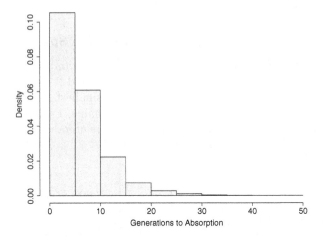

Figure 7.14. Absorption-time distribution of the brother-sister mating process. Starting at state $3 = Aa \times Aa$, absorption often occurs in less than 15 generations. See Example 7.2 (p164) and Problem 7.4.

Note: The proportion of CpGs among dinucleotides in the island model is approximately 9%; here it is only about 2%. Durbin et al. [DEKM98] discuss how, given the nucleotide sequence for a short piece of the genome, one might judge whether or not it comes from a CpG island. Further, with information about the probabilities of changing between island and "sea," one might make a Markov chain with 8 states: A′, T′, G′, C′ for CpG islands and A, T, G, C for the surrounding sea. However, when observing the nucleotides along a stretch of genome, one cannot tell A from A′, T from T′, and so on. This is an example of a **hidden Markov model**.

7.3   *Brother-sister mating (continued).*

a) In Example 7.2 (p164) verify the entries in the transition matrix $\mathbf{P}$.
b) Evaluate the products $(1/2, 0, 0, 0, 0, 1/2) \cdot \mathbf{P}$ and $(1/3, 0, 0, 0, 0, 2/3) \cdot \mathbf{P}$ by hand and comment.
c) Make several simulation runs similar to the one at the end of Example 7.2 and report the number of steps before absorption in each.

7.4   *Distribution of absorption times in brother-sister mating (continuation of Problem 7.3).* The code below simulates 10 000 runs of the brother-sister mating process starting at state 3. Each run is terminated at absorption, and the step and state at absorption for that run are recorded. The histogram from one run is shown in Figure 7.14.

Run the program several times for yourself, each time with a different starting state. Summarize your findings, comparing appropriate results with those from $\mathbf{P}^{128}$ in Example 7.2 and saying what additional information is gained by simulation.

```
m = 10000                # number of runs
step.a = numeric(m)      # steps when absorption occurs
state.a = numeric(m)     # states where absorbed

for (j in 1:m)
{
   x = 3  # initial state; inside the loop the length
          #   of x increases to record all states visited
   a = 0  # changed to upon absorption
   while(a==0)
   {
      i = length(x)   # current step; state found below
      if (x[i]==1)  x = c(x, 1)
      if (x[i]==2)
         x = c(x, sample(1:6, 1, prob=c(4,8,4,0,0,0)))
      if (x[i]==3)
         x = c(x, sample(1:6, 1, prob=c(1,4,4,4,2,1)))
      if (x[i]==4)
         x = c(x, sample(1:6, 1, prob=c(0,0,4,8,0,4)))
      if (x[i]==5)  x = c(x, 3)
      if (x[i]==6)  x = c(x, 6)
      # condition below checks for absorption
      if (length(x[x==1 | x==6]) > 0) a = i + 1
   }
   step.a[j] = a  # absorption step for jth run
   state.a[j] = x[length(x)]   # absorption state for jth run
}

hist(step.a)  # simulated distribution of absorption times
mean(step.a)  # mean time to absorption
quantile(step.a, .95)  # 95% of runs absorbed by this step
summary(as.factor(state.a))/m  # dist'n of absorption states
```

7.5 *Doubly stochastic matrix.* Consider states $S = \{0, 1, 2, 3, 4\}$ arranged clockwise around a circle with 0 adjacent to 4. A fair coin is tossed. A Markov chain moves clockwise by one number if the coin shows Heads, otherwise it does not move.

a) Write the 1-step transition matrix $\mathbf{P}$ for this chain. Is it ergodic?

b) What is the average length of time this chain spends in any one state before moving to the next? What is the average length of time to go around the circle once? From these results, deduce the long-run distribution of this chain. (In many chains with more than 2 states, the possible transitions among states are too complex for this kind of analysis to be tractable.)

c) Show that the vector $\boldsymbol{\sigma} = [1/5, 1/5, 1/5, 1/5, 1/5]$ satisfies the matrix equation $\boldsymbol{\sigma}\mathbf{P} = \boldsymbol{\sigma}$ and thus is a steady-state distribution of this chain. Is $\boldsymbol{\sigma}$ also the unique long-run distribution?

d) Transition matrices for Markov chains are sometimes called **stochastic**, meaning that each *row* sums to 1. In a **doubly stochastic** matrix, each *column* also sums to 1. Show that the limiting distribution of a K-state chain with an ergodic, doubly stochastic transition matrix $\mathbf{P}$ is uniform on the K states.

e) Consider a similar process with state space $S = \{0, 1, 2, 3\}$, but with 0 adjacent to 3, and with clockwise or counterclockwise movement at each step determined by the toss of a fair coin. (This process moves at every step.) Show that the resulting doubly stochastic matrix is not ergodic.

7.6   *An Ehrenfest Urn model.* A permeable membrane separates two compartments, Boxes A and B. There are seven molecules altogether in the two boxes. On each step of a process, the probability is 1/2 that no molecules move. If there is movement, then one of the seven molecules is chosen at random and it "diffuses" (moves) from the box it is in to the other one.

a) The number of molecules in Box A can be modeled as an 8-state Markov chain with state space $S = \{0, 1, \ldots, 7\}$. For example, if the process is currently in state 5, then the chances are 7 in 14 that it will stay in state 5 at the next step, 5 in 14 that it will go to state 4, and 2 in 14 that it will go to state 6. The more unequal the apportionment of the molecules, the stronger the tendency to equalize it. Write the 1-step transition matrix.

b) Show that the steady-state distribution of this chain is $\mathsf{BINOM}(7, \frac{1}{2})$. That is, show that it satisfies $\boldsymbol{\lambda}\mathbf{P} = \boldsymbol{\lambda}$. This is also the long-run distribution.

c) More generally, show that if there are M molecules, the long-run distribution is $\mathsf{BINOM}(M, \frac{1}{2})$.

d) If there are 10 000 molecules at steady state, what is the probability that between 4900 and 5100 are in Box A?

Note: This is a variant of the famous Ehrenfest model, modified to have probability 1/2 of no movement at any one step and thus to have an ergodic transition matrix. (See [CM65], Chapter 3, for a more advanced mathematical treatment.)

7.7   *A Gambler's Ruin problem.* As Chris and Kim begin the following gambling game, Chris has $4 and Kim has $3. At each step of the game, both players toss fair coins. If both coins show Heads, Chris pays Kim $1; if both show Tails, Kim pays Chris $1; otherwise, no money changes hands. The game continues until one of the players has $0. Model this as a Markov chain in which the state is the number of dollars Chris currently has. What is the probability that Kim wins (that is, Chris goes broke)?

Note: This is a version of the classic gambler's ruin problem. Many books on stochastic processes derive general formulas for the probability of the ruin of each player and the expected time until ruin. Approximations of these results can be obtained by adapting the simulation program of Problem 7.4.

7.8   Suppose weather records for a particular region show that 1/4 of Dry (0) days are followed by Wet (1) days. Also, 1/3 of the Wet days *that are* *immediately preceded by a Dry day* are followed by a Dry day, but there can never be three Wet days in a row.

a) Show that this situation cannot be modeled as a 2-state Markov chain.
b) However, this situation can be modeled as a 4-state Markov chain by the device of considering overlapping paired-day states: $S = \{00, 01, 10, 11\}$. For example, 00 can be followed by 00 (three dry days in a row) or 01, but it would contradict the definition of states for 00 to be followed by 10 or 11; logically, half of the entries in the 1-step transition matrix must be 0. The prohibition on three Wet days in a row dictates an additional 0 entry. Write the 4×4 transition matrix, show that it is ergodic, and find the long-run distribution.

Hints and answers: (a) The transition probability p_{11} would have to take two different values depending on the weather two days back. State two relevant conditional probabilities with different values. (b) Over the long run, about 29% of the days are Wet; give a more accurate value.

7.9   *Hardy-Weinberg Equilibrium.* In a certain large population, a gene has two alleles a and A, with respective proportions θ and $1 - \theta$. Assume these same proportions hold for both males and females. Also assume there is no migration in or out and no selective advantage for either a or A, so these proportions of alleles are stable in the population over time. Let the genotypes $aa = 1$, $Aa = 2$, and $AA = 3$ be the states of a process. At step 1, a female is of genotype aa, so that $X_1 = 1$. At step 2, she selects a mate at random and produces one or more daughters, of whom the eldest is of genotype X_2. At step 3, this daughter selects a mate at random and produces an eldest daughter of genotype X_3, and so on.

a) The X-process is a Markov chain. Find its transition matrix. For example, here is the argument that $p_{12} = 1 - \theta$: A mother of type $aa = 1$ surely contributes the allele a to her daughter, and so her mate must contribute an A-allele in order for the daughter to be of type $Aa = 2$. Under random mating, the probability of acquiring an A-allele from the father is $1 - \theta$.
b) Show that this chain is ergodic. What is the smallest N that gives $\mathbf{P}^N > \mathbf{0}$?
c) According to the Hardy-Weinberg Law, this Markov chain has the "equilibrium" (steady-state) distribution $\boldsymbol{\sigma} = [\theta^2, 2\theta(1 - \theta), (1 - \theta)^2]$. Verify that this is true.
d) For $\theta = 0.2$, simulate this chain for $m = 50\,000$ iterations and verify that the sampling distribution of the simulated states approximates the Hardy-Weinberg vector.

Hints and partial answers: (a) In deriving p_{12}, notice that it makes no difference how the A-alleles in the population may currently be apportioned among males of types AA and Aa. For example, suppose $\theta = 20\%$ in a male population with 200 alleles (100 individuals), so that there are 40 a-alleles and 160 As. If only genotypes AA and aa exist, then there are 80 AAs to choose from, any of them would contribute an

A-allele upon mating, and the probability of an Aa offspring is $80\% = 1 - \theta$. If there are only 70 AAs among the males, then there must be 20 Aas. The probability that an Aa mate contributes an A-allele is $1/2$, so that the total probability of an Aa offspring is again $1(0.70) + (1/2)(0.20) = 80\% = 1 - \theta$. Other apportionments of genotypes AA and Aa among males yield the same result. The first row of the matrix $\mathbf{P}$ is $[\theta, 1 - \theta, 0]$; its second row is $[\theta/2, 1/2, (1 - \theta)/2]$. (b) For the given σ, show that $\sigma\mathbf{P} = \sigma$. (d) Use a program similar to the one in Example 7.1.

7.10   *Algebraic approach.* For a K-state ergodic transition matrix $\mathbf{P}$, the long-run distribution is proportional to the unique row eigenvector λ corresponding to eigenvalue 1. In R, `g = eigen(t(P))$vectors[,1]; g/sum(g)`, where the transpose function `t` is needed to obtain a *row* eigenvector, `$vectors[,1]` to isolate the relevant part of the eigenvalue-eigenvector display, and the division by `sum(g)` to give a distribution. Use this method to find the long-run distributions of two of the chains in Problems 7.2, 7.5, 7.6, and 7.8—your choice, unless your instructor directs otherwise. (See [CM65] for the theory.)

Problems for Section 7.3 (Countably Infinite State Spaces)

7.11   *Reflecting barrier.* Consider a random walk on the nonnegative integers with $p_{i,i-1} = 1/2$, $p_{i,i+1} = 1/4$, and $p_{ii} = 1/4$, for $i = 1, 2, 3, \ldots$, but with $p_{00} = 1/4$ and $p_{01} = 3/4$. There is a negative drift, but negative values are impossible because the particle gets "reflected" to 1 whenever the usual leftward displacement would have taken it to -1.

a) Argue that the following R script simulates this process, run the program, and comment on whether there appears to be a long-run distribution.

```
# set.seed(1237)
m = 10000
d = sample(c(-1,0,1), m, replace=T, c(1/2,1/4,1/4))
x = numeric(m); x[1] = 0
for (i in 2:m)   {x[i] = abs(x[i-1] + d[i])}
summary(as.factor(x))
cutp=0:(max(x)+1) - .5
hist(x, breaks=cutp, prob=T)
```

b) Show that the steady-state distribution of this chain is given by $\lambda_0 = 1/4$ and $\lambda_i = \frac{3}{4}(\frac{1}{2})^i$, for $i = 1, 2, \ldots$, by verifying that these values of λ_i satisfy the equations $\lambda_j = \sum_i \lambda_i p_{ij}$, for $j = 0, 1, \ldots$. For this chain, the steady-state distribution is unique and is also the long-run distribution. Do these values agree reasonably well with those simulated in part (a)?

7.12   *Attraction toward the origin.* Consider the random walk simulated by the R script below. There is a negative drift when X_{n-1} is positive and a positive drift when it is negative, so that there is always drift towards 0. (The R function `sign` returns values -1, 0, and 1 depending on the sign of the argument.)

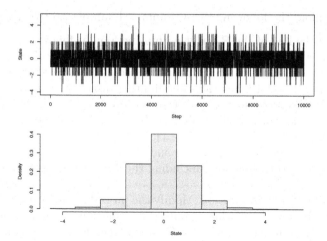

Figure 7.15. Ergodic chain with countable state space. Although the state space $S = \{0, \pm1, \pm2, \ldots\}$ of the chain in Problem 7.12 is not finite, activity is concentrated near 0, so probability does not "escape" to $\pm\infty$.

```
# set.seed(1212)
m = 10000;   x = numeric(m);   x[1] = 0
for (i in 2:m)
{
   drift = (2/8)*sign(x[i-1]);   p = c(3/8+drift, 2/8, 3/8-drift)
   x[i] = x[i-1] + sample(c(-1,0,1), 1, replace=T, prob=p)
}
summary(as.factor(x))
par(mfrow=c(2,1))  # prints two graphs on one page
   plot(x, type="l")
   cutp = seq(min(x), max(x)+1)-.5;   hist(x, breaks=cutp, prob=T)
par(mfrow=c(1,1))
```

a) Write the transition probabilities p_{ij} of the chain simulated by this program. Run the program, followed by `acf(x)`, and comment on the resulting graphs. (See Figure 7.15.)

b) Use the method of Problem 7.11 to show that the long-run distribution is given by $\lambda_0 = 2/5$ and $\lambda_i = \frac{6}{5}(\frac{1}{5})^{|i|}$ for positive and negative integer values of i. Do these values agree with your results in part (a)?

Problems for Section 7.4 (Continuous State Spaces)

7.13 *Random walk on a circle.* In Example 7.5 (p170), the displacements of the random walk on the circle are UNIF$(-0.1, 0.1)$ and the long-run distribution is UNIF$(0, 1)$. Modify the program of the example to explore the long-run behavior of such a random walk when the displacements are NORM$(0, 0.1)$. Compare the two chains.

7.14   Explore the following two variants of Example 7.6 (p171).

a) Change the line of the R script within the loop as follows:
   ```
   x[i] = rbeta(1, 2.001 - 2*x[i-1], 0.001 + 2*x[i-1]).
   ```
   You will also need to omit the plot parameter type="l". (Why?) Do
   you think this chain has the same limiting distribution as the one in the
   example? Compare histograms from several runs of each chain. Explain.
b) Simulate the Markov chain defined by the initial value $X_1 = 0.5$ and
   $X_n \sim \text{BETA}(5 + 2X_{n-1}, 7 - 2X_{n-1})$, for $n = 2, 3, \ldots$.
c) Compare the chains of parts (a) and (b) with the chain of Example 7.6.
   Discuss long-run behaviors, including the output from acf, in each case.

7.15   *Monte Carlo integration.* Modify the procedure in Example 7.7 (p174)
to make a Markov chain whose limiting distribution is uniform on the first
quadrant of the unit circle. If Z_1 and Z_2 are independent and standard normal,
use your modified chain to approximate $P\{Z_1 > 0, Z_2 > 0, Z_1^2 + Z_2^2 < 1\}$.

7.16   *Sierpinski Triangle.* Consider S, obtained by successive deletions from
a (closed) triangle $\triangle ABC$ of area 1/2 with vertices at $A = (0,0)$, $B = (0,1)$,
and $C = (1,0)$. Successively delete open subtriangles of $\triangle ABC$ as follows. At
stage 1, delete the triangle of area 1/8 with vertices at the center points of the
sides of $\triangle ABC$, leaving the union of three triangles, each of area 1/8. At stage
2, delete three more triangles, each of area 1/32 with vertices at the center
points of the sides of triangles remaining after stage 1, leaving the union of
nine triangles, each of area 1/32. Iterate this process forever. The result is the
Sierpinski Triangle.

a) Show that the area of S is 0. That is, the infinite sum of the areas of all
   the triangles removed is 1/2.
b) S is the state space of a Markov chain. Starting with $(X_1, Y_1) = (1/2, 1/2)$,
   choose a vertex of the triangle at random (probability 1/3 each) and let
   (X_2, Y_2) be the point halfway to the chosen vertex. At step 3, choose a vertex, and let (X_3, Y_3) be halfway between (X_2, Y_2) and the chosen vertex.
   Iterate. Suppose the first seven vertices chosen are A, A, C, B, B, A, A.
   (These were taken from the run in part (c).) Find the coordinates of
   (X_n, Y_n), for $n = 2, 3, \ldots, 8$, and plot them by hand.
c) As shown in Figure 7.12 (p176), the R script below generates enough points
   of S to suggest the shape of the state space. (The default distribution of
   the sample function assigns equal probabilities to the values sampled, so
   the prob parameter is not needed here.)

```
# set.seed(1212)
m = 5000
e = c(0, 1, 0);  f = c(0, 0, 1)
k = sample(1:3, m, replace=T)
x = y = numeric(m);  x[1] = 1/2;  y[1] = 1/2
```

```
for (i in 2:m)
{
  x[i] = .5*(x[i-1] + e[k[i-1]])
  y[i] = .5*(y[i-1] + f[k[i-1]])
}
plot(x,y,pch=20)
```

Within the limits of your patience and available computing speed, increase the number m of iterations in this simulation. Why do very large values of m give *less*-informative plots? Then try `plot` parameter `pch="."`. Also, make a plot of the first 100 states visited, similar to Figure 7.10. Do you think such plots would enable you to distinguish between the Sierpinski chain and the chain of Example 7.7?

d) As n increases, the number of possible values of X_n increases. For example, 3 points for $n = 2$ and 9 points for $n = 3$, considering all possible paths. Points available at earlier stages become unavailable at later stages. For example, it is possible to have $(X_2, Y_2) = (1/4, 1/4)$, but explain why this point cannot be visited at any higher numbered step. By a similar argument, no state can be visited more than once. Barnsley ([Bar88], p372) shows that the limiting distribution can be regarded as a "uniform" distribution on Sierpinski's Triangle.

Note: Properties of S related to complex analysis, chaos theory and fractal geometry have been widely studied. Type `sierpinski triangle` into your favorite search engine to list hundreds of web pages on these topics. (Those from educational and governmental sites may have the highest probability of being correct.)

7.17 *Continuation of Problem 7.16: Fractals.* Each subtriangle in Figure 7.12 is a miniature version of the entire Sierpinski set. Similarly, here each "petal" of the frond in Figure 7.16 is a miniature version of the frond itself, as is each "lobe" of each petal, and so on to ever finer detail beyond the resolution of the figure. This sort of self-similarity of subparts to the whole characterizes one type of **fractal**.

By selecting at random among more general kinds of movement, one can obtain a wide range of such fractals. Figure 7.16 resembles a frond of the black spleenwort fern. This image was made with the R script shown below. It is remarkable that such a simple algorithm can realistically imitate the appearance of a complex living thing.

a) In the fern process, the choices at each step have unequal probabilities, as specified by the vector p. For an attractive image, these "weights" are chosen to give roughly even apparent densities of points over various parts of the fern. Run the script once as shown. Then vary p in several ways to observe the role played by these weights.

```
m = 30000
a = c(0, .85, .2, -.15);    b = c(0, .04, -.26, .28)
c = c(0, -.04, .23, .26);   d = c(.16, .85, .22, .24)
```

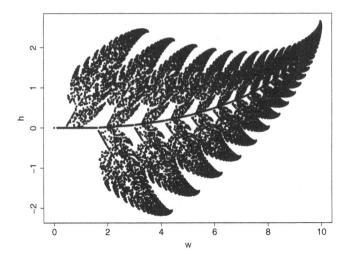

Figure 7.16. Barnsley's fractal fern. A relatively simple Markov chain has a very intricate state space. See Problem 7.17.

```
e = c(0, 0, 0, 0);        f = c(0, 1.6, 1.6, .44)
p = c(.01, .85, .07, .07)
k = sample(1:4, m, repl=T, p)
h = numeric(m);  w = numeric(m);  h[1] = 0;  w[1] = 0
for (i in 2:m)
{
  h[i] = a[k[i]]*h[i-1] + b[k[i]]*w[i-1] + e[k[i]]
  w[i] = c[k[i]]*h[i-1] + d[k[i]]*w[i-1] + f[k[i]]
}
plot(w, h, pch=20, col="darkgreen")
```

b) How can the vectors of parameters (a, b, etc.) of this script be changed to display points of Sierpinski's Triangle?

Note: See [Bar88] for a detailed discussion of fractal objects with many illustrations, some in color. Our script is adapted from pages 87–89; its numerical constants can be changed to produce additional fractal objects described there.

Problems for Section 7.5 (Metropolis and Gibbs Chains)

7.18   A bivariate normal distribution for (X, Y) with zero means, unit standard deviations, and correlation 0.8, as in Section 7.5, can be obtained as a linear transformation of independent random variables. Specifically, if U and V are independently distributed as $\mathsf{NORM}(0, 2/\sqrt{5})$, then let $X = U + V/2$ and $Y = U/2 + V$.

a) Verify analytically that the means, standard deviations, and correlation are as expected. Then use the following program to simulate and plot this

bivariate distribution. Compare your results with the results obtained in
Examples 7.8 and 7.9.

```
m = 10000
u = rnorm(m,0,2/sqrt(5));   v = rnorm(m,0,2/sqrt(5))
x = u + v/2;   y = u/2 + v
round(c(mean(x), mean(y), sd(x), sd(y), cor(x,y)), 4)
best = pmax(x, y)
mean(best >= 1.25)
plot(x, y, pch=".", xlim=c(-4,4), ylim=c(-4,4))
```

b) What linear transformation of independent normal random variables could
you use to sample from a bivariate normal distribution with zero means,
unit standard deviations, and correlation $\rho = 0.6$? Modify the code of
part (a) accordingly, run it, and report your results.

7.19 *Metropolis and Metropolis-Hastings algorithms.* Consider the following
modifications of the program of Example 7.8.

a) Explore the consequences of incorrectly using an asymmetric jump distribution in the Metropolis algorithm. Let jl = 1.25; jr = .75.
b) The Metropolis-Hastings algorithm permits and adjusts for use of an
asymmetric jump distribution by modifying the acceptance criterion.
Specifically, the ratio of variances is multiplied by a factor that corrects
for the bias due to the asymmetric jump function. In our example, this
"symmetrization" amounts to restricting jumps in X and Y to 0.75 units
in either direction. The program below modifies the one in Example 7.8
to implement the Metropolis-Hastings algorithm; the crucial change is the
use of the adjustment factor adj inside the loop. Interpret the numerical
results, the scatterplot (as in Figure 7.17, upper right), and the histogram.

```
set.seed(2008)
m = 100000;   xc = yc = numeric(m);   xc[1] = 3;   yc[1] = -3
rho = .8;   sgm = sqrt(1 - rho^2);   jl = 1.25;   jr = .75
for (i in 2:m)
{
    xc[i] = xc[i-1];   yc[i] = yc[i-1]   # if no jump
    xp = runif(1, xc[i-1]-jl, xc[i-1]+jr)
    yp = runif(1, yc[i-1]-jl, yc[i-1]+jr)
    nmtr.r = dnorm(xp)*dnorm(yp, rho*xp, sgm)
    dntr.r = dnorm(xc[i-1])*dnorm(yc[i-1], rho*xc[i-1], sgm)
    nmtr.adj = dunif(xc[i-1], xp-jl, xp+jr)*
            dunif(yc[i-1], yp-jl, yp+jr)
    dntr.adj = dunif(xp, xc[i-1]-jl, xc[i-1]+jr)*
            dunif(yp, yc[i-1]-jl, yc[i-1]+jr)
    r = nmtr.r/dntr.r;   adj = nmtr.adj/dntr.adj
    acc = (min(r*adj, 1) > runif(1))
    if (acc) {xc[i] = xp;   yc[i] = yp}
}
```

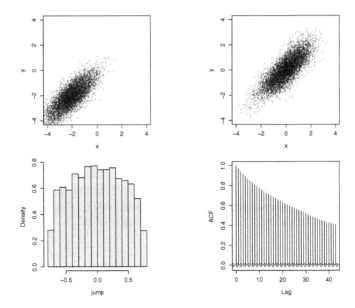

Figure 7.17. The Metropolis-Hastings algorithm. With an asymmetrical jump distribution, the Metropolis algorithm gives an incorrect result (upper left), while the Metropolis-Hastings algorithm corrects the "southwesterly" bias to sample from the intended distribution (upper right). The histogram shows that longer negative jumps responsible for asymmetry are being rejected. This relatively sluggish movement results in the very high autocorrelations seen in the ACF plot. See Problem 7.19.

```
x = xc[(m/2+1):m]; y = yc[(m/2+1):m]
round(c(mean(x), mean(y), sd(x), sd(y), cor(x,y)) ,4)
mean(diff(xc)==0);  mean(pmax(x, y) > 1.25)

par(mfrow=c(1,2), pty="s")
    jump = diff(unique(x)); hist(jump, prob=T, col="wheat")
    plot(x, y, xlim=c(-4,4), ylim=c(-4,4), pch=".")
par(mfrow=c(1,1), pty="m")
```

c) After a run of the program in part (b), make and interpret autocorrelation function plots of x and of x[thinned], where the latter is defined by thinned = seq(1, m/2, by=100). Repeat for realizations of Y.

Notes: (a) The acceptance criterion still has valid information about the shape of the target distribution, but the now-asymmetrical jump function is biased towards jumps downward and to the left. The approximated percentage of subjects awarded certificates is very far from correct. (c) Not surprisingly for output from a Markov chain, the successive pairs (X, Y) sampled by the Metropolis-Hastings algorithm after burn-in are far from independent. "Thinning" helps. To obtain the desired degree of accuracy, we need to sample more values than would be necessary in a simulation with independent realizations as in Problem 7.18. It is important to distinguish the association between X_i and Y_i on the one hand from the association among the X_i on the other hand. The first is an essential property of the target distribution, whereas the second is an artifact of the method of simulation.

7.20   We revisit the Gibbs sampler of Example 7.9.

a) Modify this program to sample from a bivariate normal distribution with zero means, unit standard deviations, and $\rho = 0.6$. Report your results. If you worked Problem 7.18, compare with those results.
b) Run the original program (with $\rho = 0.8$) and make an autocorrelation plot of X-values from $m/2$ on, as in part (c) of Problem 7.19. If you worked that problem, compare the two autocorrelation functions.
c) In the Gibbs sampler of Example 7.9, replace the second statement inside the loop by yc[i] = rnorm(1, rho*xc[i-1], sgm) and run the resulting program. Why is this change a mistake?

Note: (b) In the Metropolis-Hastings chain, a proposed new value is sometimes rejected so that there is no change in state. The Gibbs sampler never rejects.

8

Introduction to Bayesian Estimation

The rest of this book deals with Bayesian estimation. This chapter uses examples to illustrate the fundamental concepts of Bayesian point and interval estimation. It also provides an introduction to Chapters 9 and 10, where more advanced examples require computationally intensive methods.

Bayesian and frequentist statistical inference take very different approaches to statistical decision making.

- The frequentist view of probability, and thus of statistical inference, is based on the idea of an experiment that can be repeated many times.
- The Bayesian view of probability and of inference is based on a personal assessment of probability and on observations from a single performance of an experiment.

These different views lead to fundamentally different procedures of estimation, and the interpretations of the resulting estimates are also fundamentally different. In practical applications, both ways of thinking have advantages and disadvantages, some of which we will explore here.

Statistics is a relatively young science. For example, interval estimation has gradually become common in scientific research and business decision making only within the past 75 years. On this time scale it seems strange to talk about "traditional" approaches. However, frequentist viewpoints are currently much better established, particularly in scientific research, than Bayesian ones. Recently, the use of Bayesian methods has been increasing, partly because the Bayesian approach seems to be able to get more useful solutions than frequentist ones in some applications and partly because improvements in computation have made Bayesian methods increasingly convenient to apply in practice. The Gibbs sampler is one computationally intensive method that is broadly applicable in Bayesian estimation.

For some of the very simple examples considered here, Bayesian and frequentist methods give similar results. But that is not the main point. We hope you will gain some appreciation that Bayesian methods are sometimes the most natural and useful ones in practice. Also, we hope you will begin to appreciate the essential role of computation in Bayesian estimation.

For most people, the starkest contrast between frequentist and Bayesian approaches to analyzing an experiment or study is that Bayesian inference provides the opportunity—even imposes the requirement—to take explicit notice of "information" that is available before any data are collected. That is where we begin.

8.1 Prior Distributions

The Bayesian approach to statistical inference treats population parameters as random variables (not as fixed, unknown constants). The distributions of these parameters are called **prior distributions**. Often both expert knowledge and mathematical convenience play a role in selecting a particular type of prior distribution. This is easiest to explain and to understand in terms of examples. Here we introduce four examples that we carry throughout this chapter.

Example 8.1. Election Polling. Suppose Proposition A is on the ballot for an upcoming statewide election, and a political consultant has been hired to help manage the campaign for its adoption. The proportion π of prospective voters who currently favor Proposition A is the population parameter of interest here. Based on her knowledge of the politics of the state, the consultant's judgment is that the proposition is almost sure to pass, but not by a large margin. She believes that the most likely proportion π of voters in favor is 55% and that the proportion is not likely to be below 51% or above 59%.

It is reasonable to try to use the beta family of distributions to model the expert's opinion of the proportion in favor because distributions in this family take values in the interval $(0, 1)$, as do proportions. Beta distributions have density functions of the form

$$p(\pi) = K\pi^{\alpha-1}(1-\pi)^{\beta-1}$$
$$\propto \pi^{\alpha-1}(1-\pi)^{\beta-1},$$

for $0 < \pi < 1$, where α, $\beta > 0$ and K is the constant such that $\int_0^1 p(\pi)\, d\pi = 1$. Here we adopt two conventions that are common in Bayesian discussions: the use of the letter p instead of f to denote a density function, and the use of the symbol $\propto$ (read *proportional to*) instead of $=$, so that we can avoid specifying a constant whose exact value is unimportant to the discussion. The essential factor of the density function that remains when the constant is suppressed is called the **kernel** of the density function (or of its distribution).

A member of the beta family that corresponds reasonably well to the expert's opinion has $\alpha_0 = 330$ and $\beta_0 = 270$. (Its density is the fine-line curve in Figure 8.1.) This is a reasonable choice of parameters for several reasons.

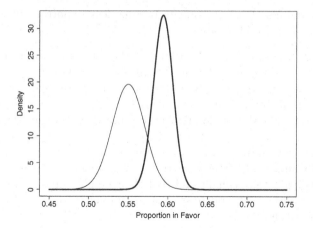

Figure 8.1. Prior and posterior densities for the proportion π of the population in favor of ballot Proposition A (see Examples 8.1 and 8.5). The prior (fine line) is BETA(330, 270) with mean 55.0%. Based on a poll of 1000 subjects with 62.0% in favor, the more concentrated posterior (heavy) is BETA(950, 650) with mean 59.5%.

- This beta distribution is centered near 55% by any of the common measures of centrality. Using analytic methods one can show that the *mean* of this distribution is $\alpha_0/(\alpha_0 + \beta_0) = 330/600 = 55.00\%$ and that its *mode* is $(\alpha_0 - 1)/(\alpha_0 + \beta_0 - 2) = 329/598 = 55.02\%$. Computational methods show the *median* to be 55.01%. (The R function qbeta(.5, 330, 270) returns 0.5500556.) The mean is the most commonly used measure of centrality. Here the mean, median, and mode are so nearly the same that it doesn't make any practical difference which is used.
- Numerical integration shows that these parameters match the expert's prior probability interval reasonably well: $P\{0.51 < \pi < 0.59\} \approx 0.95$. (The R code pbeta(.59, 330, 270) - pbeta(.51, 330, 270) returns 0.9513758.)

Of course, slightly different choices for α_0 and β_0 would match the expert's opinion about as well. It is not necessary to be any fussier in choosing the parameters than the expert was in specifying her hunches. Also, distributional shapes other than the beta might match the expert's opinion just as well. But we choose a member of the beta family because it makes the mathematics in what comes later relatively easy and because we have no reason to believe that the shape of our beta distribution is inappropriate here. (See Problems 8.2 and 8.3.)

If the consultant's judgments about the political situation are correct, then they may be helpful in managing the campaign. If she too often brings bad judgment to her clients, her reputation will suffer and she may be out of the political consulting business before long. Fortunately, as we see in the next section, the details of her judgments become less important if we also have some polling data to rely upon. ◇

Example 8.2. Counting Mice. An island in the middle of a river is one of the last known habitats of an endangered kind of mouse. The mice rove about the island in ways that are not fully understood and so are taken as random.

Ecologists are interested in the average number of mice to be found in particular regions of the island. To do the counting in a region, they set many traps there at night, using bait that is irresistible to mice at close range. In the morning they count and release the mice caught. It seems reasonable to suppose that almost all of the mice in the region where traps were set during the previous night were caught and that their number on any one night has a Poisson distribution. The purpose of the trapping is to estimate the mean λ of this distribution.

Even before the trapping is done, the ecologists doing this study have some information about λ. For example, although the mice are quite shy, there have been occasional sightings of them in almost all regions of the island, so it seems likely that $\lambda > 1$. On the other hand, from what is known of the habits of the mice and the food supply in the regions, it seems unlikely that there would be as many as 25 of them in any one region at a given time.

In these circumstances, it seems reasonable to use a gamma distribution as a prior distribution for λ. This gamma distribution has the density

$$p(\lambda) \propto \lambda^{\alpha-1} e^{-\kappa \lambda},$$

for $\lambda > 0$, where the shape parameter α and the rate parameter κ must both be positive. First, we choose a gamma distribution because it puts all of its probability on the positive half line, and λ must surely have a positive value. Second, we choose a member of the gamma family because it simplifies some important computations that we need to do later.

Using straightforward calculus, one can show that a distribution in the gamma family has mean α/κ, mode $(\alpha - 1)/\kappa$, and variance α/κ^2. These distributions are right-skewed, with the skewness decreasing as α increases.

One reasonable choice for a prior distribution on λ is a gamma distribution with $\alpha_0 = 4$ and $\kappa_0 = 1/3$. Reflecting the skewness, the mean 12, median 11.02, and mode 9 are noticeably different. (We obtained the median using R: qgamma(.5, 4, 1/3) returns 11.01618. Also, see Problem 8.9.) Numerical methods also show that $P\{\lambda < 25\} = 0.97$. (In R, pgamma(25, 4, 1/3) returns 0.9662266.) All of these values are consistent with the expert opinions of the ecologists.

It is clear that the prior experience of the ecologists with the island and its endangered mice will influence the course of this investigation in many ways: dividing the island into meaningful regions, modeling the randomness of mouse movement as Poisson, deciding how many traps to use and where to place them, choosing a kind of bait that will attract mice from a region of interest but not from all over the island, and so on. The expression of some of their background knowledge as a prior distribution is perhaps a relatively small use of their expertise. But a prior distribution is a necessary starting place for Bayesian inference, and it is perhaps the only aspect of expert opinion that will be explicitly tempered by the data that are collected. ◇

Example 8.3. Weighing an Object. A construction company buys reinforced concrete beams with a nominal weight of 700 lb. Experience with a particular supplier of these beams has shown that their beams very seldom weigh less than 680 or more than 720 lb. In these circumstances it may be convenient and reasonable to use $\mathsf{NORM}(700, 10)$ as the prior distribution of the weight of a randomly chosen beam from this supplier.

Usually, the exact weight of a beam is not especially important, but there are some situations in which it is crucial to know the weight of a beam more precisely. Then a particular beam is selected and weighed several times on a scale in order to determine its weight more exactly.

Theoretically, a frequentist statistician would ignore "prior" or background experience in doing statistical inference, basing statistical decisions only on the data collected when a beam is weighed. In real life, it is not so simple. For example, the design of the weighing experiment will very likely take past experience into account in one way or another. (For example, if you are going to weigh things, then you need to know whether you will be using a laboratory balance, a truck scale, or some intermediate kind of scale. And if you need more precision than the scale will give in a single measurement, you may need to weigh each object several times and take the average.) For the Bayesian statistician the explicit codification of some kinds of background information into a prior distribution is a required first step. ◇

Example 8.4. Precision of Hemoglobin Measurements. A hospital has just purchased a device for the assay of hemoglobin (Hgb) in the blood of newborn babies (in g/dl). Considering the claims of the manufacturer and experience with competing methods of measuring Hgb, it seems reasonable to suppose the machine gives unbiased normally distributed results X with a standard deviation σ somewhere between 0.25g/dl and 1g/dl.

For mathematical convenience in Bayesian inference, it is customary to express a prior distribution for the variability of a normal distribution in terms of a gamma distribution on the **precision** $\tau = 1/\sigma^2$. In our example, we might seek a prior distribution on τ with $P\{1/4 < \sigma < 1\} = P\{1/16 < \sigma^2 < 1\} = P\{1 < \tau < 16\} \approx 0.95$. One reasonable choice, under which this interval has probability 0.96, is $\tau \sim \mathsf{GAMMA}(\alpha_0 = 3, \kappa_0 = 0.75)$.

When τ has a gamma prior $\mathsf{GAMMA}(\alpha, \kappa)$, we say that $\theta = 1/\tau = \sigma^2$ has an **inverse gamma** prior distribution $\mathsf{IG}(\alpha, \kappa)$. This distribution family has density

$$p(\theta) = \frac{\kappa^\alpha}{\Gamma(\alpha)} \theta^{-(\alpha+1)} e^{-\kappa/\theta} \propto \theta^{-(\alpha+1)} e^{-\kappa/\theta},$$

for $\theta > 0$. The mode of this distribution is $\kappa/(\alpha + 1)$, and when $\alpha > 1$, its mean is $\kappa/(\alpha - 1)$.

In R, simulated values and quantiles of IG can be found as reciprocals of **rgamma** and **qgamma**, respectively. Cumulative probabilities can be found by using reciprocal arguments in **pgamma**. For example, with $\alpha_0 = 3$, $\kappa_0 = .75$, we find $\mathrm{Med}(\theta) = 1/\mathrm{Med}(\tau) = 0.28$ with the code **1/qgamma(.5, 3, .75)**, and we get 0.50 from **pgamma(1/0.28047, 3, .75)**. $\diamondsuit$

8.2 Data and Posterior Distributions

The second step in Bayesian inference is to collect data and combine the information in the data with the expert opinion represented by the prior distribution. The result is a **posterior distribution** that can be used for inference.

Once the data are available, we can use Bayes' Theorem to compute the posterior distribution $\pi|x$. Equation (5.7), repeated here as (8.1), states an elementary version of Bayes' Theorem for an observed event E and a partition $\{A_1, A_2, \ldots, A_k\}$ of the sample space S,

$$P(A_j|E) = \frac{P(A_j)P(E|A_j)}{\sum_{i=1}^{k} P(A_i)P(E|A_i)}. \tag{8.1}$$

This equation expresses a posterior probability $P(A_j|E)$ in terms of the prior probabilities $P(A_i)$ and the conditional probabilities $P(E|A_i)$.

Here we use a more general version of Bayes' Theorem involving data x and a parameter π,

$$p(\pi|x) = \frac{p(\pi)p(x|\pi)}{\int p(\pi)p(x|\pi)\,d\pi} \propto p(\pi)p(x|\pi), \tag{8.2}$$

where the integral is taken over all values of π for which the integrand is positive. The proportionality symbol $\propto$ is appropriate because the integral is a constant. (In case the distribution of π is discrete, the integral is interpreted as a sum.)

Thus the posterior distribution of $\pi|x$ is found from the prior distribution of π and the distribution of the data x given π. If π is a known constant, $p(x|\pi)$ is the density function of x; we might integrate it with respect to x to evaluate the probability $P(x \in A) = \int_A p(x)\,dx$. However, when we use (8.2) to find a posterior, we know the data x, and we view $p(x|\pi)$ as a function of π. When viewed in this way, $p(x|\pi)$ is called the **likelihood function** of π. (Technically, the likelihood function is defined only up to a positive constant.)

A convenient summary of our procedure for finding the posterior distribution with relationship (8.2) is to say

POSTERIOR $\propto$ PRIOR $\times$ LIKELIHOOD.

We now illustrate this procedure for each of the examples of the previous section.

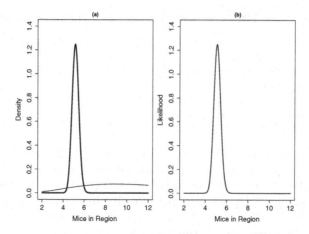

Figure 8.2. (a) Prior and posterior densities for the number of mice in a region. (b) Likelihood function of the mouse data: 50 nights with a total of 256 mice trapped. Because the prior density (fine line) is relatively flat, the data largely determine the mode 5.166 of the posterior (heavy). The MLE $\hat{\lambda} = 256/50 = 5.120$ (mode of the likelihood) is not far from the posterior mode. (See Examples 8.2 and 8.6.)

Example 8.5. Election Polling (continued). Suppose n randomly selected registered voters express opinions on Proposition A. What is the likelihood function, and how do we use it to find the posterior distribution?

If the value of π were known, then the number x of the respondents in favor of Proposition A would be a random variable with the binomial distribution: $\binom{n}{x}\pi^x(1 - \pi)^{n-x}$, for $x = 0, 1, 2, \ldots, n$. Now that we have data x, the likelihood function of π becomes $p(x|\pi) \propto \pi^x(1 - \pi)^{n-x}$.

Furthermore, display (8.2) shows how to find the posterior distribution

$$p(\pi|x) \propto \pi^{\alpha_0-1}(1 - \pi)^{\beta_0-1} \times \pi^x(1 - \pi)^{n-x}$$
$$= \pi^{\alpha_0+x-1}(1 - \pi)^{\beta_0+n-x-1} = \pi^{\alpha_n-1}(1 - \pi)^{\beta_n-1},$$

where we recognize the last line as the kernel of a beta distribution with parameters $\alpha_n = \alpha_0 + x$ and $\beta_n = \beta_0 + n - x$. It is easy to find the posterior in this case because the (beta) prior distribution we selected has a functional form that is similar to that of the (binomial) distribution of the data, yielding a (beta) posterior. In this case, we say that the beta is a **conjugate prior** for binomial data. (When nonconjugate priors are used, special computational methods are often necessary; see Problems 8.5 and 8.6.)

Recall that the parameters of the prior beta distribution are $\alpha_0 = 330$ and $\beta_0 = 270$. If $x = 620$ of the $n = 1000$ respondents favor Proposition A, then the posterior has a beta distribution with parameters $\alpha_n = \alpha_0 + x = 950$ and $\beta_n = \beta_0 + n - x = 650$. Look at Figure 8.1 for a visual comparison of the prior and posterior distributions. The density curves were plotted with the following

R script. (By using **lines**, we can plot the prior curve on the same axes as the posterior.)

```
x = seq(.45, .7, .001)
prior = dbeta(x, 330, 270)
post = dbeta(x, 950, 650)
plot(x, post, type="l", ylim=c(0, 35), lwd=2,
    xlab="Proportion in Favor", ylab="Density")
lines(x, prior)
```

The posterior mean is $950/(950+650) = 59.4\%$, a Bayesian point estimate of the actual proportion of the population currently in favor of Proposition A. Also, according to the posterior distribution, $P\{0.570 < \pi < 0.618\} = 0.95$, so that a Bayesian 95% **posterior probability interval** for the proportion π in favor is $(57.0\%, 61.8\%)$. (In R, **qbeta(.025, 950, 650)** returns 0.5695848, and **qbeta(.975, 950, 650)** returns 0.6176932.)

This probability interval resulting from Bayesian estimation is a straightforward probability statement. Based on the combined information from her prior distribution and from the polling data, the political consultant now believes it is very likely that between 57% and 62% of the population currently favors Proposition A. In contrast to a frequentist "confidence" interval, the consultant can use the probability interval without the need to view the poll as a repeatable experiment. ◊

Example 8.6. Counting Mice (continued). Suppose a region of the island is selected where the gamma distribution with parameters $\alpha_0 = 4$ and $\kappa_0 = 1/3$ is a reasonable prior for λ. The prior density is $p(\lambda) \propto \lambda^{\alpha_0-1}e^{-\kappa_0\lambda}$.

Over a period of about a year, traps are set out on $n = 50$ nights with the total number of captures $t = \sum_{i=1}^{50} x_i = 256$ for an average of 5.12 mice captured per night. Thus, the Poisson likelihood function of the data is

$$p(\mathbf{x}|\lambda) \propto \prod_{i=1}^{n} \lambda^{x_i}e^{-\lambda} = \lambda^t e^{-n\lambda}$$

and the posterior distribution is

$$p(\lambda|\mathbf{x}) \propto \lambda^{\alpha_0-1}e^{-\kappa_0\lambda} \times \lambda^t e^{-n\lambda}$$
$$= \lambda^{\alpha_0+t-1}e^{-(\kappa_0+n)\lambda},$$

in which we recognize the kernel of the gamma distribution with parameters $\alpha_n = \alpha_0 + t$ and $\kappa_n = \kappa_0 + n$. Thus, the posterior mean for our particular prior and data is

$$\frac{\alpha_n}{\kappa_n} = \frac{\alpha_0 + t}{\kappa_0 + n} = \frac{4 + 256}{1/3 + 50} = \frac{260}{50.33} = 5.166,$$

the posterior mode is $(\alpha_n - 1)/\kappa_n = 259/50.33 = 5.146$, and the posterior median is 5.159. Based on this posterior distribution, a 95% probability interval for λ is $(4.56, 5.81)$. (In R, **qgamma(.025, 260, 50.33)** returns 4.557005

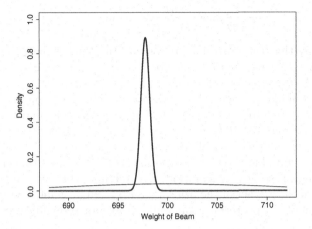

Figure 8.3. Prior density and posterior density for the weight of a beam. The normal prior (fine line) is so flat that the normal posterior (heavy) is overwhelmingly influenced by the data, obtained by repeated weighing of the beam on a scale of relatively high precision. (See Examples 8.3 and 8.7, and Problem 8.12.)

and qgamma(.975, 260, 50.33) returns 5.812432.) The prior and posterior densities are shown in Figure 8.2. ◊

Example 8.7. Weighing a Beam (continued). Suppose that a particular beam is selected from among the beams available. Recall that, according to our prior distribution, the weight of beams in this population is NORM(700, 10), so $\mu_0 = 700$ pounds and $\sigma_0 = 10$ pounds. The beam is weighed $n = 5$ times on a balance that gives unbiased, normally distributed readings with a standard deviation of $\sigma = 1$ pound. Denote the data by $\mathbf{x} = (x_1, \ldots, x_n)$, where the x_i are independent NORM(μ, σ) and μ is the parameter to be estimated. Such data have the likelihood function

$$p(\mathbf{x}|\mu) \propto \exp\left[-\frac{1}{2\sigma^2}\sum_{i=1}^{n}(x_i - \mu)^2\right],$$

where the distribution of μ is determined by the prior and $\sigma = 1$ is known. Then, after some algebra (see Problem 8.13), the posterior is seen to be

$$p(\mu|\mathbf{x}) \propto p(\mu)p(\mathbf{x}|\mu) \propto \exp[-(\mu - \mu_n)^2/2\sigma_n^2],$$

which is the kernel of NORM(μ_n, σ_n), where

$$\mu_n = \frac{\frac{1}{\sigma_0^2}\mu_0 + \frac{n}{\sigma^2}\bar{x}}{\frac{1}{\sigma_0^2} + \frac{n}{\sigma^2}} \quad \text{and} \quad \sigma_n^2 = \frac{1}{\frac{1}{\sigma_0^2} + \frac{n}{\sigma^2}}.$$

It is common to use the term **precision** to refer to the reciprocal of a variance. If we define $\tau_0 = 1/\sigma_0^2$, $\tau = 1/\sigma^2$, and $\tau_n = 1/\sigma_n^2$, then we have

$$\mu_n = \frac{\tau_0}{\tau_0 + n\tau}\mu_0 + \frac{n\tau}{\tau_0 + n\tau}\bar{x} \quad \text{and} \quad \tau_n = \tau_0 + n\tau.$$

Thus, we say that the posterior precision is the sum of the precisions of the prior and the data and that the posterior mean is a precision-weighted average of the means of the prior and the data.

In our example, $\tau_0 = 0.01$, $\tau = 1$, and $\tau_n = 5.01$. Thus the weights used in computing μ_n are $0.01/5.01 \approx 0.002$ for the prior mean μ_0 and $5/5.01 \approx 0.998$ for the mean $\bar{x}$ of the data. We see that the posterior precision is almost entirely due to the precision of the data, and the value of the posterior mean is almost entirely due to their sample mean. In this case, the sample of five relatively high-precision observations is enough to concentrate the posterior and diminish the impact of the prior. (See Problem 8.12 and Figure 8.3 for the computation of the posterior mean and a posterior probability interval.) ◊

In the previous example, we considered the situation in which the mean of a normal distribution is to be estimated but its standard deviation is known. In the following example, data are again normal, but the mean is known and its standard deviation is to be estimated.

Example 8.8. Precision of Hemoglobin Measurements (continued). Suppose researchers use the new device to make Hgb determinations v_i on blood samples from $n = 42$ randomly chosen newborns and also make extremely precise corresponding laboratory determinations w_i on the same samples. Based in part on assumptions in Example 8.4, we assume $x_i = v_i - w_i \sim \text{NORM}(0, \sigma)$. Assuming the laboratory measurements to be of "gold standard" quality, we ignore their errors and take $\tau = 1/\sigma^2$ to be a useful measure of the precision of the new device.

If we observe $s = (\sum_i x_i^2/n)^{1/2} = 0.34$ and use the prior distribution $\tau \sim \text{GAMMA}(3, 0.75)$ of Example 8.4, then what Bayesian posterior probability intervals can we give for τ and for σ? The likelihood function of the data $\mathbf{x} = (x_1, \ldots, x_n)$ is

$$p(\mathbf{x}|\theta) \propto \prod_{i=1}^{n} \theta^{-1/2} \exp\left(-\frac{x_i^2}{2\theta}\right) = \theta^{-n/2} \exp\left(-\frac{ns^2}{2\theta}\right),$$

where we denote $\sigma^2 = \theta$, and the posterior distribution of θ is

$$p(\theta|\mathbf{x}) \propto \theta^{-(\alpha_0+1)} \exp\left(-\frac{\kappa_0}{\theta}\right) \times \theta^{-n/2} \exp\left(-\frac{ns^2}{2\theta}\right)$$
$$= \theta^{-(\alpha_n+1)} \exp\left(-\frac{\kappa_n}{\theta}\right),$$

where $\alpha_n = \alpha_0 + n/2$ and $\kappa_n = \kappa_0 + ns^2/2$. We recognize this as the kernel of the $\text{IG}(\alpha_n, \kappa_n)$ density function. Notice that the posterior has a relatively simple form because θ appears in the denominator of the exponential factor of the inverse-gamma prior. If we had used a gamma prior for the variance

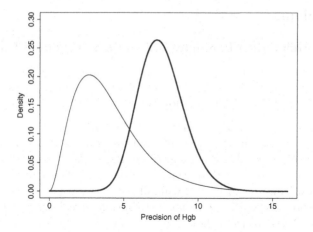

Figure 8.4. Prior density and posterior density for the precision of hemoglobin measurements. The gamma prior (fine line) contributes information corresponding to six measurements. The posterior (heavy line) combines this information with data on 42 subjects to give greater precision. (See Examples 8.4 and 8.8.)

θ (instead of the precision $\tau = 1/\theta$), then θ would have appeared in the numerator of the exponential factor, making the posterior density unwieldy.

For our data, $\alpha_n = 3 + 42/2 = 24$ and $\kappa_n = 0.75 + 42(0.34)^2/2 = 3.178$, so that a 95% posterior probability interval for τ is $(4.84, 10.86)$, computed in R as qgamma(c(.025, .975), 24, 3.18). The corresponding interval for σ is $(0.303, 0.455)$. The frequentist 95% confidence interval for $\sigma = \sqrt{\theta}$ based on $ns^2/\theta \sim \mathsf{CHISQ}(n)$ is $(0.280, 0.432)$, and can be computed in R as sqrt(42*(.34)^2/qchisq(c(.975,.025), 42)). The gamma prior and posterior distributions for the precision τ are shown in Figure 8.4 for τ in the interval $(1, 16)$.

> Notes: (1) Because the normal mean is assumed known, $\mu = 0$, we have $ns^2/\sigma^2 = \sum(x_i - \mu)^2/\sigma^2 = \sum x_i^2/\sigma^2$ distributed as chi-squared with n (not $n-1$) degrees of freedom. (2) This example is loosely based on a real situation reported in [HF94] and used as an extended example in Unit 14 of [Tru02]. In this study, $s = 0.34$ based on $n = 42$ subjects. Complications in practice are that readings from the new device appear to be slightly biased and that the laboratory determinations, while more precise than those from the new device, are hardly free of measurement error. Fortunately, in this clinical setting the precision of both kinds of measurements is much better than it needs to be. ◇

In the next two chapters, we look at Bayesian estimation problems where computationally intensive methods are required to find posterior distributions. Specifically, the concepts of continuous Markov chains from Chapter 7 are used to implement Gibbs samplers.

8.3 Problems

Problems Related to Examples 8.1 and 8.5 (Binomial Data)

8.1   In a situation similar to Example 8.1, suppose a political consultant chooses the prior BETA(380, 220) to reflect his assessment of the proportion of the electorate favoring Proposition B.

a) In terms of a most likely value for π and a 95% probability interval for π, describe this consultant's view of the prospects for Proposition B.
b) If a poll of 100 randomly chosen registered voters shows 62% *opposed* to Proposition B, do you think the consultant (a believer in Bayesian inference) now fears Proposition B will fail? Quantify your answer with specific information about the posterior distribution. Recall that in Example 8.5 a poll of 1000 subjects showed 62% in favor of Proposition A. Contrast that situation with the current one.
c) Modify the R code of Example 8.5 to make a version of Figure 8.5 (p207) that describes this problem.
d) Pollsters sometimes report the margin of sampling error for a poll with n subjects as being roughly given by the formula $100/\sqrt{n}\,\%$. According to this formula, what is the (frequentist's) margin of error for the poll in part (b)? How do you suppose the formula is derived?

Hints: (a) Use R code `qbeta(c(.025,.975), 380, 220)` to find one 95% prior probability interval. (b) One response: $P\{\pi < 0.55\} < 1\%$. (d) A standard formula for an interval with roughly 95% confidence is $p \pm 1.96\sqrt{p(1-p)/n}$, where n is "large" and p is the sample proportion in favor (see Example 1.6). What value of π maximizes $\pi(1-\pi)$? What if $\pi = 0.4$ or 0.6?

8.2   In Example 8.1, we require a prior distribution with $E(\pi) \approx 0.55$ and $P\{0.51 < \pi < 0.59\} \approx 0.95$. Here we explore how one might find suitable parameters α and β for such a beta-distributed prior.

a) For a beta distribution, the mean is $\mu = \alpha/(\alpha + \beta)$ and the variance is $\sigma^2 = \alpha\beta/[(\alpha+\beta)^2(\alpha+\beta+1)]$. Also, a beta distribution with large enough values of α and β is roughly normal, so that $P\{\mu - 2\sigma < \pi < \mu + 2\sigma\} \approx 0.95$. Use these facts to find values of α and β that approximately satisfy the requirements. (Theoretically, this normal distribution would need to be truncated to have support $(0, 1)$.)
b) The following R script finds values of α and β that may come close to satisfying the requirements and then checks to see how well they succeed.

```
alpha = 1:2000        # trial values of alpha
beta = .818*alpha     # corresponding values of beta

# Vector of probabilities for interval (.51, .59)
prob = pbeta(.59, alpha, beta) - pbeta(.51, alpha, beta)
prob.err = abs(.95 - prob)  # errors for probabilities
```

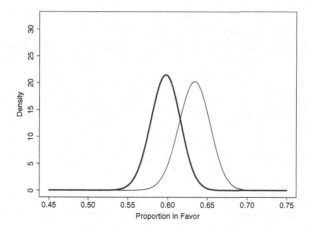

Figure 8.5. Prior and posterior densities for the population proportion π favoring Proposition B (see Problem 8.1). Here the prior (fine line) reflects strong optimism that the proposition is leading. The posterior (heavy line), taking into account results of a relatively small poll with 62% *opposed*, does little to dampen the optimism.

```
# Results: Target parameter values
t.al = alpha[prob.err==min(prob.err)]
t.be = round(.818*t.al)
t.al; t.be

# Checking: Achieved mean and probability
a.mean = t.al/(t.al + t.be)
a.mean
a.prob = pbeta(.59, t.al, t.be) - pbeta(.51, t.al, t.be)
a.prob
```

What assumptions about α and β are inherent in the script? Why do we use $\beta = 0.818\alpha$? What values of α and β are returned? For the values of the parameters considered, how close do we get to the desired values of $E(\pi)$ and $P\{0.51 < \pi < 0.59\}$?

c) If the desired mean is 0.56 and the desired probability in the interval $(0, 51, 0.59)$ is 90%, what values of the parameters are returned by a suitably modified script?

8.3   In practice, the beta family of distributions offers a rich variety of shapes for modeling priors to match expert opinion.

a) Beta densities $p(\pi)$ are defined on the *open* unit interval. Observe that parameter α controls behavior of the density function near 0. In particular, find the value $p(0^+)$ and the slope $p'(0^+)$ in each of the following five cases: $\alpha < 1, \alpha = 1, 1 < \alpha < 2, \alpha = 2$, and $\alpha > 2$. Evaluate each limit as being 0, positive and finite, ∞, or $-\infty$. (As usual, 0^+ means to take the limit as the argument approaches 0 through positive values.)

b) By symmetry, parameter β controls behavior of the density function near 1. Thus, combinations of the parameters yield 25 cases, each with its own "shape" of density. In which of these 25 cases does the density have a unique mode in $(0, 1)$? The number of possible inflection points of a beta density curve is 0, 1, or 2. For each of the 25 cases, give the number of inflection points.

c) The R script below plots examples of each of the 25 cases, scaled vertically (with top) to show the properties in parts (a) and (b) about as well as can be done and yet show most of each curve.

```
alpha = c(.5, 1, 1.2, 2, 5);  beta = alpha
op = par(no.readonly = TRUE)  # records existing parameters
par(mfrow=c(5, 5))            # formats 5 x 5 matrix of plots
par(mar=rep(2, 4), pty="m")   # sets margins
x = seq(.001, .999, .001)

for (i in 1:5)  {
  for (j in 1:5)  {
    top = .2 + 1.2 * max(dbeta(c(.05, .2, .5, .8, .95),
      alpha[j], beta[i]))
    plot(x,dbeta(x, alpha[i], beta[j]),
      type="l", ylim=c(0, top), xlab="", ylab="",
      main=paste("BETA(",alpha[j],",", beta[i],")", sep=""))  }
}
par(op)                       # restores former parameters
```

Run the code and compare the resulting matrix of plots with your results above (α-cases are rows, β columns). What symmetries within and among the 25 plots are lost if we choose beta = c(.7, 1, 1.7, 2, 7)? (See Figure 8.6.)

8.4   In Example 8.1, we require a prior distribution with $E(\pi) \approx 0.55$ and $P\{0.51 < \pi < 0.59\} \approx 0.95$. If we are willing to use nonbeta priors, how might we find ones that meet these requirements?

a) If we use a normal distribution, what parameters μ and σ would satisfy the requirements?

b) If we use a density function in the shape of an isosceles triangle, show that it should have vertices at $(0.4985, 0)$, $(0.55, 19.43)$, and $(0.6015, 0)$.

c) Plot three priors on the same axes: BETA$(330, 270)$ of Example 8.1 and the results of parts (a) and (b).

d) Do you think the expert would object to any of these priors as an expression of her feelings about the distribution of π?

Notes: (c) Plot: Your result should be similar to Figure 8.7. Use the method in Example 8.5 to put several plots on the same axes. Experiment: If v = c(.51, .55, .59)

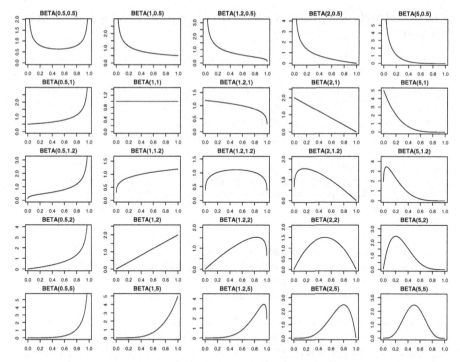

Figure 8.6. Shapes of beta density functions. Shape parameters α and β control the behavior of the density near 0 and 1, respectively; 25 fundamentally different shapes are shown here. (See Problem 8.3.)

and $w = c(0, 10, 0)$, then what does `lines(v, w)` add to an existing `plot`? (d) The triangular prior would be agreeable only if she thinks values of π below 0.4985 or above 0.6015 are *absolutely* impossible.

8.5 Computational methods are often necessary if we multiply the kernels of the prior and likelihood and then can't recognize the result as the kernel of a known distribution. This can occur, for example, when we don't use a conjugate prior. We illustrate several computational methods using the polling situation of Examples 8.1 and 8.5 where we seek to estimate the parameter π.

To begin, suppose we are aware of the beta prior $p(\pi)$ (with $\alpha = 330$ and $\beta = 270$) and the binomial likelihood $p(x|\pi)$ (for $x = 620$ subjects in favor out of $n = 1000$ responding). But we have *not* been clever enough to notice the convenient beta form of the posterior $p(\pi|x)$.

We wish to compute the posterior estimate of centrality $E(\pi|x)$ and the posterior probability $P\{\pi > .6|x\}$ of a potential "big win" for the ballot proposition. From the *equation* in (8.2), we have $E(\pi|x) = \int_0^1 \pi p(\pi)p(x|\pi)\,d\pi/D$ and $P(\pi > 0.6|x) = \int_{0.6}^1 p(\pi)p(x|\pi)\,d\pi/D$, where the denominator of the posterior is $D = \int_0^1 p(\pi)p(x|\pi)\,d\pi$. You should verify these equations for yourself before going on.

a) The following R script uses Riemann approximation to obtain the desired posterior information. Match key quantities in the program with those in the equations above. Also, interpret the last two lines of code. Run the program and compare your results with those obtainable directly from the known beta posterior of Example 8.5. (In R, pi means 3.1416, so we use pp, population proportion, for the grid points of parameter π.)

```
x = 620;  n = 1000                              # data
m = 10000;  pp = seq(0, 1, length=m)            # grid points
igd = dbeta(pp, 330, 270) * dbinom(x, n, pp)    # integrand
d = mean(igd);  d                               # denominator

# Results
post.mean = mean(pp*igd)/d;  post.mean
post.pr.bigwin = (1/m)*sum(igd[pp > .6])/d;  post.pr.bigwin
post.cum = cumsum((igd/d)/m)
min(pp[post.cum > .025]);  min(pp[post.cum > .975])
```

b) Now suppose we choose the prior $\mathsf{NORM}(0.55, 0.02)$ to match the expert's impression that the prior should be centered at $\pi = 55\%$ and put 95% of its probability in the interval $51\% < \pi < 59\%$. The shape of this distribution is very similar to $\mathsf{BETA}(330, 270)$ (see Problem 8.4). However, the normal prior is *not a conjugate prior*. Write the kernel of the posterior, and say why the method of Example 8.5 is intractable. Modify the program above to use the normal prior (substituting the function dnorm for dbeta). Run the modified program. Compare the results with those in part (a).

c) The scripts in parts (a) and (b) above are "wasteful" because grid values of π are generated throughout $(0, 1)$, but both prior densities are very nearly 0 outside of $(0.45, 0.65)$. Modify the program in part (b) to integrate over this shorter interval.

Strictly speaking, you need to divide d, post.mean, and so on, by 5 because you are integrating over a region of length 1/5. (Observe the change in b if you shorten the interval without dividing by 5.) Nevertheless, show that this correction factor cancels out in the main results. Compare your results with those obtained above.

d) Modify the R script of part (c) to do the computation for a normal prior using Monte Carlo integration. Increase the number of iterations to $m \geq 100\,000$, and use pp = sort(runif(m, .45, .65)). Part of the program depends on having the π-values sorted in order. Which part? Why? Compare your results with those obtained by Riemann approximation. (If this were a multidimensional integration, some sort of Monte Carlo integration would probably be the method of choice.)

e) *(Advanced)* Modify part (d) to generate normally distributed values of pp (with sorted rnorm(m, .55, .02)), removing the dnorm factor from the

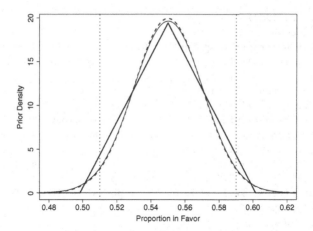

Figure 8.7. Two nonbeta priors. One (thick lines) has an isosceles triangle as its density. The other, NORM(.55, .02) (dashed), is hardly distinguishable from BETA(330, 270) of Example 8.1 (thin). For all three priors, $P\{.51 < \pi < .59\} \approx 95\%$. Only the beta prior is conjugate with binomial data. Bayesian inference using the nonbeta priors requires special numerical methods. (See Problems 8.4, 8.5, and 8.6.)

integrand. Explain why this works, and compare the results with those above.

This method is efficient because it concentrates values of π in the "important" part of $(0, 1)$, where computed quantities are largest. So there would be no point in restricting the range of integration as in parts (c) and (d). This is an elementary example of **importance sampling**.

8.6   *Metropolis algorithm.* In Section 7.5, we illustrated the Metropolis algorithm as a way to sample from a bivariate normal distribution having a known density function. In Problem 8.5, we considered some methods of computing posterior probabilities that arise from nonconjugate prior distributions. Here we use the Metropolis algorithm in a more serious way than before to sample from posterior distributions arising from the nonconjugate prior distributions of Problem 8.4.

a) Use the Metropolis algorithm to sample from the posterior distribution of π arising from the prior NORM(0.55, 0.02) and a binomial sample of size $n = 1000$ with $x = 620$ respondents in favor. Simulate $m = 100\,000$ observations from the posterior to find a 95% Bayesian probability interval for π. Also, if you did Problem 8.5, find the posterior probability $P\{\pi > 0.6|x\}$. The R code below implements this computation using a symmetrical uniform jump function and compares results with those from the very similar conjugate prior BETA(330, 270). See the top panel in Figure 8.8.

```
set.seed(1234)
m = 100000
piec = numeric(m);   piec[1] = 0.7                   # states of chain
for (i in 2:m)  {
  piec[i] = piec[i-1]                                # if no jump
  piep = runif(1, piec[i-1]-.05, piec[i-1]+.05) # proposal
  nmtr = dnorm(piep, .55, .02)*dbinom(620, 1000, piep) %%1
  dmtr = dnorm(piec[i-1], .55, .02)*dbinom(620, 1000, piec[i-1])
  r = nmtr/dmtr;   acc = (min(r,1) > runif(1))   # accept prop.?
  if(acc) {piec[i] = piep}  }
pp = piec[(m/2+1):m]                                 # after burn-in
quantile(pp, c(.025,.975));   mean(pp > .6)
qbeta(c(.025,.975), 950, 650);   1-pbeta(.6, 950, 650)
hist(pp, prob=T, col="wheat", main="")
xx = seq(.5, .7, len=1000)
lines(xx, dbeta(xx, 950, 650), lty="dashed", lwd=2)
```

b) Modify the program of part (a) to find the posterior corresponding to the "isosceles" prior of Problem 8.4. Make sure your initial value is within the support of this prior, and use the following lines of code for the numerator and denominator of the ratio of densities. Notice that, in this ratio, the constant of integration cancels, so it is not necessary to know the height of the triangle. In some more advanced applications of the Metropolis algorithm, the ability to ignore the constant of integration is an important advantage. Explain why results here differ considerably from those in part (a). See the bottom panel in Figure 8.8.

```
nmtr = max(.0515-abs(piep-.55), 0)*dbinom(620, 1000, piep)
dmtr = max(.0515-abs(piec[i-1]-.55), 0)*
          dbinom(620, 1000, piec[i-1])
```

Notes: (a) In the program, the code %%1 (mod 1) restricts the value of nmtr to $(0,1)$. This might be necessary if you experiment with parameters different from those in this problem. (b) Even though the isosceles prior may seem superficially similar to the beta and normal priors, it puts no probability above 0.615, so the posterior can put no probability there either. In contrast, the data show 620 out of 1000 respondents are in favor.

8.7   A commonly used frequentist principle of estimation provides a point estimate of a parameter by finding the value of the parameter that maximizes the likelihood function. The result is called a **maximum likelihood estimate** (MLE). Here we explore one example of an MLE and its similarity to a particular Bayesian estimate.

Suppose we observe $x = 620$ successes in $n = 1000$ binomial trials and wish to estimate the probability π of success. The likelihood function is $p(x|\pi) \propto \pi^x(1 - \pi)^{n-x}$ taken as a function of π.

a) Find the MLE $\hat{\pi}$. A common way to maximize $p(x|\pi)$ in π is to maximize $\ell(\pi) = \ln p(x|\pi)$. Solve $d\ell(\pi)/d\pi = 0$ for π, and verify that you have found

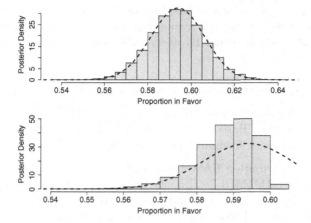

Figure 8.8. Posteriors from nonconjugate priors. Data: 620 subjects in favor out of 1000. Top: The simulated posterior distribution from the prior NORM(.55, .02) is nearly the same as the posterior BETA(950, 650) (dashed) from the conjugate prior BETA(330, 270). Bottom: In contrast, support of the posterior from the "isosceles" prior in Figure 8.7 cannot extend beyond (0.485, 0.615). (See Problem 8.6.)

an absolute maximum. State the general formula for $\hat{\pi}$ and then its value for $x = 620$ and $n = 1000$.

b) Plot the likelihood function for $n = 1000$ and $x = 620$. Approximate its maximum value from the graph. Then do a numerical maximization with the R script below. Compare it with the answer in part (a).

```
pp = seq(.001, .999, .001)    # avoid 'pi' (3.1416)
like = dbinom(620, 1000, pp)
plot(like, type="l");  pp[like==max(like)]
```

c) *Agresti-Coull confidence interval.* The interval $\tilde{\pi} \pm 1.96\sqrt{\tilde{\pi}(1 - \tilde{\pi})/(n + 4)}$, where $\tilde{\pi} = (x + 2)/(n + 4)$, has approximately 95% confidence for estimating π. (This interval is based on the normal approximation to the binomial; see Example 1.6 on p13 and Problems 1.16 and 1.17.) Evaluate its endpoints for 620 successes in 1000 trials.

d) Now we return to Bayesian estimation. A prior distribution that provides little, if any, definite information about the parameter to be estimated is called a **noninformative prior** or **flat prior**. A commonly used noninformative beta prior has $\alpha_0 = \beta_0 = 1$, which is the same as UNIF(0, 1). For this prior and data consisting of x successes in n trials, find the posterior distribution and its mode.

e) For the particular case with $n = 1000$ and $x = 620$, find the posterior mode and a 95% probability interval.

Note: In many estimation problems, the MLE is in close numerical agreement with the Bayesian point estimate based on a noninformative prior and on the posterior

mode. Also, a confidence interval based on the MLE may be numerically similar
to a Bayesian probability interval from a noninformative prior. But the underlying
philosophies of frequentists and Bayesians differ, and so the ways they interpret
results in practice may also differ.

Problems Related to Examples 8.2 and 8.6 (Poisson Data)

8.8   Recall that in Example 8.6 researchers counted a total of $t = 256$ mice
on $n = 50$ occasions. Based on these data, find the interval estimate for λ
described in each part. Comment on similarities and differences.

a) The prior distribution GAMMA(α_0, κ_0) has least effect on the posterior
   distribution GAMMA$(\alpha_0 + t, \kappa_0 + n)$ when α_0 and κ_0 are both small. So
   prior parameters $\alpha_0 = 1/2$ and $\kappa_0 = 0$ give a Bayesian 95% posterior
   probability interval based on little prior information.
b) Assuming that $n\lambda$ is large, an approximate 95% frequentist confidence
   interval for λ is obtained by dividing $t \pm 1.96\sqrt{t}$ by n.
c) A frequentist confidence interval guaranteed to have at least 95% coverage
   has lower and upper endpoints computable in R as `qgamma(.025, t, n)`
   and `gamma(.975, t+1, n)`, respectively.

Notes: (a) Actually, using $\kappa_0 = 0$ gives an improper prior. See the discussion in
Problem 8.12. (b) This style of CI has coverage inaccuracies similar to those of the
traditional CIs for binomial π (see Section 1.2). (c) See [Sta08], Chapter 12.

8.9   In a situation similar to that in Examples 8.2 and 8.6, suppose that we
want to begin with a prior distribution on the parameter λ that has $E(\lambda) \approx 8$
and $P\{\lambda < 12\} \approx 0.95$. Subsequently, we count a total of $t = 158$ mice in
$n = 12$ trappings.

a) To find the parameters of a gamma prior that satisfy the requirements
   above, **write a program** analogous to the one in Problem 8.2. (You can
   come very close with α_0 an integer, but don't restrict κ_0 to integer values.)
b) Find the gamma posterior that results from the prior in part (a) and the
   data given above. Find the posterior mean and a 95% posterior probability
   interval for λ.
c) As in Figure 8.2(a), plot the prior and the posterior. Why is the posterior
   here less concentrated than the one in Figure 8.2(a)?
d) The ultimate *noninformative* gamma prior is the *improper* prior having
   $\alpha_0 = \kappa_0 = 0$ (see Problems 8.7 and 8.12 for definitions). Using this prior
   and the data above, find the posterior mean and a 95% posterior probability interval for λ. Compare the interval with the interval in part (c).

Partial answers: In (a) you can use a prior with $\alpha_0 = 13$. Our posterior intervals
in (c) and (d) agree when rounded to integer endpoints: $(11, 15)$, but not when
expressed to one- or two-place accuracy—as you should do.

8.10   In this chapter, we have computed 95% posterior probability intervals
by finding values that cut off 2.5% from each tail. This method is computationally relatively simple and gives satisfactory intervals for most purposes.
However, for skewed posterior densities, it does not give the *shortest* interval
with 95% probability.

The following R script finds the shortest interval for a gamma posterior.
(The vectors p.low and p.up show endpoints of enough 95% intervals that
we can come very close to finding the one for which the length, long, is a
minimum.)

```
alp = 5; kap = 1
p.lo = seq(.001,.05, .00001);   p.up = .95 + p.lo
q.lo = qgamma(p.lo, alp, kap);   q.up = qgamma(p.up, alp, kap)
long = q.up - q.lo            # avoid confusion with function 'length'
c(q.lo[long==min(long)], q.up[long==min(long)])
```

a) Compare the length of the shortest interval with that of the usual
(probability-symmetric) interval. What probability does the shortest interval put in each tail?
b) Use the same method to find the shortest 95% posterior probability interval in Example 8.6. Compare it with the probability interval given there.
Repeat, using suitably modified code, for 99% intervals.
c) Suppose a posterior density function has a single mode and decreases
monotonically as the distance away from the mode increases (for example,
a gamma density with $\alpha > 1$). Then the shortest 95% posterior probability
interval is also the 95% probability interval corresponding to the highest
values of the posterior: a **highest posterior density** interval. Explain
why this is true. For the 95% intervals in parts (a) and (b), verify that the
heights of the posterior density curve are indeed the same at each end of
the interval (as far as allowed by the spacing 0.00001 of the probability
values used in the script).

8.11   *Mark-recapture estimation of population size.* In order to estimate the
number ν of fish in a lake, investigators capture r of these fish at random,
tag them, and then release them. Later (leaving time for mixing but not for
significant population change), they capture s fish at random from the lake
and observe the number x of tagged fish among them. Suppose $r = 900$,
$s = 1100$, and we observe $x = 103$. (This is similar to the situation described
in Problem 4.27 (p116), partially reprised here in parts (a) and (b).)

a) Method of moments estimate (MME). At recapture, an unbiased estimate
of the true proportion r/ν of tagged fish in the lake is x/s. That is,
$E(x/s) = r/\nu$. To find the MME of ν, equate the observed value x/s to its
expectation and solve for ν. (It is customary to truncate to an integer.)
b) Maximum likelihood estimate (MLE). For known r, s, and ν, the **hyper-
geometric** distribution function $p_{r,s}(x|\nu) = \binom{r}{x}\binom{\nu-r}{s-x}/\binom{\nu}{s}$ gives the prob-
ability of observing x tagged fish at recapture. With known r and s and

observed data x, the likelihood function of ν is $p_{r,s}(x|\nu)$. Find the MLE; that is, the value of ν that maximizes $p_{r,s}(x|\nu)$.

c) Bayesian interval estimate. Suppose we believe ν lies in $(6000, 14\,000)$ and are willing to take the prior distribution of ν as uniform on this interval. Use the R code below to find the cumulative posterior distribution of $\nu|x$ and thence a 95% Bayesian interval estimate of ν. Explain the code.

```
r = 900;  s = 1100;  x = 103;  nu = 6000:14000;  n = length(nu)
prior = rep(1/n, n);  like = dhyper(x, r, nu-r, s)
denom = sum(prior*like)
post = prior*like/denom;  cumpost = cumsum(post)
c(min(nu[cumpost >= .025]), max(nu[cumpost <= .975]))
```

d) Use the negative binomial prior: `prior = dnbinom(nu-150, 150, .014)`. Compare the resulting Bayesian interval with that of part (c) and with a bootstrap confidence interval obtained as in Problem 4.27.

Problems Related to Examples 8.3 and 8.7 (Normal Data, σ Known)

8.12   In Example 8.7, we show formulas for the mean and precision of the posterior distribution. Suppose five measurements of the weight of the beam, using a scale known to have precision $\tau = 1$, are: 698.54, 698.45, 696.09, 697.14, 698.62 ($\bar{x} = 697.76$).

a) Based on these data and the prior distribution of Example 8.3, what is the posterior mean of μ? Does it matter whether we choose the mean, the median, or the mode of the posterior distribution as our point estimate? (Explain.) Find a 95% posterior probability interval for μ. Also, suppose we are unwilling to use this beam if it weighs more than 699 pounds; what are the chances of that?

b) Modify the R script shown in Example 8.5 to plot the prior and posterior densities on the same axes. (Your result should be similar to Figure 8.3.)

c) Taking a frequentist point of view, use the five observations given above and the known variance of measurements produced by our scale to give a 95% confidence interval for the true weight of the beam. Compare it with the results of part (a) and comment.

d) The prior distribution in this example is very "flat" compared with the posterior: its precision is small. A practically noninformative normal prior is one with precision τ_0 that is much smaller than the precision of the data. As τ_0 decreases, the effect of μ_0 diminishes. Specifically, $\lim_{\tau_0 \to 0} \mu_n = \bar{x}$ and $\lim_{\tau_0 \to 0} \tau_n = n\tau$. The *effect* is as if we had used $p(\mu) \propto 1$ as the prior. Of course, such a prior distribution is not strictly possible because $\int_{-\infty}^{\infty} p(\mu)\,d\mu$ would be ∞. But it is convenient to use such an **improper prior** as shorthand for understanding what happens to a posterior as the prior gets less and less informative. What posterior mean and 95% probability interval result from using an improper prior with our data? Compare with the results of part (c).

e) Now change the example: Suppose that our vendor supplies us with a more consistent product so that the prior $\text{NORM}(701, 5)$ is realistic and that our data above come from a scale with known precision $\tau = 0.4$. Repeat parts (a) and (b) for this situation.

8.13   *(Theoretical)* The purpose of this problem is to derive the posterior distribution $p(\mu|\mathbf{x})$ resulting from the prior $\text{NORM}(\mu_0, \sigma_0)$ and n independent observations $x_i \sim \text{NORM}(\mu, \sigma)$. (See Example 8.7.)

a) Show that the *likelihood* is

$$f(\mathbf{x}|\mu) \propto \prod_{i=1}^{n} \exp\left[-\frac{1}{2\sigma^2}(x_i - \mu)^2\right] \propto \exp\left[-\sum_{i=1}^{n}\frac{1}{2\sigma^2}(\bar{x} - \mu)^2\right].$$

To obtain the first expression above, recall that the likelihood function is the joint density function of $\mathbf{x} = (x_1, \ldots, x_n)|\mu$. To obtain the second, write $(x_i - \mu)^2 = [(x_i - \bar{x}) + (\bar{x} - \mu)]^2$, expand the square, and sum over i. On distributing the sum, you should obtain three terms. One of them provides the desired result, another is 0, and the third is irrelevant because it does not contain the variable μ. (A constant term in the exponential is a constant factor of the likelihood, which is not included in the kernel.)

b) To derive the expression for the kernel of the *posterior,* multiply the kernels of the prior and the likelihood, and expand the squares in each. Then put everything in the exponential over a common denominator, and collect terms in μ^2 and μ. Terms in the exponent that do not involve μ are constant factors of the posterior density that may be adjusted as required in completing the square to obtain the desired posterior kernel.

Problems Related to Examples 8.3 and 8.7 (Normal Data, $\mu = 0$)

8.14   For a pending American football game, the "point spread" is established by experts as a measure of the difference in ability of the two teams. The point spread is often of interest to gamblers. Roughly speaking, the favored team is thought to be just as likely to win by more than the point spread as to win by less or to lose. So ideally a fair bet that the favored team "beats the spread" could be made at even odds. Here we are interested in the difference $x = v - w$ between the point spread v, which might be viewed as the favored team's predicted lead, and the actual point difference w (the favored team's score minus its opponent's) when the game is played.

a) Suppose an amateur gambler, perhaps interested in bets that would not have even odds, is interested in the precision of x and is willing to assume $x \sim \text{NORM}(0, \sigma)$. Also, recalling relatively few instances with $|x| > 30$, he decides to use a prior distribution on σ that satisfies $P\{10 < \sigma < 20\} = P\{100 < \sigma^2 = 1/\tau < 400\} = P\{1/400 < \tau < 1/100\} = 0.95$. Find parameters α_0 and κ_0 for a gamma-distributed prior on τ that approximately satisfy this condition. (Imitate the program in Problem 8.2.)

b) Suppose data for point spreads and scores of 146 professional football games show $s = (\sum x_i^2/n)^{1/2} = 13.3$. Under the prior distribution of part (a), what 95% posterior probability intervals for τ and σ result from these data?

c) Use the noninformative improper prior distribution with $\alpha_0 = \kappa_0 = 0$ and the data of part (b) to find 95% posterior probability intervals for τ and σ. Also, use these data to find the frequentist 95% confidence interval for σ based on the distribution CHISQ(146), and compare it with the posterior probability interval for σ.

Notes and hints: (a) Parameters $\alpha_0 = 11, \kappa_0 = 2500$ give probability 0.945, but your program should give integers that come closer to 95%. (b) The data **x** in part (b), taken from more extensive data available online [Ste92], are for 1992 NFL home games; $\bar{x} \approx 0$ and the data pass standard tests for normality. For a more detailed discussion and analysis of point spreads, see [Ste91]. (c) The two intervals for σ agree closely, roughly (12, 15). You should report results to one decimal place.

8.15   We want to know the precision of an analytic device. We believe its readings are normally distributed and unbiased. We have five standard specimens of known value to use in testing the device, so we can observe the error x_i that the device makes for each specimen. Thus we assume that the x_i are independent NORM$(0, \sigma)$, and we wish to estimate $\sigma = 1/\sqrt{\tau}$.

a) We use information from the manufacturer of the device to determine a gamma-distributed prior for τ. This information is provided in terms of σ. Specifically, we want the prior to be consistent with a median of about 0.65 for σ and with $P\{\sigma < 1\} \approx 0.95$. If a gamma prior distribution on τ has parameter $\alpha_0 = 5$, then what value of the parameter κ_0 comes close to meeting these requirements?

b) The following five errors are observed when analyzing test specimens: $-2.65, 0.52, 1.82, -1.41, 1.13$. Based on the prior distribution in part (a) and these data, find the posterior distribution, the posterior *median* value of τ, and a 95% posterior probability interval for τ. Use these to give the posterior median value of σ and a 95% posterior probability interval for σ.

c) On the same axes, make plots of the prior and posterior distributions of τ. Comment.

d) Taking a frequentist approach, find the maximum likelihood estimate (MLE) $\hat{\tau}$ of τ based on the data given in part (b). Also, find 95% confidence intervals for σ^2, σ, and τ. Use the fact that $\sum_{i=1}^{n} x_i^2/\sigma^2 \sim$ CHISQ$(n) =$ GAMMA$(n/2, 1/2)$. Compare these with the Bayesian results in part (b).

Notes: The invariance principle of MLEs states that $\hat{\tau} = 1/\widehat{\sigma^2} = 1/\hat{\sigma}^2$, where "hats" indicate MLEs of the respective parameters. Also, the median of a random variable is invariant under any monotone transformation. Thus, for the prior or posterior distribution of τ (always positive), Med$(\tau) = 1/$Med$(\sigma^2) = 1/[$Med$(\sigma)]^2$. But, in general, expectation is invariant only under *linear* transformations. For example, $E(\tau) \neq 1/E(\sigma^2)$ and $E(\sigma^2) \neq [E(\sigma)]^2$.

9

Using Gibbs Samplers to Compute Bayesian Posterior Distributions

In Chapter 8, we introduced the fundamental ideas of Bayesian inference, in which prior distributions on parameters are used together with data to obtain posterior distributions and thus interval estimates of parameters. However, in practice, Bayesian posterior distributions are often difficult to compute.

Gibbs sampling is a computational method that uses Markov chains, as discussed in Chapter 7, to approximate posterior distributions. The central idea is to use available information about a prior distribution and data to construct an ergodic Markov chain whose limiting distribution is the desired posterior distribution. Then we simulate enough steps of the chain to obtain a good approximation to the limiting distribution.

In this chapter, we consider several relatively simple Bayesian models, explicitly illustrating how to program suitable chains in R in order to approximate posterior distributions and obtain interval estimates of parameters. In Chapter 10, we show how WinBUGS software can simplify the programming to do inference for more intricate Bayesian models.

9.1 Bayesian Estimates of Disease Prevalence

In Section 5.2, we considered how one might use the properties and results of a medical screening test to estimate the prevalence of a disease. In particular, we assumed we know the sensitivity and specificity of a screening test,

$$\eta = P\{\text{Positive test}|\text{Disease present}\} = P\{T = 1|D = 1\}$$

and

$$\theta = P\{\text{Negative test}|\text{Disease absent}\} = P\{T = 0|D = 0\},$$

respectively. Based on these quantities, we sought to estimate the prevalence of the disease $\pi = P\{D = 1\}$ from the equation $\pi = (\tau + \theta - 1)/(\eta + \theta - 1)$, where $\tau = P\{T = 1\}$. If τ is estimated by $t = A/n$, which is the ratio of the

E.A. Suess and B.E. Trumbo, *Introduction to Probability Simulation and Gibbs Sampling with R*, Use R!, DOI 10.1007/978-0-387-68765-0_9, © Springer Science+Business Media, LLC 2010

number of individuals with positive tests to the sample size, then replacing τ by t in this equation gives an estimate p of π,

$$p = \frac{A/n + \theta - 1}{\eta + \theta - 1} = \frac{t + \theta - 1}{\eta + \theta - 1}. \tag{9.1}$$

For the derivation, see page 123. Endpoints of a confidence interval for τ can be plugged into this equation to obtain a confidence interval for π.

However, we have seen some circumstances in which such an estimate of π falls outside the interval $[0, 1]$. Even more often, the corresponding confidence interval for π can extend beyond this interval. Another difficulty with this method arises when the sample size is small and the proportion of "Successes" is near 0 or 1. Then binomial confidence intervals are known to be problematic and equation (9.1) may not provide a useful interval estimate of π.

In Section 6.4, we investigated the situation in which, for a particular population, we know the predictive power of a positive test $\gamma = P\{D = 1|T = 1\}$ and a negative test $\delta = P\{D = 0|T = 0\}$ in addition to η and θ. Here the relationships $\gamma = \pi\eta/[\pi\eta + (1-\pi)(1-\theta)]$ and $\delta = (1-\pi)\theta/[\pi(1-\eta) + (1-\pi)\theta]$ follow from Bayes' Theorem. So if we knew π along with η and θ, we could compute γ and δ. Accordingly, it seems reasonable that we should be able to compute π if η, θ, γ, and δ are known. In the examples of Section 6.4, we saw how this can be done either by simulation (simple Gibbs sampler) or analytically (solving for the steady state of a Markov chain). Unfortunately, these particular procedures are mainly of theoretical and pedagogical interest because data to estimate γ and (especially) δ are not typically available in practical situations.

Fortunately, in a framework with a Bayesian prior distribution, we can use a Gibbs sampler to find useful estimates of π. With a prior distribution on π having support $[0, 1]$, the following example shows how to obtain a posterior probability interval for π based on data A and n, and on known values of the sensitivity η and specificity θ, with no need to make assumptions about the predictive values γ and δ. Because values of π outside of $[0, 1]$ are not contemplated in the prior, they have zero probability under the posterior.

Example 9.1. Suppose we use a screening test with sensitivity $\eta = 99\%$ and specificity $\theta = 97\%$, and among $n = 1000$ subjects we see $A = 49$ positive results. We use a beta prior distribution for π. That is, $\pi \sim \mathsf{BETA}(\alpha, \beta)$. Not claiming to have advance information about π, we choose the flat prior with $\alpha = \beta = 1$, so that the prior is $\pi \sim \mathsf{BETA}(1, 1) = \mathsf{UNIF}(0, 1)$.

Our Gibbs sampler starts with an arbitrary initial value π_1^* of π. From π_1^* and our knowledge of η and θ, we speculate as to the number X of the A test-positive subjects that may have the disease. Based on π_1^*, the probability that any one of these A subjects has the disease is equal to the predictive value of a positive test, $\gamma_1^* = \pi_1^*\eta/[\pi_1^*\eta + (1 - \pi_1^*)(1 - \theta)]$. So we use a binomial distribution with A trials and this Success probability γ_1^* to simulate X. In much the same way, we simulate the number Y of the $B = n - A$ test-negative subjects that have the disease. Altogether, we now have $X + Y$ subjects out of n with the disease, and we can use this simulated total to update the beta distribution for π, as in the election polling examples of Chapter 8. From this updated distribution, we simulate π_2^*, and we iterate the procedure from there to get $\pi_3^*, \pi_4^*, \ldots$. In symbols, the partial conditional distributions relating the key quantities are

$$X|A, \pi \sim \mathsf{BINOM}(A, \gamma), \ \ Y|B, \pi \sim \mathsf{BINOM}(B, 1 - \delta), \ \text{ and}$$

$$\pi|X, Y \sim \mathsf{BETA}(\alpha + X + Y, \ \beta + n - X - Y),$$

where $\gamma = \pi\eta/[\pi\eta + (1 - \pi)(1 - \eta)]$, $\delta = (1 - \pi)\theta/[\pi(1 - \eta) + (1 - \pi)\theta]$, and $B = n - A$. These relationships are used in the R code below. Simulated values of π shown with asterisks (*) above are elements of the vector PI (all capitals) in the code.

Because the distribution of $\pi_i^* = $ PI[i] depends only on known parameters and the previous value $\pi_{i-1}^* = $ PI[i-1], the values in PI simulate a Markov process with a continuous state space as in Chapter 7. It can be shown that the limiting distribution of this process is the posterior distribution of π based on the prior and the data.

```
# set.seed(1237)
m = 50000                        # iterations
PI = numeric(m);  PI[1] = .5     # vector for results, initial value
alpha = 1;  beta = 1             # parameters of beta prior
eta = .99;  theta = .97          # sensitivity; specificity
n = 1000;  A = 49;  B = n - A    # data

for (i in 2:m)
{
   num.x = PI[i-1]*eta;  den.x = num.x + (1-PI[i-1])*(1 - theta)
   X = rbinom(1, A, num.x/den.x)
   num.y = PI[i-1]*(1 - eta);  den.y = num.y + (1-PI[i-1])*theta
   Y = rbinom(1, B, num.y/den.y)
   PI[i] = rbeta(1, X + Y + alpha, n - X - Y + beta)
}
aft.brn = seq(m/2 + 1,m)
mean(PI[aft.brn])
quantile(PI[aft.brn], c(.025, .975))
par(mfrow=c(2,1))
   plot(aft.brn, PI[aft.brn], type="l")
   hist(PI[aft.brn], prob=T)
par(mfrow=c(1,1))

> mean(PI[aft.brn])
[1] 0.02059591
> quantile(PI[aft.brn], c(.025, .975))
      2.5%        97.5%
0.007428221 0.035523630
```

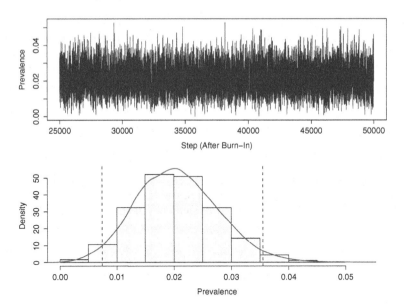

Figure 9.1. History plot (top) and histogram of 25 000 sampled prevalence values after burn-in. The plot shows good mixing of the Gibbs sampler in Example 9.1. The histogram approximates the posterior distribution $\pi|A, B$; an estimated density curve is superimposed (see Problem 9.7). Dotted lines indicate the 95% Bayesian interval estimate of π. Compare with Figure 9.10 on page 239.

The histogram in the top panel of Figure 9.1 indicates the posterior distribution of π. Taking the mean of this distribution, we have the point estimate $\pi = 0.021$, and cutting off 2.5% from each tail of this simulated distribution, we have the Bayesian interval estimate $(0.007, 0.036)$ for π. Problem 9.1 invites you to see how these results change when we use some informative prior distributions.

Essentially, Figure 9.2 is made using the following additional statements.

```
par(mfrow=c(1,2))
    acf(PI[aft.brn], ylim=c(0, .6))
    plot(1:m, cumsum(PI)/(1:m), type="l", ylim=c(.016, .024))
par(mfrow=c(1,1))
```

The three diagnostic graphs in the top panel of Figure 9.1 and in Figure 9.2 show that, in spite of some positive autocorrelation for neighboring values of $\pi_i^*|A, B$, the sampler mixes well and running averages after burn-in converge smoothly to the point estimate. (See Problem 9.5 for more about running averages and burn-in.)

It is easy to see why there is positive autocorrelation. If by chance at some step i in the iteration we obtain a rather large value of π_i^*, then it is somewhat likely that unusually large values of X or Y or both will result at step

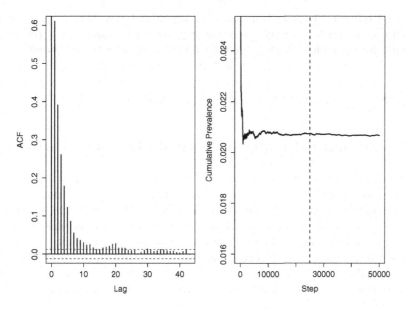

Figure 9.2. The ACF plot (left) of sampled values of π after burn-in for Example 9.1 shows autocorrelations of sampled values decaying to insignificance for lags above 25. The plot of cumulative averages of sampled values shows steady convergence to the Bayesian point estimate of π after burn-in (dotted vertical line).

$i + 1$. Consequently, the first parameter of the beta distribution may be inflated, and along with it the expectation of the next value of π_i^*. However, the data and the prior exert an overall tendency towards appropriate values of π_i^*, so the process does not consistently "run away" towards ever larger values. A mirror image of this argument holds in case we get an unusually low value of π_i^* at some step in the simulation. (See Problem 9.6 for more on autocorrelations.)

In this example, equation (9.1) provides the traditional point estimate $p = 0.020$ of π and the corresponding confidence interval $(0.006, 0.034)$. [The Agresti-Couil estimates of τ give $p = 0.022$ and $(0.008, 0.036)$.] So in this situation where equation (9.1) works well, the Gibbs sampler with a noninformative prior gives almost identical results. Moreover, Problem 9.3 illustrates that a Gibbs sampler gives reasonable point and interval estimates of π, even in situations where equation (9.1) gives problematic negative estimates. $\Diamond$

Although there are many important applications in which equation (9.1) is not useful, the benefit of the Bayesian framework is not just to avoid absurd estimates outside the range of possible parameter values. In practice, one seldom encounters a situation where there is no prior information at all about prevalence. For example, if $\pi = 93\%$— or even $\pi = 30\%$—for a serious disease, the evidence of this public health catastrophe would be evident all around us without reference to data from medical screening tests. Also, in practice one often encounters situations where there is very little data and a reasonable approach is to meld expert opinion with the bit of objective information that is available. For such reasons, it is fair to say that Gibbs sampling, as illustrated in Example 9.1, has wide applicability in estimating prevalence from the results of screening tests across a broad spectrum of applications.

9.2 Bayesian Estimates of Normal Mean and Variance

In Chapter 8, we discussed separately (i) Bayesian estimation of the mean μ of a normal population when the variance is known and (ii) Bayesian estimation of the variance $\sigma^2 = \theta$ when the mean is known. In this section, we use a Gibbs sampler to provide Bayesian estimates of the normal mean and variance simultaneously. In the following example, we see that—when relatively flat priors are used—Bayesian results are numerically very similar to those obtained by traditional methods based on Student's t and chi-squared distributions. In several problems, we explore the effect of informative priors.

Example 9.2. Changes in Students' Heights. Heights of $n = 41$ young men at a boarding school are measured in the morning and also in the evening. For each student, the difference x_i, morning height minus evening height, is found. Considering these subjects to be a random sample from an appropriate population, our main purpose is to estimate the population mean μ of the change in height.

If we take differences in height x_i, for $i = 1, 2, ..., 41$, to be normally distributed, this example is similar in some ways to Examples 8.3 and 8.7 (concrete beams) and Examples 8.4 and 8.8 (hemoglobin). Here we assume $x_i \sim \text{NORM}(\mu, \sigma)$, where both parameters are unknown. The classical unbiased point estimators are $\bar{x} = \frac{1}{n} \sum_i x_i$ for μ and $s^2 = \frac{1}{n-1} \sum_i (x_i - \bar{x})^2$ for $\sigma^2 = \theta$. We seek Bayesian point and interval estimates for μ and σ.

Prior Distributions. First, we choose a prior distribution for μ of the form $\text{NORM}(\mu_0, \sigma_0)$, where $\theta_0 = \sigma_0^2$. Specifically, we choose $\mu_0 = 0$ because we have no reason to suppose heights differ systematically between morning and evening. Also, we want a reasonably flat prior because we claim no particular expertise in the matter of height changes, and we do not really know whether students might grow or shrink a little during the day. Thus, rather arbitrarily, we choose $\sigma_0 = 20$ mm (about 3/4 of an inch), so $\theta_0 = 400$.

Next, we choose a prior distribution for θ of the form $\text{IG}(\alpha_0, \kappa_0)$, where α_0 and κ_0 are shape and rate parameters, respectively. We do not have much idea how accurately the measuring will be done, and differences involve two measurements. Also, if there are differences in height during a day, those differences may be larger for some students than for others. Accordingly, we choose $\alpha_0 = 1/2$ and $\kappa_0 = 1/5$, which means we think the standard deviation of the differences is pretty sure to be between 0.3 mm and 20 mm (computed from `sqrt(1/qgamma(c(.975, .025), 1/2, 1/5))`. This choice seems reasonable. As heights go, a millimeter is very small and it seems unlikely that measurements could be made much more precisely than that. Also, 20 mm seems an unbelievably large amount of measurement error or variability in daily changes among students.

Data. From the data, we find $n = 41$ differences x_i, for which the sample mean is $\bar{x} = 9.6$ mm and the sample variance is $s^2 = 7.48$, so that $s = 2.73$ mm. (See Problem 9.12 for the 41 differences.) A Bayesian analysis will combine these data and our priors to give posterior distributions upon which we base our inferences.

Posterior Distributions. Almost exactly as in Example 8.7, we have

$$\mu | \mathbf{x}, \theta \sim \mathsf{NORM}(\mu', \sqrt{\theta'}),$$

where the updated parameters (denoted with primes), reflecting the data, are $\mu' = \theta'(n\bar{x}/\theta + \mu_0/\theta_0)$ and $\theta' = (n/\theta + 1/\theta_0)^{-1} = \theta_0\theta/(n\theta_0 + \theta)$. Also, similar to our results in Example 8.8,

$$\theta | \mathbf{x}, \mu \sim \mathsf{IG}(\alpha', \kappa'),$$

where $\alpha' = \alpha_0 + n/2$ and $\kappa' = \kappa_0 + [(n-1)s^2 + n(\bar{x} - \mu)^2]/2$. The important change from Example 8.8 is the second term inside brackets in the expression for κ', needed here to take $\bar{x}$ into account because μ is not known. (See Problem 9.14 for some details of the derivation.)

Now, in order to find the posterior distributions of $\mu | \mathbf{x}$ and $\theta | \mathbf{x}$, we use a Gibbs sampler to perform the required integrations:

$$p(\mu | \mathbf{x}) \propto \int p(\mu | \mathbf{x}, \theta) \, p(\theta | \mathbf{x}) \, d\theta \quad \text{and} \quad p(\theta | \mathbf{x}) \propto \int p(\theta | \mathbf{x}, \mu) \, p(\mu | \mathbf{x}) \, d\mu.$$

Gibbs sampler. Using the R code below, we simulate a bivariate Markov chain with vectors denoted in the program as `MU` and `THETA`. The limiting distribution of this chain provides estimates of the posterior distributions of μ and θ, respectively, upon which Bayesian estimates are based. The simulation begins with known quantities: the parameters μ_0 and θ_0 of the normal prior distribution on μ, the parameters α_0 and κ_0 of the inverse gamma prior distribution on θ, the data $\bar{x}$ and s^2, and an arbitrary starting value `THETA[1]`.

Iteratively, at step i of the Gibbs sampler, we generate values `MU[i]` and `THETA[i]` of the Markov chain. We sample `MU[i]` from $\mathsf{NORM}(\mu', \sqrt{\theta'})$, where θ in the expressions for μ' and θ' is taken to be `THETA[i-1]`. Then we sample `THETA[i]` from $\mathsf{IG}(\alpha', \kappa')$, where μ in the expression for κ' is taken to be `MU[i]`.

We simulate $m = 50\,000$ steps, with the first half of them as burn-in values. Thus we take values of `MU` and `THETA` from steps $i = m/2 + 1 = 25\,001$ through m to represent distributions of $\mu | \mathbf{x}$ and $\theta | \mathbf{x}$, respectively. Cutting off 2.5% from the tails of these simulated distributions gives us Bayesian interval estimates of μ and θ. From the interval estimate of $\theta = \sigma^2$, we obtain an interval estimate of σ.

```
# set.seed(1237)
m = 50000                               # iterations
MU = numeric(m);   THETA = numeric(m)   # sampled values
THETA[1] = 1                            # initial value
n = 41;  x.bar = 9.6;  x.var = 2.73^2   # data
mu.0 = 0;  th.0 = 400                   # mu priors
alp.0 = 1/2;  kap.0 = 1/5               # theta priors

for (i in 2:m)
{
  th.up = 1/(n/THETA[i-1] + 1/th.0)
  mu.up = (n*x.bar/THETA[i-1] + mu.0/th.0)*th.up
  MU[i] = rnorm(1, mu.up, sqrt(th.up))

  alp.up = n/2 + alp.0
  kap.up = kap.0 + ((n-1)*x.var + n*(x.bar - MU[i])^2)/2
  THETA[i] = 1/rgamma(1, alp.up, kap.up)
}

# Bayesian point and probability interval estimates
aft.brn = (m/2 + 1):m
mean(MU[aft.brn])               # point estimate of mu
bi.MU = quantile(MU[aft.brn], c(.025,.975));  bi.MU
mean(THETA[aft.brn])            # point estimate of theta
bi.THETA = quantile(THETA[aft.brn], c(.025,.975));  bi.THETA
SIGMA = sqrt(THETA)
mean(SIGMA[aft.brn])            # point estimate of sigma
bi.SIGMA = sqrt(bi.THETA);  bi.SIGMA

par(mfrow=c(2,2))
  plot(aft.brn, MU[aft.brn], type="l")
  plot(aft.brn, SIGMA[aft.brn], type="l")
  hist(MU[aft.brn], prob=T);  abline(v=bi.MU, col="red")
  hist(SIGMA[aft.brn], prob=T);  abline(v=bi.SIGMA, col="red")
par(mfrow=c(1,1))

> mean(MU[aft.brn])               # point estimate of mu
[1] 9.594313
> bi.MU = quantile(MU[aft.brn], c(.025,.975));  bi.MU
     2.5%      97.5%
 8.753027 10.452743

> mean(THETA[aft.brn])            # point estimate of theta
[1] 7.646162
> bi.THETA = quantile(THETA[aft.brn], c(.025,.975));  bi.THETA
     2.5%      97.5%
 4.886708 11.810233
> SIGMA = sqrt(THETA)
```

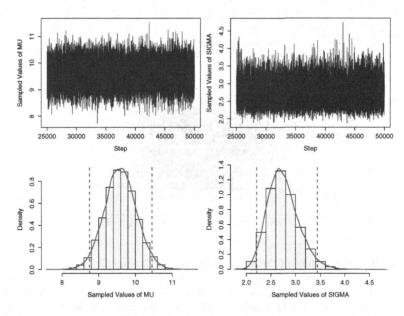

Figure 9.3. History plots (top) and histograms of sampled values approximating $\mu|\mathbf{x}$ (left) and $\sigma|\mathbf{x}$ in the Gibbs sampler of Example 9.2. All values are after burn-in. Vertical dashed lines show 95% Bayesian interval estimates.

```
> mean(SIGMA[aft.brn])        # point estimate of sigma
[1] 2.747485
> bi.SIGMA = sqrt(bi.THETA);  bi.SIGMA
    2.5%     97.5%
2.210590 3.436602
```

Diagnostic graphs in Figure 9.3 (top) show good behavior of the Gibbs sampler, so the numerical results from MU and SIGMA can be trusted to represent the posterior distributions $\mu|\mathbf{x}$ and $\sigma|\mathbf{x}$ accurately. Also, because this is a bivariate Markov chain, we show, in Figure 9.4 on page 228, a scatterplot of the last 10 000 sampled pairs approximating $(\mu, \sigma)|\mathbf{x}$. (For additional diagnostic graphs, see Problem 9.9.)

The 95% Bayesian interval estimates are $(8.73, 10.44)$ for μ and $(2.22, 3.45)$ for σ. On average, it seems that from morning to evening the students shrink in height by about a centimeter (10 mm or about 3/8 in). Other studies have found similar decreases in height. A plausible explanation is that the cartilage between vertebrae is compressed during the day and expands during sleep.

Frequentist methods that use Student's t and chi-squared distributions give a 95% confidence interval $(8.74, 10.46)$ for μ and a 95% confidence interval $(2.24, 3.49)$ for σ (see Problem 9.10). The Bayesian probability intervals are slightly shorter than the corresponding frequentist confidence intervals, possibly because our prior distributions, even though diffuse, provide some

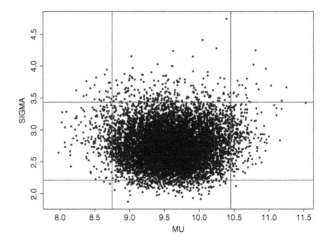

Figure 9.4. Scatterplot of the last 10 000 pairs in Example 9.2 simulating $(\mu, \sigma)|\mathbf{x}$. The prior distributions on μ and $\theta = \sigma^2$ are independent, as are the sample statistics $\bar{x}$ and s. Also, this plot shows no marked association between simulated values of $\mu|\mathbf{x}$ and $\sigma|\mathbf{x}$. Reference lines indicate 95% Bayesian interval estimates for μ and σ.

useful information about variability. But in this example the effect of our prior distributions is relatively small because there are enough data to overwhelm the effect of priors that are not strongly informative. $\Diamond$

In general, if we make the prior parameter σ_0 very large and the parameters α_0 and κ_0 very small, then neither prior affects the posterior by much, and the Bayesian intervals are nearly in numerical agreement with the corresponding frequentist confidence intervals. Specifically, one formulation of a noninformative prior gives the posterior distributions

$$t = \frac{\mu - \bar{x}}{s/\sqrt{n}} \sim \mathsf{T}(n-1) \quad \text{and} \quad (n-1)s^2/\sigma^2 \sim \mathsf{CHISQ}(n-1),$$

where μ and σ are random variables and $\bar{x}$ and s are observed values. These yield 95% Bayesian interval estimates for μ and σ that are numerically exactly the same as the respective traditional frequentist 95% confidence intervals.

Moreover, there are particular ways to formulate informative priors so that posterior distributions given $\bar{x}$ and s can be expressed in closed form. Then Bayesian interval estimates can be found for μ and σ without the need for Gibbs sampling. (For discussions of more general priors, see [BT73] and [Lee04].)

In practice, Gibbs samplers are especially important in models with many parameters, for which Example 9.2 provides an important pedagogical bridge. In the next section, we consider a three-parameter model for which traditional methods may be especially inappropriate and for which a Gibbs sampler is a practical way to compute useful Bayesian inferences.

9.3 Bayesian Estimates of Components of Variance

In Section 9.1, we saw that a Bayesian approach to estimating disease prevalence gave useful estimates in circumstances where traditional methods can give absurd results. In this section, we look at one more practical situation in which a traditional frequentist approach often does not provide useful estimates and a Bayesian framework does.

Suppose a manufacturing process has two steps. Precursors of the finished items are made in batches, and then the batches are used to produce the individual items. If a key measurement on the final items shows excessive variability, the question arises whether this variability may arise mainly at the batch level or mainly at the final stage of the overall process. A logical step towards reducing variability is to try to understand where it arises. We want to estimate the two components of variance that contribute towards overall variance of individual items.

Assuming normal distributions for errors, we can write the measured value of the jth item from the ith batch as

$$x_{ij} = \mu + A_i + e_{ij},$$

where $A_i \sim$ NORM$(0, \sqrt{\theta_A})$, $e_{ij} \sim$ NORM$(0, \sqrt{\theta})$, and all A_i and e_{ij} are mutually independent. This implies that measurements on two items from two different batches are independent but that measurements x_{ij} and $x_{ij'}$ on two items from batch i are correlated. Specifically, $V(x_{ij}) = V(x_{ij'}) = \theta_A + \theta$, $Cov(x_{ij}, x_{ij'}) = \theta_A$, and $\rho_I = \rho(x_{ij}, x_{ij'}) = \theta_A/(\theta_A + \theta)$. The ratio ρ_I, called the **intraclass correlation**, is the proportion of the total variance that arises at the batch level of the manufacturing process.

Example 9.3. Consider a pilot project to manufacture a pharmaceutical drug in two steps as just described. Technicians want to know if variability among batches makes an important contribution to product variability. They assay $r = 10$ individual items from each of $g = 12$ batches.

In this example, so we can know whether our estimates are reasonable, we generate data with known parameter values $\mu = 100$, $\theta_A = 15^2 = 225$, and $\theta = 9^2 = 81$, so that $\rho_I = 225/306 = 0.7353$—values roughly modeled after proprietary data. These $gr = 120$ observations are plotted in Figure 9.5, and the procedure for generating them is shown in Problem 9.15. Because the data are normal, it is sufficient to look at the $g = 12$ batch means $\bar{x}_{i.} = \frac{1}{r}\sum_j x_{ij}$ and variances $s_i^2 = \frac{1}{r-1}\sum_j (x_{ij} - \bar{x}_{i.})^2$. Summary data by batch are shown in the printout below.

Batch	1	2	3	4	5	6
Mean	91.9	129.0	104.1	75.7	108.7	100.2
SD	9.96	10.07	4.98	12.16	5.06	10.65

Batch	7	8	9	10	11	12
Mean	62.6	107.5	66.7	129.1	106.8	93.4
SD	6.52	11.05	9.90	8.39	8.99	8.14

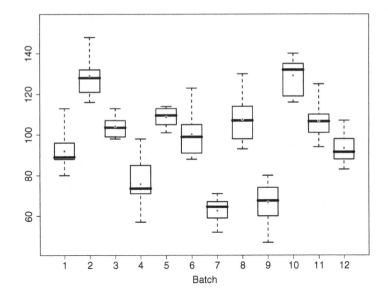

Figure 9.5. Boxplots of the data of Example 9.3. Dots show group means. Clearly, variance among batches contributes significantly to the variability of individual observations. Compare with Figure 9.6, which illustrates the data of Problem 9.18, where batch-to-batch variability is relatively much smaller.

In these circumstances, it is traditional to look at the following statistics.

$$\bar{x}_{..} = \sum_{i=1}^{g}\sum_{j=1}^{r} x_{ij} = 97.975,$$

$$\text{MS(Batch)} = \frac{r}{g-r}\sum_{i=1}^{g}(\bar{x}_{i.} - \bar{x}_{..})^2 = 4582.675,$$

$$\text{MS(Error)} = \frac{1}{g(r-1)}\sum_{i=1}^{g}\sum_{j=1}^{r}(x_{ij} - \bar{x}_{i.})^2 = \frac{1}{g}\sum_{i=1}^{g}s_i^2 = 82.68056.$$

For normal data, these three statistics are independent, and the following distributions are useful for making confidence intervals.

$$(\bar{x}_{..} - \mu)/\sqrt{\text{MS(Batch)}/gr} \sim \mathsf{T}(g-1),$$
$$(g-1)\text{MS(Error)}/(r\theta_A + \theta) \sim \mathsf{CHISQ}(g-1),$$
$$(gr-1)\text{MS(Error)}/\theta \sim \mathsf{CHISQ}(gr-1),$$
$$\text{MS(Batch)}/\text{MS(Error)} \sim \mathsf{F}(g-1, gr-1).$$

Unbiased point estimates are $\hat{\mu} = x_{..} = 97.975$ (compared with the known $\mu = 100$), $\hat{\theta} = \text{MS(Error)}$, and $\hat{\theta}_A = [\text{MS(Batch)} - \text{MS(Error)}]/r = 449.999$.

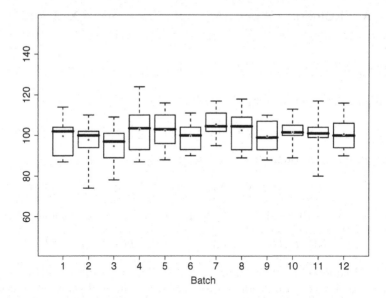

Figure 9.6. Boxplots of the data of Problem 9.18, drawn to the same scale as Figure 9.5 for easy comparison. Here batch-to-batch variance θ_A is so small that traditional methods of estimating it are problematic. A Gibbs sampler yields a useful Bayesian probability interval (see problem 9.18 on page 245).

Taking square roots, we have estimates for the standard deviations $\hat{\sigma} = 9.09$ (compared with $\sigma = 9$) and $\hat{\sigma}_A = 21.21$ (compared with $\sigma_A = 15$). In general, estimating θ_A and σ_A can be problematic. Information about batch variability is entangled with information about item variability. Because $\hat{\theta}_A$ is found by subtraction, we could potentially get $\hat{\theta}_A < 0$, even though $\theta_A = \sigma_A^2 \geq 0$.

Frequentist confidence intervals for μ, σ, and ρ_I can be obtained from t, chi-squared, and F distributions, respectively (see Problem 9.16). There is no such straightforward confidence interval for θ_A. Also, the confidence interval for ρ_I includes negative values whenever $\hat{\theta}_A < 0$. There are models in which intraclass correlation can legitimately be negative (for example, see [SC80]), but ours is not one of them.

A Bayesian framework for this model is similar to that of Example 9.2, with one additional variance parameter. Our prior distributions are

$$\mu \sim \text{NORM}(\mu_0, \sqrt{\theta_0}), \quad \theta_A \sim \text{IG}(\alpha_0, \kappa_0), \quad \text{and} \quad \theta \sim \text{IG}(\beta_0, \lambda_0).$$

In order to have noninformative priors for this example with simulated data, we select a large value of θ_0 and small values of all four inverse gamma parameters.

Partial conditional distributions used in the Gibbs sampler to compute the posterior distributions of μ, θ, and θ_A are as follows (see [GS85], page 405).

$$\mu|\theta_A, \mathbf{A} \sim \text{NORM}(\mu', \sqrt{\theta'}), \quad \theta_A|\mathbf{A}, \mu \sim \text{IG}(\alpha', \kappa'), \quad \text{and} \quad \theta|\mathbf{X}, \mathbf{A} \sim \text{IG}(\beta', \lambda'),$$

where

$$\mu' = (\mu_0\theta_A + \theta_0 \textstyle\sum_i A_i)/(\theta_A + r\sum_i A_i) \quad \text{and} \quad \theta' = \theta_0\theta_A/(\theta_A + r\sum_i A_i)$$

in the partial conditional for $\mu|\theta_A, \mathbf{A}$;

$$\alpha' = \alpha_0 + g/2 \quad \text{and} \quad \kappa' = \kappa_0 + \tfrac{1}{2}\textstyle\sum_i(A_i - \mu)^2$$

in the partial conditional for $\theta_A|\mathbf{A}, \mu$; and

$$\beta' = \beta_0 + gr/2 \quad \text{and} \quad \lambda' = \lambda_0 + \tfrac{1}{2}[(r-1)\textstyle\sum_i s_i^2 + r\sum_i(A_i - \bar{x}_{i.})^2]$$

in the partial conditional for $\theta|\mathbf{X}, \mathbf{A}$. In the above, the g elements of $\mathbf{A}$ are

$$A_i \sim \text{NORM}((r\theta_A\bar{x}_{i.} + \theta\mu)/(r\theta_A + \theta), \ [\theta\theta_A/(r\theta_A + \theta)]^{1/2}).$$

The R code below shows how these relationships can be used in a Gibbs sampler to simulate a multidimensional Markov chain of sampled values. Results include vectors denoted MU, VAR.BAT, and VAR.ERR, from which we can find Bayesian interval estimates of μ, θ_A, and θ, respectively. Each step of the sampler uses the prior distributions and the data.

- The sampler starts with an arbitrary initial value of MU[1]. We also require values of the random effects A_i, so-called **latent variables**, which are not directly observable as data. In the program, these are denoted by the g-vector a. On the first pass through the loop, we use the group means $\bar{x}_{i.}$ as initial values of a. On later passes, updated values of a are available.
- Next, the sampler uses values of a and MU[1] to sample VAR.BAT[2] and VAR.ERR[2], and then uses VAR.BAT[2] and VAR.ERR[2] to sample MU[2].
- At the end of the loop, new latent values a are sampled using MU[2], VAR.BAT[2], and VAR.ERR[2].
- The loop is iterated at each pass using values of MU[k-1], VAR.BAT[k-1], VAR.ERR[k-1], and the newest values a to sample elements of the vectors with index [k].

Finally, when all iterations are completed, Bayesian interval estimates of μ, θ_A, and θ are found from the values after burn-in of the three simulated vectors, and intervals for σ_A, σ, and ρ_I are found from information the Gibbs sampler provides about $\theta_A = \sigma_A^2$ and $\theta = \sigma^2$.

```
##Assumes matrix X with g rows (batches), r columns (reps),
##Or provide g-vectors of batch means and SDs as the 2nd line.
# set.seed(443)
X.bar = apply(X, 1, mean);   X.sd = apply(X, 1, sd)
m = 50000;  b = m/4              # iterations;  burn-in
MU = VAR.BAT = VAR.ERR = numeric(m)
```

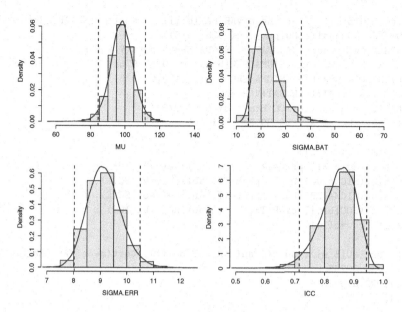

Figure 9.7. Histograms of simulated posteriors for Example 9.3. Results are from the vectors MU, SIGMA.BAT, SIGMA.ERR, and ICC of the Gibbs sampler after burn-in. Indications of the 95% Bayesian interval estimate and the estimated posterior density are superimposed on each histogram.

```
mu.0  = 0;    th.0  = 10^10      # prior parameters for MU
alp.0 = .001; kap.0 = .001       # prior parameters for VAR.BAT
bta.0 = .001; lam.0 = .001       # prior parameters for VAR.ERR
MU[1] = 150;  a = X.bar          # initial values


for (k in 2:m)  {
  alp.up = alp.0 + g/2
  kap.up = kap.0 + sum((a - MU[k-1])^2)/2
  VAR.BAT[k] = 1/rgamma(1, alp.up, kap.up)

  bta.up = bta.0 + r*g/2
  lam.up = lam.0 + (sum((r-1)*X.sd^2) + r*sum((a - X.bar)^2))/2
  VAR.ERR[k] = 1/rgamma(1, bta.up, lam.up)

  mu.up = (VAR.BAT[k]*mu.0 + th.0*sum(a))/(VAR.BAT[n] + g*th.0)
  th.up = th.0*VAR.BAT[k]/(VAR.BAT[n] + g*th.0)
  MU[k] = rnorm(1, mu.up, sqrt(th.up))

  deno = r*VAR.BAT[k] + VAR.ERR[k]
  mu.a = (r*VAR.BAT[k]*X.bar + VAR.ERR[k]*MU[k])/deno
  th.a = (VAR.BAT[k]*VAR.ERR[k])/deno
  a = rnorm(g, mu.a, sqrt(th.a))  }
```

```
mean(MU[b:m]);  sqrt(mean(VAR.BAT[b:m]));  sqrt(mean(VAR.ERR[b:m]))
bi.MU = quantile(MU[b:m], c(.025,.975))
SIGMA.BAT = sqrt(VAR.BAT);  SIGMA.ERR = sqrt(VAR.ERR)
bi.SG.B = quantile(SIGMA.BAT[b:m], c(.025,.975))
bi.SG.E = quantile(SIGMA.ERR[b:m], c(.025,.975))
ICC = VAR.BAT/(VAR.BAT+VAR.ERR);
bi.ICC = quantile(ICC[b:m], c(.025,.975))
bi.MU;  bi.SG.B;  bi.SG.E;   bi.ICC


par(mfrow=c(2,2))
  hist(MU[b:m], prob=T);          abline(v=bi.MU)
  hist(SIGMA.BAT[b:m], prob=T); abline(v=bi.SG.B)
  hist(SIGMA.ERR[b:m], prob=T); abline(v=bi.SG.E)
  hist(ICC[b:m], prob=T);        abline(v=bi.ICC)
par(mfrow=c(1,1))


> mean(MU[b:m]);  sqrt(mean(VAR.BAT[b:m]));  sqrt(mean(VAR.ERR[b:m]))
[1] 98.00235
[1] 23.41598
[1] 9.177717

> bi.MU;  bi.SG.B;  bi.SG.E;   bi.ICC
     2.5%      97.5%
 84.30412 111.59195
     2.5%      97.5%
 14.84336  36.16719
     2.5%      97.5%
 8.022159 10.488738
     2.5%      97.5%
0.7146413 0.9421721
```

From the printouts for one run of the Gibbs sampler, we see that the Bayesian point estimates (98.0 for μ, 23.4 for σ_A, and 9.2 for σ) are not much different from the traditional ones (98.0, 21.2, and 9.1, respectively). Also, the 95% Bayesian interval estimates of these parameters all happen to cover the known values we used to simulate the data (100, 15, and 9, respectively). Based on distributions stated earlier, traditional 95% confidence intervals are $(85.7, 110.2)$ for μ, $(8.0, 10.5)$ for σ, and $(0.71, 0.94)$ for ρ_I.

Figure 9.7 shows the approximate posterior distributions and 95% Bayesian interval estimates for μ, σ_A, σ, and ρ_I. Figure 9.8 shows diagnostic plots—all favorable—for the dimension of the sampler estimating σ_A, and we leave the remaining diagnostic plots to Problem 9.17.

The 95% Bayesian interval estimate of σ_A is very wide because we have information on only $g = 12$ batches. In contrast, we have much more information about σ, and that information is not entangled with other effects, so the interval for σ is shorter. If feasible, it would be appropriate to increase g. (See Problem 9.23.) But, in practice, batches are often prohibitively expensive. In our consulting experience, the number of batches has rarely exceeded 12.

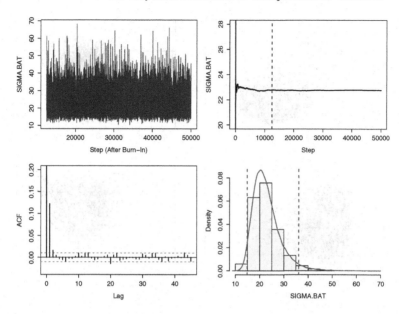

Figure 9.8. Diagnostic plots for the simulation of the posterior distribution of σ_A in Example 9.3. Evidently, the Gibbs sampler converges smoothly to its limiting distribution. Only the plot of cumulative means (upper right) shows all 50 000 steps; the others use steps after burn-in. The histogram is also shown in Figure 9.7.

Faced with long interval estimates for the batch component of variance, some authors and practitioners use 90% intervals instead. (In this particular example, that would give an interval that doesn't cover 15.) In a Bayesian context where appropriate prior information is available, an informative prior on $\theta_A = \sigma_A^2$ might give a shorter and more useful interval estimate. $\diamond$

In this example, the traditional method of estimating θ_A gives useful answers. But this method becomes problematic when the batch component of variance is relatively small. Then, as mentioned above, the estimate of θ_A can be negative and the confidence interval for ρ_I can include negative values. As we see in Problem 9.19, this happens more than occasionally.

- One standard interpretation when $\hat{\theta}_A < 0$ is to say this is an indication that θ_A must be "very small." Maybe so, but presumably we would not have chosen a model containing θ_A without reason to believe batches might make some contribution to overall variance, and this analysis leaves us with no idea how large θ_A might really be.
- A related traditional approach is to test the null hypothesis H_0: $\theta_A = 0$ against H_1: $\theta_A > 0$. What do we say if H_0 is accepted, as it surely will be when $\hat{\theta}_A < 0$? Again the interpretation is that θ_A is "very small." But then we must speculate about the power of the test, the probability of

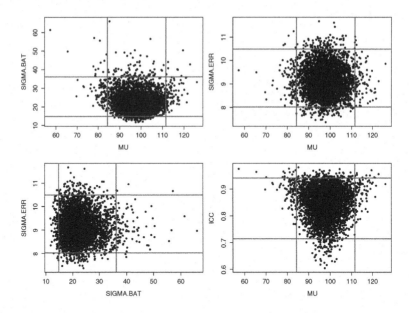

Figure 9.9. Bivariate plots from the Gibbs sampler of Example 9.3. In each of the four panels, every 10th step after burn-in is plotted. The vectors MU, SIGMA.BAT, and SIGMA.ERR are not mutually independent. In particular, values of MU far from $\bar{x}_{..}$ tend to be associated with large values of ICC. See Problem 9.17.

accepting H_0, for various possible values of $\theta_A > 0$. We still have no idea how large θ_A might actually be. Perhaps this difficulty has been made more obscure by the terminology of hypothesis testing, but it has not gone away.

In Problem 9.18, we show data for which $\hat{\theta}_A < 0$, but the Gibbs sampler of Example 9.3 gives useful Bayesian interval estimates for all parameters (see Figure 9.12 on page 244). Several additional problems show real data that result in $\hat{\theta}_A > 0$. For a comparison of the models of Examples 9.2 and 9.3, see Problem 9.24.

Technical note: Although traditional method of moments estimates (MMEs) of θ_A, essentially obtained by making $\mathrm{E}(\hat{\theta}_A) = \theta_A$, can have negative values, we take this opportunity to mention that computationally intensive methods are available to find approximate maximum likelihood estimates (MLEs) of θ_A, which are never negative. Moreover, except when the MLE of θ_A is small, these methods can also provide approximate confidence intervals. Typically, these MLE results are numerically similar to Bayesian results from a Gibbs sampler using a noninformative prior. However, when the MLE of θ_A is small, computational difficulties involving collinearity arise, so that it is not feasible to construct MLE-based confidence intervals.

In this chapter, we have seen situations in which a Bayesian approach has something to offer over a traditional one, and in which a Gibbs sampler is a useful method for computing approximate posterior distributions. An inconvenience in using a Gibbs sampler is the need to specify partial conditional distributions upon which to base the programming. In Chapter 10, we show how BUGS software can do Gibbs sampling simply by specifying the model, but without having to write and explicitly program partial conditional distributions.

9.4 Problems

Problems for Section 9.1 (Estimating Prevalence of a Disease)
In working these problems, modify the program of the example as appropriate.

9.1   Estimating prevalence π with an informative prior.

a) According to the prior distribution BETA$(1, 10)$, what is the probability that π lies in the interval $(0, 0.2)$?
b) If the prior BETA$(1, 10)$ is used with the data of Example 9.1, what is the (posterior) 95% Bayesian interval estimate of π?
c) What parameter β would you use so that BETA$(1, \beta)$ puts about 95% probability in the interval $(0, 0.05)$?
d) If the beta distribution of part (c) is used with the data of Example 9.1, what is the 95% Bayesian interval estimate of π?

Hints: c) Use beta = seq(1:100); x = pbeta(.05, 1, beta); min(beta[x>=.95]).
Explain. d) The mean of the posterior distribution $\pi|X, Y$ is about 1.8%.

9.2   Run the program of Example 9.1 and use your simulated posterior distribution of π to find Bayesian point and interval estimates of the predictive power of a positive test in the population from which the data are sampled. How many of the 49 subjects observed to test positive do you expect are actually infected?

9.3   In Example 5.2 on p124, the test has $\eta = 99\%$ and $\theta = 97\%$, the data are $n = 250$ and $A = 6$, and equation (9.1) on p220 gives an absurd negative estimate of prevalence, $\pi = -0.62\%$.

a) In this situation, with a uniform prior, what are the Bayesian point estimate and (two-sided) 95% interval estimate of prevalence? Also, find a one-sided 95% interval estimate that provides an upper bound on π.
b) In part (a), what estimates result from using the prior BETA$(1, 30)$?

Comment: a) See Figure 9.10. Two-sided 95% Bayesian interval: $(0.03\%, 2.9\%)$. Certainly, this is more useful than a negative estimate, but don't expect a narrow interval with only $n = 250$ observations. Consider that a flat-prior 95% Bayesian interval estimate of τ based directly on $t = 6/250$ is roughly $(1\%, 5\%)$.

9.4   In each part below, use the uniform prior distribution on π and suppose the test procedure described results in $A = 24$ positive results out of $n = 1000$ subjects.

a) Assume the test used is not a screening test but a gold-standard test, so that $\eta = \theta = 1$. Follow through the code for the Gibbs sampler in Example 9.1, and determine what values of X and Y must always occur. Run the sampler. What Bayesian interval estimate do you get? Explain why the result is essentially the same as the Bayesian interval estimate you would get from a uniform prior and data indicating 24 *infected* subjects in 1000, using the code qbeta(c(.025,.975), 25, 977).

b) Screening tests exist because it is not feasible to administer a gold-standard test to a large group of subjects. So the situation in part (a) is not likely to occur in the real world. But it does often happen that everyone who gets a positive result on the screening test is given a gold-standard test, and no gold-standard tests are given to subjects with negative screening test results. Thus, in the end, we have $\eta = 99\%$ and $\theta = 1$. In this case, what part of the Gibbs sampler becomes deterministic? Run the Gibbs sampler with these values and report the result.

c) Why are the results from parts (a) and (b) not much different?

Hints: a) The Gibbs sampler simulates a large sample precisely from BETA$(25, 977)$ and cuts off appropriate tails. Why these parameters? Run the additional code:

    set.seed(1237); pp=c(.5, rbeta(m-1, 25, 977)); mean(pp[(m/2):m])

c) Why no false positives among the 24 in either (a) or (b)? Consider false negatives.

9.5   *Running averages and burn-in periods.* In simulating successive steps of a Markov chain, we know that it may take a number of steps before the running averages of the resulting values begin to stabilize to the mean value of the limiting distribution. In a Gibbs sampler, it is customary to disregard values of the chain during an initial burn-in period. Throughout this chapter, we rather arbitrarily choose to use $m = 50\,000$ iterations and take the burn-in period to extend for the first $m/4$ or $m/2$ steps. These choices have to do with the appearance of stability in the running average plot and how much simulation error we are willing to tolerate. For example, the running averages in the right-hand panel of Figure 9.2 (p223) seem to indicate smooth convergence of the mean of the π-process to the posterior mean after $25\,000$ iterations. The parts below provide an opportunity to explore the perception of stability and variations in the length of the burn-in period. Use $m = 50\,000$ iterations throughout.

a) Rerun the Gibbs sampler of Example 9.1 three times with different seeds, which you select and record. How much difference does this make in the Bayesian point and interval estimates of π? Use one of the same seeds in parts (b) and (c) below.

b) Redraw the running averages plot of Figure 9.2 so that the vertical plotting interval is $(0, 0.5)$. (Change the plot parameter ylim.) Does this affect

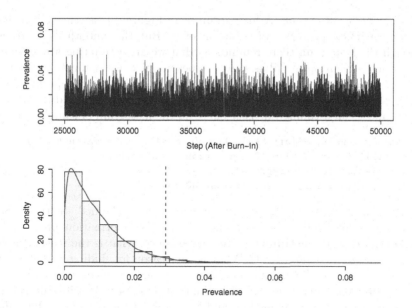

Figure 9.10. History plot (top) and histogram of 25 000 sampled prevalence values after burn-in for Problem 9.3. Here, traditional methods give a nonsensical negative point estimate of prevalence π. But a one-sided 95% Bayesian interval provides a useful upper bound on π (dotted line). Compare this with Figure 9.1 on page 222.

your perception of when the process "becomes stable"? Repeat, letting the vertical interval be $(0.20, 0.22)$, and comment.

c) Change the code of the Gibbs sampler in the example so that the burn-in period extends for 15 000 steps. Compared with the results of the example, what change does this make in the Bayesian point and interval estimates of π? Repeat for a burn-in of 30 000 steps and comment.

9.6   *Thinning.* From the ACF plot in Figure 9.2 on p223, we see that the autocorrelation is near 0 for lags of 25 steps or more. Also, from the right-hand plot in this figure, it seems that the process of Example 9.1 stabilizes after about 15 000 iterations. One method suggested to mitigate effects of autocorrelation, called **thinning,** is to consider observations after burn-in located sufficiently far apart that autocorrelation is not an important issue.

a) Use the data and prior of Example 9.1. What Bayesian point estimate and probability interval do you get by using every 25th step, starting with step 15 000? Make a histogram of the relevant values of PI. Does thinning in this way have an important effect on the inferences?

b) Use the statement `acf(PI[seq(15000, m, by=25)])` to make the ACF plot of these observations. Explain what you see.

9.7   *Density estimation.* A histogram, as in Figure 9.1, is one way to show
the approximate posterior distribution of π. But the smooth curve drawn
through the histogram there reminds us that we are estimating a *continuous*
posterior distribution. A Gibbs sampler does not give us the functional form of
the posterior density function, but the smooth curve is a good approximation.
After the Gibbs sampler of Example 9.1 is run, the following additional code
superimposes an estimated density curve on the histogram of sampled values.

```
est.d = density(PI[aft.brn], from=0, to=1);   mx = max(est.d$y)
hist(PI[aft.brn], ylim=(0, mx), prob=T, col="wheat")
lines(est.d, col="darkgreen")
median(PI[aft.brn]);   est.d$x[est.d$y==mx]
```

a) Run the code to verify that it gives the result claimed. In the R Ses-
sion window, type ?density and browse the information provided on
kernel density estimation. In this instance, what is the reason for the
parameters from=0, to=1? What is the reason for finding mx before the
histogram is made? In this book, we have used the mean of sampled values after burn-in as the Bayesian point estimate of π. Possible alternative
estimates of π are the median and the mode of the sampled values after
burn-in. Explain how the last statement in the code roughly approximates
the mode.

b) To verify how well kernel density estimation works in one example, do
the following: Generate 50 000 observations from BETA(2, 3), make a
histogram of these observations, superimpose a kernel density-estimated
curve in one color, and finally superimpose the true density function of
BETA(2, 3) as a dotted curve in a different color. Also, find the estimated
mode and compare it with the exact mode 1/3 of this distribution.

9.8   So far as is known, a very large herd of livestock is entirely free of a
certain disease ($\pi = 0$). However, in a recent routine random sample of $n = 100$
of these animals, two have tested positive on a screening test with sensitivity
95% and specificity 98%. One "expert" argues that the two positive tests
warrant slaughtering all of the animals in the herd. Based on the specificity
of the test, another "expert" argues that seeing two positive tests out of 100
is just what one would expect by chance in a disease-free herd, and so mass
slaughter is not warranted by the evidence.

a) Use a Gibbs sampler with a flat prior to make a one-sided 95% probability
interval that puts an upper bound on the prevalence. Based on this result,
what recommendation might you make?

b) How does the posterior mean compare with the estimate from equation (9.1) on p220?

c) Explain what it means to believe the prior BETA(1, 40). Would your recommendation in part (a) change if you believed this prior?

d) What Bayesian estimates would you get with the prior of part (c) if there are no test-positive animals among 100? In this case, what part of the Gibbs sampling process becomes deterministic?

Comments: In (a) and (b), the Bayesian point estimate and the estimate from equation (9.1) are about the same. If there are a few thousand animals in the herd, these results indicate there might indeed be at least one infected animal. Then, if the disease is one that may be highly contagious beyond the herd or if diseased animals pose a danger to humans, we could be in for serious trouble. If possible, first steps might be to quarantine this herd for now, find the two animals that tested positive, and quickly subject them to a gold-standard diagnostic test for the disease. That would provide more reliable information than the Gibbs sampler based on the screening test results. d) Used alone, a screening test with $\eta = 95\%$ and $\theta = 98\%$ applied to a relatively small proportion of the herd seems a very blunt instrument for trying to say whether the herd is free of a disease.

Problems for Section 9.2 (Estimating Normal Mean and Variance)

9.9   Write and execute R code to make diagnostic graphs for the Gibbs sampler of Example 9.2 showing ACFs and traces (similar to the plots in Figure 9.2). Comment on the results.

9.10   Run the code below. Explain step-by-step what each line (beyond the first) computes. How do you account for the difference between diff(a) and diff(b)?

```
x.bar = 9.60;   x.sd = 2.73;   n = 41
x.bar + qt(c(.025, .975), n-1)*x.sd/sqrt(n)
a = sqrt((n-1)*x.sd^2 / qchisq(c(.975,.025), n-1));   a;   diff(a)
b = sqrt((n-1)*x.sd^2 / qchisq(c(.98,.03), n-1));   b;   diff(b)
```

9.11   Suppose we have $n = 5$ observations from a normal population that can be summarized as $\bar{x} = 28.31$ and $s = 5.234$.

a) Use traditional methods based on Student's t and chi-squared distributions to find 95% confidence intervals for μ and σ.
b) In the notation of Example 9.2, use prior distributions with parameters $\mu_0 = 25$, $\sigma_0 = \sqrt{\theta_0} = 2$, $\alpha_0 = 30$, and $\kappa_0 = 1000$, and use a Gibbs sampler to find 95% Bayesian interval estimates for μ and σ. Discuss the priors. Make diagnostic plots. Compare with the results of part (a) and comment.
c) Repeat part (b), but with $\mu_0 = 0$, $\sigma_0 = 1000$, $\alpha_0 = 0.01$, and $\kappa_0 = 0.01$. Compare with the results of parts (a) and (b) and comment.

Hints: In (a)–(c), the sample size is small, so an informative prior is influential. In (a) and (c): (21.8, 34.8) for μ; (3, 15) for σ. Roughly.

9.12   Before drawing inferences, one should always look at the data to see whether assumptions are met. The vector x in the code below contains the $n = 41$ observations summarized in Example 9.2.

```
x = c( 8.50,   9.75,   9.75,   6.00,   4.00, 10.75,   9.25, 13.25,
      10.50, 12.00, 11.25, 14.50, 12.75,   9.25, 11.00, 11.00,
       8.75,   5.75,   9.25, 11.50, 11.75,   7.75,   7.25, 10.75,
       7.00,   8.00, 13.75,   5.50,   8.25,   8.75, 10.25, 12.50,
       4.50, 10.75,   6.75, 13.25, 14.75,   9.00,   6.25, 11.75,  6.25)

mean(x)
var(x)
shapiro.test(x)

par(mfrow=c(1,2))
  boxplot(x, at=.9, notch=T, ylab="x",
    xlab = "Boxplot and Stripchart")
  stripchart(x, vert=T, method="stack", add=T, offset=.75, at = 1.2)
  qqnorm(x)
par(mfrow=c(1,1))
```

a) Describe briefly what each statement in the code does.
b) Comment on the graphical output in Figure 9.11. (The angular sides of the box in the boxplot, called **notches**, indicate a nonparametric confidence interval for the population median.) Also comment on the result of the test. Give several reasons why it is reasonable to assume these data come from a normal population.

Note: Data are from [MR58], also listed and discussed in [Rao89] and [Tru02]. Each data value in x is the difference between a morning and an evening height value. Each height value is the average of four measurements on the same subject.

9.13   Modify the code for the Gibbs sampler of Example 9.2 as follows to reverse the order of the two key sampling steps at each passage through the loop. Use the starting value MU[1] = 5. At each step i, first generate THETA[i] from the data, the prior on θ, and the value MU[i-1]. Then generate MU[i] from the data, the prior on μ, and the value THETA[i]. Compare your results with those in the example, and comment.

9.14   (*Theoretical*) In Example 9.2, the prior distribution of the parameter $\theta = \sigma^2$ is of the form $\theta \sim \mathsf{IG}(\alpha_0, \kappa_0)$, so that $p(\theta) \propto \theta^{-(\alpha_0+1)} \exp(-\kappa_0/\theta)$. Also, the data $\mathbf{x}$ are normal with x_i randomly sampled from $\mathsf{NORM}(\mu, \sigma)$, so that the likelihood function is

$$p(\mathbf{x}|\mu, \theta) \propto \theta^{n/2} \exp\left\{-\frac{1}{2\theta} \sum_{i=1}^{n} (x_i - \mu)^2\right\}.$$

a) By subtracting and adding $\bar{x}$, show that the exponential in the likelihood function can be written as $\exp\{-\frac{1}{2\theta}[(n-1)s^2 + n(\bar{x} - \mu)^2]\}$.
b) The distribution of $\theta|\mathbf{x}, \mu$ used in the Gibbs sampler is based on the product $p(\theta|\mathbf{x}, \mu) \propto p(\theta)\, p(\mathbf{x}|\mu, \theta)$. Expand and then simplify this product to verify that $\theta|\mathbf{x}, \mu \sim \mathsf{IG}(\alpha_n, \kappa_n)$, where α_n and κ_n are as defined in the example.

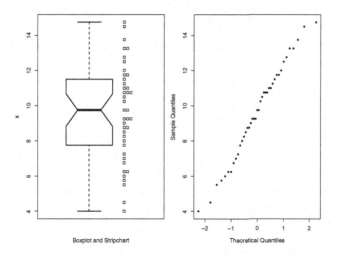

Figure 9.11. A boxplot, stripchart, and normal quantile plot of the height differences in Example 9.2. The overall impression is that the data are consistent with the assumption of a normally distributed population. See Problem 9.12.

Problems for Section 9.3 (Estimating Variance Components)

9.15 The R code below was used to generate the data used in Example 9.3. If you run the code using the same (default) random number generator in R we used and the seed shown, you will get the same data.

```
set.seed(1212)
g = 12                                   # number of batches
r = 10                                   # replications per batch
mu = 100;  sg.a = 15;  sg.e = 9          # model parameters
a.dat = matrix(rnorm(g, 0, sg.a), nrow=g, ncol=r)
          # ith batch effect across ith row
e.dat = matrix(rnorm(g*r, 0, sg.e), nrow=g, ncol=r)
          # g x r random item variations
X = round(mu + a.dat + e.dat)            # integer data
X
```

```
> X
      [,1] [,2] [,3] [,4] [,5] [,6] [,7] [,8] [,9] [,10]
 [1,]  103  113   88   96   89   88   80   92   89    81
 [2,]  143  116  126  127  132  121  129  148  129   119
 [3,]  107  107   98  103  113  104   99  103   98   109
 [4,]   71   72   89   63   85   71   75   76   98    57
 [5,]  105  101  113  110  109  101  114  114  113   107
 [6,]   88   93  100   91   98  105  103   91  123   110
 [7,]   71   52   67   59   67   67   60   68   62    53
 [8,]  115  102   93  111  130  114   97  103  112    98
 [9,]   58   70   65   78   67   60   74   80   47    68
```

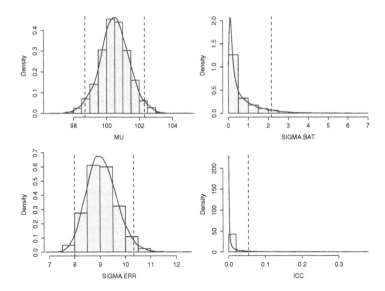

Figure 9.12. Histograms of simulated posteriors with estimated posterior densities for Problem 9.18. Results are from the vectors MU, SIGMA.BAT, SIGMA.ERR, and ICC of the Gibbs sampler after burn-in. Indications of the 95% Bayesian interval estimate (one-sided for SIGMA.BAT and ICC). Compare this with Figure 9.7 (p233).

```
[10,]   133   119   130   136   133   116   131   118   140     135
[11,]   103   101    97   110   125   107   115   106   110      94
[12,]    83   106    86    91    88   107    92    98    88      95
```

a) Run the code and verify whether you get the same data. Explain the results of the statements a.dat, var(a.dat[1,]), var(a.dat[,1]), and var(as.vector(e.dat)). How do the results of the first and second statements arise? What theoretical values are approximated (not very well because of the small sample size) by the last two statements.

b) Explain why the following additional code computes MS(Batch) and MS(Error). How would you use these quantities to find the unbiased estimate of θ_A shown in the example?

```
X.bar = apply(X, 1, mean);   X.sd = apply(X, 1, sd)
MS.Bat = r*var(X.bar);   MS.Err = mean(X.sd^2)
```

Hints: a) By default, matrices are filled by columns; shorter vectors recycle. The variance components of the model are estimated.

9.16   (*Continuation of Problem 9.15*) *Computation and derivation of frequentist confidence intervals related to Example 9.3.*

a) The code below shows how to find the 95% confidence intervals for μ, θ, and ρ_I based on information in Problem 9.15 and Example 9.3.

```
mean(X.bar) + qt(c(.025,.975), g-1)*sqrt(MS.Bat/(g*r))
df.Err*MS.Err/qchisq(c(.975,.025), df.Err)
R = MS.Bat/MS.Err;   q.f = qf(c(.975,.025), g-1, g*r-g)
(R - q.f)/(R + (r-1)*q.f)
```

b) (*Intermediate*) Derive the confidence intervals in part (b) from the distributions of the quantities involved.

Hint: b) For ρ_I, start by deriving a confidence interval for $\psi = \theta_A/\theta$. What multiple of R is distributed as $\mathsf{F}(g - 1, g(r - 1))$?

9.17   Figure 9.8 on page 235 shows four diagnostic plots for the simulated posterior distribution of σ_A in the Gibbs sampler of Example 9.3. Make similar diagnostic plots for the posterior distributions of μ, σ, and ρ_I.

9.18   *Small contribution of batches to the overall variance.* Suppose the researchers who did the experiment in Example 9.3 find a way to reduce the batch component of variance. For the commercial purpose at hand, that would be important progress. But when they try to analyze a second experiment, there is a good chance that standard frequentist analysis will run into trouble. The code below is essentially the same as in Problem 9.15, but with the parameters and the seed changed. Group means and standard deviations, sufficient for running the Gibbs sampler of Example 9.3 are shown as output.

```
set.seed(1237)
g = 12;  r = 10
mu = 100;  sg.a = 1;  sg.e = 9
a.dat = matrix(rnorm(g, 0, sg.a), nrow=g, ncol=r)
e.dat = matrix(rnorm(g*r, 0, sg.e), nrow=g, ncol=r)
X = round(mu + a.dat + e.dat)
X.bar = apply(X, 1, mean);  X.sd = apply(X, 1, sd)
round(rbind(X.bar, X.sd), 3)

> round(rbind(X.bar, X.sd), 3)
        [,1]    [,2]    [,3]    [,4]    [,5]    [,6]
X.bar 96.90 103.700 97.300 100.900 95.100 95.900
X.sd  11.77  12.781  8.693  10.418  6.244  7.505

        [,7]   [,8]    [,9]   [,10]   [,11]   [,12]
X.bar 94.900 99.00 98.200 98.200 98.700 102.400
X.sd   9.871 10.76 10.304  6.356 11.146   8.289
```

a) Figure 9.6 shows boxplots for each of the 12 batches simulated above. Compare it with Figure 9.5 (p230). How can you judge from these two figures that the batch component of variance is smaller here than in Example 9.3?

b) Run the Gibbs sampler of Section 9.3 for these data using the same uninformative priors as shown in the code there. You should obtain 95% Bayesian interval estimates for $\mu, \sigma, \sigma_A = \sqrt{\theta_A}$, and ρ_I that cover the values used to generate the data X. See Figure 9.12, where one-sided intervals are used for σ_A and ρ_I.

9.19   *Continuation of Problem 9.18. Negative estimates of θ_A and ρ_I.*

a) Refer to results stated in Problems 9.15 and 9.16. Show that the unbiased estimate of θ_A is negative. Also, show that the 95% confidence interval for ρ_I includes negative values. Finally, find 95% confidence intervals for μ and $\sigma = \sqrt{\theta}$ and compare them with corresponding results from the Gibbs sampler in Problem 9.18.
b) Whenever $R = \text{MS(Batch)}/\text{MS(Error)} < 1$, the unbiased estimate $\hat{\theta}_A$ of θ_A is negative. When the batch component of variance is relatively small, this has a good chance of occurring. Evaluate $P\{R < 1\}$ when $\sigma_A = 1$, $\sigma = 9$, $g = 12$, and $r = 10$, as in this problem.
c) The null hypothesis $H_0: \theta_A = 0$ is accepted (against $H_1: \theta_A > 0$) when R is smaller than the 95th quantile of the F distribution with $g - 1$ and $g(r - 1)$ degrees of freedom. Explain why this null hypothesis is always accepted when $\hat{\theta}_A < 0$.

Hints: b) Exceeds 1/2. c) The code qf(.95, 11, 108) gives a result exceeding 1.

9.20   Calcium concentration in turnip leaves (% dry weight) is assayed for four samples from each of four leaves. Consider leaves as "batches." The data are shown below as R code for the matrix X in the program of Example 9.3; that is, each row of X corresponds to a batch.

```
X = matrix(c(3.28, 3.09, 3.03, 3.03,
             3.52, 3.48, 3.38, 3.38,
             2.88, 2.80, 2.81, 2.76,
             3.34, 3.38, 3.23, 3.26), nrow=4, ncol=4, byrow=T)
```

a) Run the program, using the same noninformative prior distributions as specified there, to find 95% Bayesian interval estimates for μ, σ_A, σ, and ρ_I from these data.
b) Suppose the researchers have previous experience making calcium determinations from such leaves. While calcium content and variability from leaf to leaf can change from one crop to the next, they have observed that the standard deviation σ of measurements from the same leaf is usually between 0.075 and 0.100. So instead of a flat prior for σ, they choose $\text{IG}(\alpha_0 = 35, \lambda_0 = 0.25)$. In these circumstances, explain why this is a reasonable prior.
c) With flat priors for μ and θ_A, but the prior of part (b) for θ, run the Gibbs sampler to find 95% Bayesian interval estimates for μ, σ_A, σ, and ρ_I from the data given above. Compare these intervals with your answers in part (a) and comment.

Note: Data are from page 239 of [SC80]. The unbiased estimate of $\theta_A = \sigma_A^2$ is positive here. Estimation of σ_A by any method is problematic because there are so few batches.

9.21 In order to assess components of variance in the two-stage manufacture of a dye, researchers obtain measurements on five samples from each of six batches. The data are shown below as R code for the matrix X in the program of Example 9.3; that is, each row of X corresponds to a batch.

```
X = matrix(c(1545, 1440, 1440, 1520, 1580,
             1540, 1555, 1490, 1560, 1495,
             1595, 1550, 1605, 1510, 1560,
             1445, 1440, 1595, 1465, 1545,
             1595, 1630, 1515, 1635, 1625,
             1520, 1455, 1450, 1480, 1445), 6, 5, byrow=T)
```

a) Use these data to find unbiased point estimates of μ, σ_A, and σ. Also find 95% confidence intervals for μ, σ, and ρ_I (see Problem 9.16).
b) Use a Gibbs sampler to find 95% Bayesian interval estimates for μ, σ_A, σ, and ρ_I from these data. Specify noninformative prior distributions as in Example 9.3. Make diagnostic plots.

Answers: b) Roughly: (1478, 1578) for μ; (15, 115) for σ_A. See [BT73] for a discussion of these data, reported in [Dav57].

9.22 In order to assess components of variance in the two-stage manufacture of a kind of plastic, researchers obtain measurements on four samples from each of 22 batches. Computations show that MS(Error) = 23.394. Also, *sums* of the four measurements from each of the 22 batches are as follows:

218	182	177	174	208	186
206	192	187	154	208	176
196	179	181	158	158	198
160	178	148	194		

a) Compute the batch means, and thus $\bar{x}_{..}$ and MS(Batch). Use your results to find the unbiased point estimates of μ, θ_A, and θ.
b) Notice that the batch standard deviations s_i, $i = 1, \ldots, 12$, enter into the program of Example 9.3 only as $\sum_i (r-1)s_i^2$. Make minor changes in the program so that you can use the information provided to find 90% Bayesian interval estimates of μ, σ_A, σ, and ρ_I based on the same noninformative prior distributions as in the example.

Note: Data are taken from [Bro65], p325. Along with other inferences from these data, the following traditional 90% confidence intervals are given there: (43.9, 47.4) for μ; (17.95, 31.97) for θ; and (0.32, 1.62) for $\psi = \theta_A/\theta$. (See Problem 9.16.)

9.23 *Design considerations.* Here we explore briefly how to allocate resources to get a narrower probability interval for σ_A than we got in Example 9.3 with $g = 12$ batches and $r = 10$ replications. Suppose access to additional randomly chosen batches comes at negligible cost, so that the main expenditure for the experiment is based on handling $gr = 120$ items. Then an experiment with $g = 60$ and $r = 2$ would cost about the same as the one in the example.

Modify the code shown in Problem 9.15 to generate data from such a 60-batch experiment, but still with $\mu = 100$, $\sigma_A = 15$, and $\sigma = 9$ as in the example. Then run the Gibbs sampler of the example with these data. Compare the lengths of the probability intervals for μ, σ_A, σ, and ρ_I from this 60-batch experiment with those obtained for the 12-batch experiment of the example. Comment on the trade-offs.

Note: In several runs with different seeds for generating the data, we got much shorter intervals for σ_A based on the larger number of batches, but intervals for σ were a little longer. What about the lengths of probability intervals for μ and ρ_I? In designing an experiment, one must keep its goal in mind. For the goal of getting the shortest frequentist confidence interval for μ within a given budget, [SC80] shows an optimization based on relative costs of batches and items. Additional explorations: (i) For these same parameters, investigate a design with $g = 30$ and $r = 4$. (ii) Investigate the effect of increasing the number of batches when σ_A is small, as for the data generated in Problem 9.18.

9.24 *Using the correct model.* To assess the variability of a process for making a pharmaceutical drug, measurements of potency were made on one pill from each of 50 bottles. These results are entered into a spreadsheet as 10 rows of 5 observations each. Row means and standard deviations are shown below.

Row	1	2	3	4	5	6	7	8	9	10
Mean	124.2	127.8	119.4	123.4	110.6	130.4	128.4	127.6	122.0	124.4
SD	10.57	14.89	11.55	10.14	12.82	9.99	12.97	12.82	16.72	8.53

a) Understanding from a telephone conversation with the researchers that the rows correspond to different batches of the drug made on different days, a statistician uses the Gibbs sampler of Example 9.3 to analyze the data. Perform this analysis for yourself.

b) The truth is that all 50 observations come from the same batch. Recording the data in the spreadsheet by rows was just someone's idea of a convenience. Thus, the data would properly be analyzed without regard to bogus "batches" according to a Gibbs sampler as in Example 9.2. (Of course, this requires summarizing the data in a different way. Use $s^2 = [9\text{MS(Batch)} + 40\text{MS(Error)}]/49$, where s is the standard deviation of all 50 observations.) Perform this analysis, compare it with the results of part (a), and comment.

Note: Essentially a true story, but with data simulated from $\mathsf{NORM}(125, 12)$ replacing unavailable original data. The most important "prior" of all is to get the model right.

10

Using WinBUGS for Bayesian Estimation

Historically, an important roadblock to using Bayesian inference has been the difficulty of computing posterior distributions of parameters. Thus, a major focus of this book is to show how such computations can be done using modern hardware and software.

- We introduced Bayesian inference in Chapter 8, showing a few situations in which the exact posterior distributions can be found analytically. For example, in Example 8.5 (on election polling), the posterior distribution is in the beta family. Printed tables of beta distributions were never very widely available, but nowadays it is easy to use the R function qbeta, or similar functions in other software, to find interval estimates of π, the population proportion in favor of a particular candidate.

- In general, however, it may be difficult or impossible to find the posterior distribution analytically, and then simulation methods are useful. For example, in Section 9.3 (on variance components), a Gibbs sampler is used to find posterior interval estimates of batch and error variances and of other parameters in the model. This requires finding partial conditional distributions analytically and using them to program the simulation of a multidimensional Markov chain whose stationary distribution provides the desired interval estimates.

- The discovery that the Gibbs sampler can be used to solve a wide variety of Bayesian estimation problems was a major advance in the accessibility of Bayesian inference in practice. However, it may not be trivial to find the partial conditional distributions or to do the programming necessary to implement a Gibbs sampler. For example, the program of Example 9.3 is about a page long when printed in the format of this book.

In this chapter, we introduce WinBUGS, showing some instances in which this software greatly simplifies the implementation of Gibbs sampling and other computational methods for Bayesian estimation. We begin with some basic information about WinBUGS software. Then we use this software to

E.A. Suess and B.E. Trumbo, *Introduction to Probability Simulation and Gibbs Sampling with R*,   249
Use R!, DOI 10.1007/978-0-387-68765-0_10, © Springer Science+Business Media, LLC 2010

analyze some of the models illustrated in Chapters 8 and 9 and also some
additional Bayesian inference models we have not previously considered.

10.1 What Is BUGS?

The BUGS program is designed for Bayesian modeling. The term *BUGS*
stands for **B**ayesian **I**nference **U**sing **G**ibbs **S**ampling. WinBUGS is an implementation of BUGS for Microsoft Windows. As this is being written, the
latest version of WinBUGS is available from OpenBUGS website:

www.openbugs.info

This website also has links to prior versions of OpenBUGS, information about
running BUGS software on other operating systems, manuals, information
about the authors of the software, and so on. In this chapter, we use Open-
BUGS 3.0.3, a version of WinBUGS from this site. (See [TOLS06].)

To begin a specific analysis, the user provides a relatively brief WinBUGS
program, written in code that has syntax that is somewhat similar to R. The
program includes a statement of the model (including prior distributions), the
data, and initial values of the parameters to be estimated (as the first step in
a simulated Markov chain).

At the core of the BUGS software is the Gibbs sampler, which is used
to sample from the conditional posterior distributions of the parameters.
These distributions are provided by the program, based on a listing of the
model, the prior distributions, and the data. When nonconjugate models
are used, WinBUGS may employ various computational methods to sample
from these conditional posterior distributions. For example, the Metropolis-
Hastings algorithm and other adaptive rejection methods may also be implemented as appropriate (see Chapters 7 and 8).

The posterior analysis is performed using the simulated Monte Carlo
Markov chain output produced by the program. Posterior statistics and posterior densities can be calculated to produce posterior estimates of the parameters in the model. Much as in Chapter 9, diagnostic graphical displays such
as trace plots, mean plots, and autocorrelation plots can be made as an aid
in determining whether the simulated chain converges satisfactorily.

In earlier chapters of this book, we have stressed that some Markov chains
either do not have stationary distributions or that it may be difficult to approximate a stationary distribution using simulation. Although WinBUGS
is relatively easy to use, that does not diminish the responsibility of users
to determine whether the results are trustworthy. Here we quote an explicit
warning on this issue from the creators of WinBUGS [STBL07]:

> Potential users are reminded to be extremely careful if using this program
> for serious statistical analysis. We have tested the program on quite a wide
> set of examples, but be particularly careful with types of model that are

currently not featured. If there is a problem, WinBUGS might just crash, which is not very good, but it might well carry on and produce answers that are wrong, which is even worse. Please let us know of any successes or failures.

Beware: MCMC sampling can be dangerous!

10.2 Running WinBUGS: The Binomial Proportion

In this section, we show how to use WinBUGS to make Bayesian inferences about a population proportion, assuming binomial data. In Chapter 8, we found exact values based on beta-distributed prior and posterior distributions, so it is not necessary to use WinBUGS in this situation. But this example provides an easy introduction to WinBUGS.

Example 10.1. Election Polling Revisited. Recall Example 8.1 on election polling. This model denotes by x the number of potential voters out of a sample of size n in favor of Proposition A. We assume that $x \sim \text{BINOM}(n, \pi)$, where π is the population proportion of potential voters who favor Proposition A, and we use the prior distribution $\pi \sim \text{BETA}(\alpha, \beta)$, with $\alpha_0 = 330$ and $\beta_0 = 270$. Note that in the syntax of WinBUGS the parameterization of the binomial distribution is with the population proportion listed first and the sample size listed second. These parameters center the prior at $\pi = 55\%$ with high probability that π is between 51% and 59%. The data are $x = 620$ favorable responses out of $n = 1000$ subjects sampled.

Because this model and prior are conjugate, the posterior distribution can be derived analytically, so WinBUGS implements Monte Carlo simulation directly from the posterior and does not implement a Gibbs sampler. We show the steps necessary to program this situation into WinBUGS. We use only 2000 iterations for this first example so we can see the simulated history plots more clearly. Longer runs of the program would produce more accurate results and would match the exact answers from Chapter 8 to any desired number of decimal places.

We begin by typing the code below into a new *Program window*, available by selecting New in the File menu. There are three parts to the program: the Model and two lists, one for the Data and one for the Inits (initial values).

```
# Model
model; {
    x ~ dbin(pp, n)              # Note syntax differs from R
    pp ~ dbeta(330, 270)         # Prior
}
# Data
list(x = 620, n = 1000)          # observed data
# Inits
list(pp = 0.25)                  # starting values for pp
```

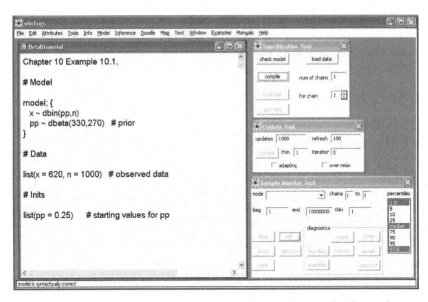

Figure 10.1. Polling data: WinBUGS screenshot with code for Example 10.1.

Notice that we use **pp** for π and that the name and syntax for the binomial density differ from that of R.

To run this program, begin by opening the *Specification Tool* from the *Model* pull-down menu. See Figures 10.1 and 10.2 for a view of the screen at this point and an enlargement of the *Specification Tool*, respectively.

- To specify the model, highlight the code **model;** under the comment **# Model**, and then click the *check model* button in the *Specification Tool*. The response in the lower-left corner of the *Program window* should be "model is syntactically correct."

- To load the data, highlight the code **list** under the comment **# Data**, and then click the *load data* button in the *Specification Tool*. The response in the lower-left corner of the *Program window* should be "data loaded."

- To compile the model, click the *compile* button in the *Specification Tool*. The response in the lower-left corner of the *Program window* should be "model compiled."

- To load the initial values, highlight the code **list** under the comment **# Inits**, and then click the *load inits* button in the *Specification Tool*. The response in the lower-left corner of the *Program window* should be "model is initialized."

Next, we open two new tools: the *Update Tool* from the *Model* pull-down menu (shown in Figure 10.3) and the *Sample Monitor Tool* from the *Inference* pull-down menu (shown in Figure 10.4). The screen should now look like Figure 10.1.

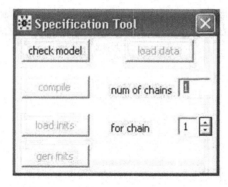

Figure 10.2. Polling data: Use of the *Specification Tool* for Example 10.1.

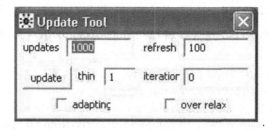

Figure 10.3. Polling data: Use of the *Update Tool* for Example 10.1

Once the *Sample Monitor Tool* is open, we can enter the *node* of the model. (WinBUGS uses the word *node* to mean a parameter to be estimated.) Since we have one model parameter, we need to enter only one node. In the *Sample Monitor Tool*, type the node **pp** and click the *set* button. To indicate that no more nodes are to be entered, type the star symbol ∗ in the *node* box. Then you should see additional buttons appear in the *Sample Monitor Tool*.

Now we run the model for 2000 iterations. In the *Update Tool*, click the *update* button twice (once for each 1000 iterations). To view the resulting sampled values, click the *history* button in the *Sample Monitor Tool*. Then, to view the sampled values of the estimated parameter **pp**, its estimated posterior density, and the ACF of the simulated values, click buttons *history*, *stats*, *density*, and *auto cor* in the *Sample Monitor Tool*. At this point, the screen should look like Figure 10.5.

With 2000 iterations from one run of this program in WinBUGS, we obtained posterior statistics for π that are similar to the exact values obtained in Example 8.6 on election polling. Our posterior mean was 59.39% in favor of Proposition A. And the posterior probability interval for the proportion in favor was $(56.9\%, 61.79\%)$. These results are very close to the exact values 59.4% for the mean and $(57.0\%, 61.8\%)$ for the interval. With a longer run of the WinBUGS program, we could come closer to the actual values, but

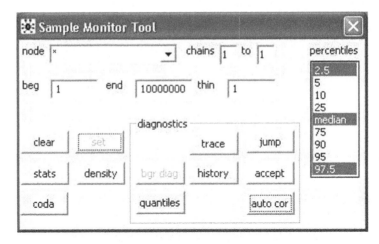

Figure 10.4. Polling data: Use of the *Sample Monitor Tool* for Example 10.1.

with 2000 values we come very close. We chose a relatively small number of iterations to give a clear view of the degree of mixing.

As seen in Chapter 8, this problem can be solved directly because of the conjugate prior, and so we are sampling directly from the posterior density and convergence is achieved very near the beginning of the Markov chain simulation. This is Monte Carlo integration of the posterior density, similar to what we presented in Chapter 3.

It should be noted that the summary results from the simulated values from the posterior distribution of **pp** shown in Figure 10.5 indicate that the sampling is independent, as demonstrated in the ACF. And we present the summary statistics below (slightly reformatted to fit on our printed page).

```
        mean     sd      MC_error
pp  0.5939  0.01222  2.389E-4

val2.5pc  median  val97.5pc  start  sample
0.569     0.5939  0.6179      1      2000
```

Also, we tried using the first 1000 iterations as a burn-in period by changing the starting value *beg* from 1 to 1001 in the *Sample Monitor Tool*. Then our summary results changed to:

```
        mean     sd      MC_error
pp  0.5935  0.01221  3.51E-4

    val2.5pc  median  val97.5pc  start  sample
pp  0.5684    0.5935  0.6178      1001   1000
```

Problems 10.1 through 10.3 provide some basic practice using WinBugs and also additional insight into estimation in this situation. ◇

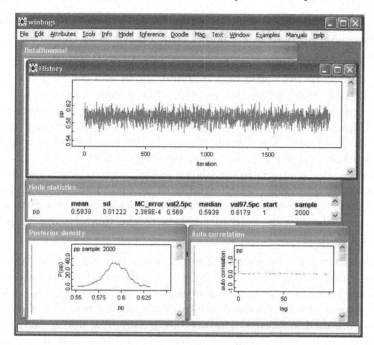

Figure 10.5. Polling data: WinBUGS screenshot for Example 10.1

In the next section, we use WinBugs to solve an elementary inferential problem involving normal data that is not included in Chapters 8 or 9. Unlike this section, where the sampling for one parameter was independent, it should be clear in the next section how the Gibbs sampler is a form of dependent sampling when there are two or more parameters in the model.

10.3 Running WinBUGS with a Script: Two-Sample Problem

This section deals with a two-sample model in which two independent samples are assumed to be normal with the same population variance and in which we seek an interval estimate for the difference between the population means. We use flat prior distributions, so the numerical results should be nearly the same as we would get from a frequentist two-sample t procedure.

Example 10.2. The Speed of a Mental Process. One way to quantify the speed of a mental process is by repeated measurements of reaction times. In one study, 48 student subjects are divided at random into two groups, called Choice and Simple. Subjects in both groups are asked to memorize three nonsense words, each starting with a different letter. In repeated trials, each subject says a particular word as soon as possible after a cue appears on a computer monitor, but the method of cueing depends on the group.

- On each trial, a subject in the Simple group is reminded of a specific word by showing its first letter on a computer monitor. Then he or she is asked to say this word as soon as possible after a subsequent cue appears on the screen. Thus the reaction time involves only the time to see the cue and "program" the voice to say the word into a microphone connected to the computer. The computer records the reaction time between the appearance of the cue and the utterance of the word.

- On each trial, a subject in the Choice group is alerted that a cue is coming, but without revealing which word is to be said. The cue itself gives the first letter of the word. Thus the reaction time includes the time it takes to recall the specific word in addition to the time it takes to program saying it.

The idea is that, on average, the difference between Choice and Simple reaction times should be the time it takes a person to recall a nonsense word after being told its first letter. Similar reaction-time experiments for determining the speeds of mental processes have been done for over a century. (Early experiments used stopwatches rather than computers, and so also involved components of error due to experimenter reaction times.) The data presented here are from an experiment by Stuart Klapp at California State University, East Bay, reported and discussed in [Tru02].

A practical complication is that reaction times—even for a given subject— are quite variable and distinctly right-skewed. So each subject did many trials, and the data in the program below show the median reaction time (in milliseconds) for each subject. For a demonstration that such medians are nearly normally distributed, see Problem 10.5.

In classical statistics, a pooled two-sample confidence interval could be found for the difference between population means for the Simple and Choice methods (see Problem 10.4). Here we show a Bayesian model for estimating the difference in means between the two groups.

The WinBUGS program below specifies that observations in the Choice group are randomly sampled from $N(\mu_1, \tau)$ and that observations from the Simple group are from $N(\mu_2, \tau)$. As usual in Bayesian modeling, we specify variability in terms of precision τ (reciprocal of variance). This matches the syntax for the normal distribution in WinBUGS, which is different from that of R. In WinBUGS, the second parameter of the function dnorm is the precision, not the standard deviation. Notice that we assume the same precision τ in both populations.

Our model uses prior distributions that are conjugate and nearly flat. The priors for the means μ_1 and μ_2 are normal with very small precision, and the prior for the precision τ has a gamma distribution with very small shape and rate parameters. Two additional parameters, useful in interpreting results,

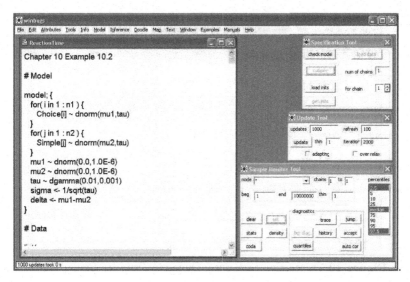

Figure 10.6. Reaction-time data: WinBUGS screenshot for Example 10.2.

are defined: the common population standard deviation $\sigma = 1/\sqrt{\tau}$ and the difference $\delta = \mu_1 - \mu_2$ between the two population means.

```
# Model
model; {
    for(i in 1:n1) {
        Choice[i] ~ dnorm(mu1, tau)
    }
    for(j in 1:n2) {
        Simple[j] ~ dnorm(mu2, tau)
    }
    mu1 ~ dnorm(0.0, 1.0E-6)
    mu2 ~ dnorm(0.0, 1.0E-6)
    tau ~ dgamma(0.01, 0.001)
    sigma <- 1/sqrt(tau)
    delta <- mu1 - mu2
}
# Data
list(
Choice = c(462, 397, 523, 481, 494, 430, 516, 472,
           521, 397, 441, 474, 468, 503, 492, 383,
           432, 569, 444, 534, 437, 553, 434, 435),
Simple = c(237, 260, 340, 322, 255, 273, 252, 316,
           276, 339, 304, 268, 291, 355, 292, 225,
           381, 334, 256, 325, 413, 307, 308, 312), n1=24, n2=24)
# Inits
list(mu1=0, mu2=0, tau=1)
```

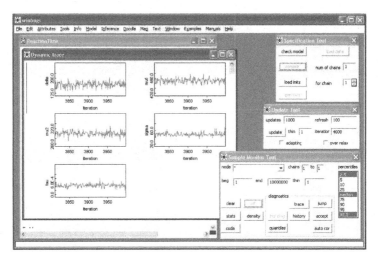

Figure 10.7. Reaction-time data: Dynamic Trace for Example 10.2.

Open this program in WinBUGS and pull down the *Specification Tool* from *Model* on the menu bar. To specify the model, highlight the code word `model;` under the comment `#Model`, and then click the *check model* button in the *Specification Tool*.

To load the data, highlight the code `list` under the comment `#Data` and click the *load data* button. Then compile the model by clicking the *compile* button.

To load the initial values, highlight the code `list` under the comment `#Inits` and click the *load inits* button. The response in the lower-left corner should be "model is initialized."

Next, we select the *Update Tool* from the *Model* menu and the *Sample Monitor Tool* from the *Inference* menu. Once the *Sample Monitor Tool* is active, we can enter the five nodes in the model. In turn, type `mu1`, `mu2`, `tau`, `sigma`, and `delta` into the *node* box—clicking the *set* button immediately after typing each one. Then enter * into the *node* box. At this point, all of the buttons on the *Sample Monitor Tool* should appear (see Figure 10.6).

Now run the simulation for 2000 iterations as a burn-in period by clicking the *update* button twice. Select the *trace* button in the *Sample Monitor Tool* to watch the sample values as they are simulated. Then run the simulation for 2000 more iterations by clicking the *update* button two more times to produce the results we regard as sampled values from the posterior distributions. Note that in the *Dynamic trace window* the simulated values are automatically updated (see Figure 10.7).

To view all 4000 of the sampled values, click the *history* button. To see the posterior analysis only for values sampled after burn-in, change the *beg* value to 2001 and click the *stats* and *density* buttons. At this point, your screen should look like Figure 10.8. The numerical results (in msec) are shown below.

There is a clear difference in reaction times between the two groups. To see this, we can examine the model parameter $\delta = \mu_1 - \mu_2$. Its posterior mean estimate is 168.8 msec, with a posterior 95% interval estimate of (141.1, 196.7). There is a 95% probability that the difference $\mu_1 - \mu_2$ is in this interval, which does not include 0, indicating that the difference between the two group means is significantly different from 0.

	mean	sd	MC_error		
delta	168.9	13.76	0.2404		
	val2.5pc	median	val97.5pc	start	sample
delta	141.3	169.3	197.0	2001	2000

	mean	sd	MC_error		
mu1	470.4	9.69	0.2021		
	val2.5pc	median	val97.5pc	start	sample
mu1	451.7	470.6	489.4	2001	2000

	mean	sd	MC_error		
mu2	301.6	10.01	0.2161		
	val2.5pc	median	val97.5pc	start	sample
mu2	281.5	301.8	321.0	2001	2000

	mean	sd	MC_error		
sigma	48.54	5.168	0.1007		
	val2.5pc	median	val97.5pc	start	sample
sigma	39.5	48.09	59.62	2001	2000

	mean	sd	MC_error		
tau	4.387E-4	9.195E-5	1.75E-6		
	val2.5pc	median	val97.5pc	start	sample
tau	2.826E-4	4.325E-4	6.413E-4	2001	2000

Theoretically, a more precise experimental design might use both methods of giving cues on each subject (paired data) so that we could measure each person's individual recall time. But it takes a while for subjects to get familiar with the equipment and cueing procedure. Experience has shown that subjects get confused when asked to switch between methods, so it is best to have two groups. Problems 10.7 and 10.8 illustrate situations where it is feasible to use paired data to estimate a difference between means. ◇

An alternative way to run a WinBUGS program is by writing a script. Here is an example of the script needed to estimate parameters from the reaction-time data.

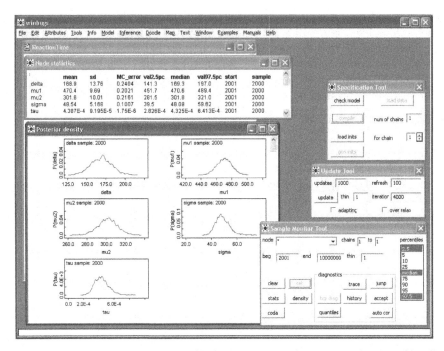

Figure 10.8. Reaction-time data: Node statistics and posterior density plots for Example 10.2.

```
modelCheck('C:/ReactionTime/ReactionTimeModel.odc')
modelData('C:/ReactionTime/ReactionTimeData.odc')
modelCompile(1)
modelInits('C:/ReactionTimeInit.odc')
modelUpdate(1000)
samplesSet(mu1)
samplesSet(mu2)
samplesSet(tau)
samplesSet(sigma)
samplesSet(delta)
modelUpdate(1000)
samplesStats('*')
samplesDensity('*')
samplesHistory('*')
samplesAutoC('*')
```

This script calls three files, which need to be prepared in advance: a model file, a data file, and an initial values file. The files contain the three parts of the WinBUGS program for reaction times as shown above. However, notice that the WinBUGS script uses forwardslashes / instead of backslashes \ in specifying MS Windows directory locations.

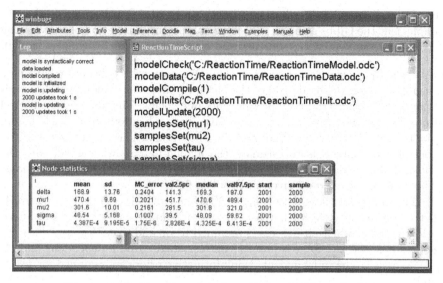

Figure 10.9. Reaction-time data: Script for Example 10.2.

To run the script, open the *Log* window, where you will see the script as it runs. From the *Info* pull-down menu, select *Open Log*. Then click on the window in WinBUGS containing the script, and from the *Model* pull-down menu select *Script*. In the *Log* window, you should see each command in the script. When the script finishes, the screen should look similar to Figure 10.9.

10.4 Running WinBUGS and MCMC Diagnostics: Variance Components

WinBUGS includes a large library of example programs. In this section, we show one of them. These examples can be found in the *Examples* pull-down menu, where there are three volumes of general examples. The one we illustrate here is from *Example Volume 1, Dyes: variance components model* (see Problem 9.21). This example gives the WinBUGS implementation of the analysis of the data presented by Box and Tiao [BT73]. This is a very useful example because it has more parameters than the examples given earlier in the chapter and shows the ease of using WinBUGS to implement Bayesian hierarchical models. It also provides an opportunity to use the convergence diagnostics available in WinBUGS and thus to illustrate some of the issues with convergence of MCMC methods.

In particular, it is a nice example of how different prior distributions (possibly nonconjugate) may influence the convergence of the Markov chains produced. In Problem 10.9, we also illustrate that the Gibbs sampler implemented by WinBUGS may produce slightly different results than if one derives the conditional posterior distributions directly and writes R code based on them to implement the Gibbs sampler. This is an important illustration of the warning given with the WinBUGS software and quoted earlier in this chapter (p250).

Example 10.3. Dye Data Revisited. The Dye data were collected to examine the influence on the final yield of the batch-to-batch (or "among batch") variation in a raw material used in a production process. The data were collected from six batches, $i = 1, ..., g = 6$, with five replications from each batch, $j = 1, ..., r = 5$, and thus there are $gr = 30$ observations y_{ij} altogether. The goal of the analysis is to estimate the among-batch variation σ_A^2 and the within-batch variation σ^2, assuming independent batches and replications. These are the same data given in Problem 9.21.

The model for these data is $y_{ij} \sim N(A_i, \tau)$, with $A_i \sim N(\mu, \tau_A)$, where A_i is the true mean yield for batch i, $\tau_A = 1/\sigma_A^2$, and $\tau = 1/\sigma^2$. The total variation in the production process is assumed to be additive, $V(y_{ij}) = \sigma_A^2 + \sigma^2$. For each i, the y_{ij}, for $j = 1, ..., r$, are assumed to be independent and A_i are assumed to be independent. Noninformative priors are assumed for μ, τ_A, and τ.

We present the code for one of the recommended priors given in the example WinBUGS program. It is suggested that the prior $\mathsf{UNIF}(0, 1)$ be placed on the intraclass correlation. This is suggested to improve the convergence of the Gibbs sampler in WinBUGS for this model. The intraclass correlation ρ_I is

$$\rho_I = \frac{\sigma_A^2}{\sigma_A^2 + \sigma^2} = \frac{\theta_A}{\theta_A + \theta}.$$

This leads to a derived formula for the between variance among batches:

$$\sigma_A^2 = \frac{\rho_I}{1 - \rho_I} \sigma^2.$$

The WinBUGS model with this uniform prior on the intraclass correlation is shown below along with the data and some arbitrary initial parameter values.

```
# Model
model; {
    for(i in 1:g) {
        A[i] ~ dnorm(mu, tau.A)
        for(j in 1:r) {
            y[i,j] ~ dnorm(A[i], tau)
        }
    }
    mu ~ dnorm(0.0, 1.0E-10)        # flat prior on grand mean
    tau ~ dgamma(0.001, 0.001)      # flat prior on within-variation
    theta <- 1/tau
    sigma <- sqrt(theta)
    rho.I ~ dunif(0,1)              # Uniform prior on ICC
    theta.A <- theta * rho.I/(1-rho.I)
```

```
sigma.A <- sqrt(theta.A)
tau.A <- 1/theta.A
}

# Data
list(g = 6, r = 5,
y = structure(
   .Data = c(1545, 1440, 1440, 1520, 1580,
             1540, 1555, 1490, 1560, 1495,
             1595, 1550, 1605, 1510, 1560,
             1445, 1440, 1595, 1465, 1545,
             1595, 1630, 1515, 1635, 1625,
             1520, 1455, 1450, 1480, 1445), .Dim = c(6,5)))

# Inits
list(mu=1500, tau=1, rho.I=0.5)
list(mu=1000, tau=0.9, rho.I=0.2)
list(mu=2000, tau=10, rho.I=0.8)
```

For this model, we run three chains simulatively to examine the convergence of the Gibbs sampler. To simulate the three chains with different starting values, we run WinBUGS in the same way as the last two examples in this chapter, but before clicking the *compile* button we change the *num of chains* from 1 to 3.

When running multiple chains, it is usual to provide different *inits* for each chain. Accordingly, we have listed three sets of starting values in the WinBUGS program above. To implement them, highlight each *list* separately and click *load inits*. Our *inits* lists do not include all of the model parameters requiring initial values, so after loading the three sets of initial values there are still some uninitialized variables. The rest of the initial values can be randomly generated by clicking the *gen inits* button at the bottom of the *Specification Tool*.

All of the model parameters should be examined for convergence, but for illustration here we focus on the four parameters, $\mu, \rho_I, \sigma,$ and σ_A, that were examined in Chapter 9. In the *Sample Monitor Tool*, we enter the nodes mu, rho.I, sigma, and sigma.A, clicking the *set* button after entering each one. We run the program for 500 000 iterations, with a burn-in of 200 000 iterations.

To assess the convergence of the simulated chains, we examine the *history plots, acf plots, quantiles,* and *BGR diagnostic* (see Figure 10.10).

- First, we set the beginning value to *beg* 200 001, and then we click *history* to see the history plots of the three chains. Notice that the three chains overlap.
- We view the acf plots by clicking the *auto cor* button. The acf plots indicate high autocorrelation within the chains. We could also try selecting the *over relax* option on the *Update Tool* or increasing the *thin* options to 10 or 20. These options would reduce the autocorrelation within the Markov chains.

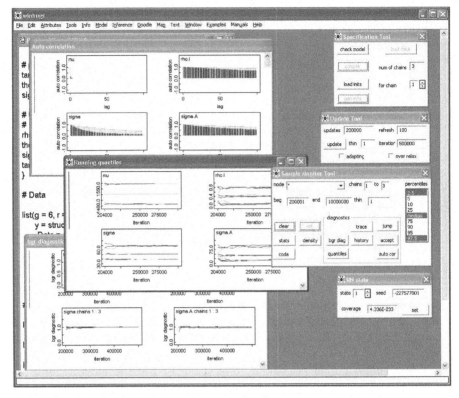

Figure 10.10. Dye data: Convergence diagnostics for Example 10.3.

- We click on the *quantiles* button to see the running 2.5, 50, and 97.5 quantiles. Also, Notice that the quantile plots overlap. Both plots indicate that the chains converge.
- Finally, we click on the *bgr diag* to see the output of BGR convergence diagnostic. The plot shows that all of the chains converge around 1 in the plot. This is also an indication of convergence.

Below, we give the posterior statistics from the three chains. Notice that they all produce similar posterior estimates of each model parameter. (In each case, we sample 300 000 values, starting at burn-in value 200 000. These values are omitted from the printouts below. See Figure 10.11.)

```
Chain 1 results:
            mean      sd  MC_error  val2.5pc   median  val97.5pc
mu        1527.0   22.79  0.05223    1482.0   1527.0     1573.0
rho.I     0.4337  0.1817  0.005604   0.09642  0.4303     0.7916
sigma       51.3   7.538   0.1239     38.91    50.46      68.16
sigma.A    46.94   19.77   0.5591      18.1    43.36      95.93
```

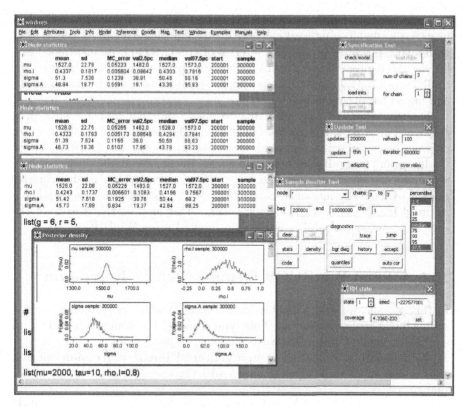

Figure 10.11. Dye data: Posterior statistics from three chains for Example 10.3.

```
Chain 2 results:
            mean      sd   MC_error   val2.5pc   median   val97.5pc
mu         1528.0   22.75   0.05265    1482.0    1528.0     1573.0
rho.I      0.4323   0.1793  0.005173   0.09546   0.4294     0.7841
sigma      51.38    7.624   0.1165     39.0      50.58      68.83
sigma.A    46.73    19.36   0.5107     17.95     43.78      93.23

Chain 3 results:
            mean      sd   MC_error   val2.5pc   median   val97.5pc
mu         1528.0   22.08   0.05226    1483.0    1527.0     1572.0
rho.I      0.4243   0.1737  0.006601   0.1083    0.4196     0.7567
sigma      51.42    7.618   0.1925     38.76     50.44      68.2
sigma.A    45.73    17.89   0.634      19.37     42.84      88.25
```

As a summary, we average over all three chains, that is, 900 000 values after burn-in.

```
            mean      sd   MC_error   val2.5pc   median   val97.5pc
mu         1528.0   22.54   0.03025    1482.0    1527.0     1573.0
rho.I      0.4301   0.1783  0.003364   0.09916   0.4265     0.7798
sigma      51.37    7.594   0.08563    38.87     50.49      68.38
sigma.A    46.47    19.03   0.3295     18.62     43.31      92.81
```

For comparison with Problem 9.21, the estimate for μ is 15.28 with a 95% posterior probability interval of $(1482, 1573)$ and the estimate for $\sigma_A = 46.47$ with a 95% posterior probability interval of $(18.62, 92.81)$. As mentioned earlier in this chapter, the results of WinBUGS may be slightly different from the results produced by directly computing the conditional posterior distributions and implementing them in R directly.

For this example, also note that we use a different model with a nonconjugate prior on the intraclass correlation. The model from Chapter 9 uses different priors; in particular, noninformative inverse gamma distributions are used as priors for both variance components. In the new model presented here a noninformative prior is used on one variance component, a noninformative prior is used on the intraclass correlation, and from this some information is assumed for the other variance component. This may be the reason for the shorter posterior probability intervals computed using WinBUGS. $\diamond$

10.5 A Final WinBUGS Example: Linear Regression

In this section, we give a final example using WinBUGS to implement the simple linear regression model. This type of model is commonly used in data analysis in general and in Bayesian modeling in particular.

We estimate the parameters in a **centered linear regression model**. This is also an example given in the WinBUGS User Manual. The model is

$$Y_i = \alpha + \beta(x_i - \bar{x}) + \varepsilon_i, \tag{10.1}$$

where $\varepsilon_i \sim N(0, \tau)$ and $\tau = 1/\sigma^2$. This is the same model as the simple linear regression model

$$Y_i = \alpha_0 + \beta x_i + \varepsilon_i \tag{10.2}$$

with $\alpha_0 = \alpha - \beta\bar{x}$. In order to get better behavior of the simulated Markov chains within the Gibbs sampler, we use equation (10.1) for the model and then estimate α_0 in equation (10.2). This centering of the predictor variable, x_i, makes the estimate of α independent of the estimate of β, which leads to much less correlation of the simulated Markov chains for α and β. See Problem 10.10.

Example 10.4. Linear Regression Model Using the Old Faithful Data. Recall the data on eruptions of the Old Faithful geyser discussed in Chapter 6, Example 6.1. These data were collected in an attempt to predict the waiting time Y_i until the next eruption given the length x_i of an eruption just finished. Values of Y_i and x_i for 107 eruptions are shown in the following WinBUGS code.

```
# Model      WaitNext = alpha + beta ( DurLast - mean(DurLast) )
model; {
   for( i in 1 : N ) {
      mu[i] <- alpha + beta * (x[i] - xbar)
   }
   for( i in 1 : N ) {
      Y[i] ~ dnorm(mu[i],tau)
   }
   alpha ~ dnorm(0.0, 1.0E-6)
   beta ~ dnorm(0.0, 1.0E-6)
   tau ~ dgamma(0.001, 0.001)
   sigma <- sqrt(1 / tau)
}


# Data
list(x = c(4.4, 3.9, 4.0, 4.0, 3.5, 4.1, 2.3, 4.7, 1.7, 4.9,
           1.7, 4.6, 3.4, 4.3, 1.7, 3.9, 3.7, 3.1, 4.0, 1.8,
           4.1, 1.8, 3.2, 1.9, 4.6, 2.0, 4.5, 3.9, 4.3, 2.3,
           3.8, 1.9, 4.6, 1.8, 4.7, 1.8, 4.6, 1.9, 3.5, 4.0,
           3.7, 3.7, 4.3, 3.6, 3.8, 3.8, 3.8, 2.5, 4.5, 4.1,
           3.7, 3.8, 3.4, 4.0, 2.3, 4.4, 4.1, 4.3, 3.3, 2.0,
           4.3, 2.9, 4.6, 1.9, 3.6, 3.7, 3.7, 1.8, 4.6, 3.5,
           4.0, 3.7, 1.7, 4.6, 1.7, 4.0, 1.8, 4.4, 1.9, 4.6,
           2.9, 3.5, 2.0, 4.3, 1.8, 4.1, 1.8, 4.7, 4.2, 3.9,
           4.3, 1.8, 4.5, 2.0, 4.2, 4.4, 4.1, 4.1, 4.0, 4.1,
           2.7, 4.6, 1.9, 4.5, 2.0, 4.8, 4.1),
       Y = c(78,  74,  68,  76,  80,  84,  50,  93,  55,  76,
             58,  74,  75,  80,  56,  80,  69,  57,  90,  42,
             91,  51,  79,  53,  82,  51,  76,  82,  84,  53,
             86,  51,  85,  45,  88,  51,  80,  49,  82,  75,
             73,  67,  68,  86,  72,  75,  75,  66,  84,  70,
             79,  60,  86,  71,  67,  81,  76,  83,  76,  55,
             73,  56,  83,  57,  71,  72,  77,  55,  75,  73,
             70,  83,  50,  95,  51,  82,  54,  83,  51,  80,
             78,  81,  53,  89,  44,  78,  61,  73,  75,  73,
             76,  55,  86,  48,  77,  73,  70,  88,  75,  83,
             61,  78,  61,  81,  51,  80,  79), xbar = 3.461, N = 107)

# Inits
list(alpha = -1, beta = -1, tau = 0.1)
list(alpha = 0, beta = 0, tau = 1)
list(alpha = 1, beta = 1, tau = 10)
```

These data can also be saved in a column format and imported into WinBUGS as shown in Problem 10.11.

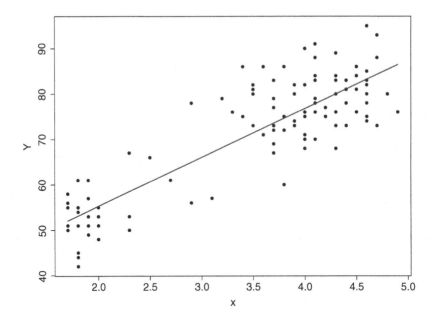

Figure 10.12. Old Faithful data: Fitted line plot for Example 10.4.

Here is the WinBUGS script file to run the regression model.

```
modelCheck('C:/Regr/RegressionModel.odc')
modelData('C:/Regr/RegressionData.odc')
modelCompile(1)
modelInits('C:/Regr/RegressionInits.odc')
modelUpdate(2000)
samplesSet(alpha)
samplesSet(beta)
samplesSet(tau)
samplesSet(sigma)
modelUpdate(2000)
samplesStats('*')
samplesDensities('*')
samplesHistory('*')
samplesAutoC('*')
```

The resulting history plots show that the model converges quickly and the autocorrelation plots show good mixing (see Problem 10.10). The estimated parameters are shown below.

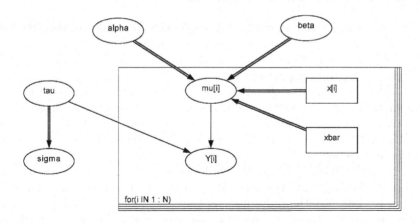

Figure 10.13. Old Faithful data: Graphical model for Example 10.4.

	mean	sd	MC_error	val2.5pc	median	val97.5pc
alpha	70.99000	0.651800	1.227e-02	69.72000	71.0000	72.26000
alpha0	33.78000	2.257000	4.353e-02	29.40000	33.8100	38.31000
beta	10.75000	0.625200	1.177e-02	9.51400	10.7500	11.95000
sigma	6.72000	0.473500	8.717e-03	5.88000	6.6960	7.70600
sigma2	45.39000	6.438000	1.192e-01	34.58000	44.8400	59.38000
tau	0.02247	0.003147	5.692e-05	0.01684	0.0223	0.02892

The fitted regression model from equation (10.1) for the fitted values $\hat{Y}$ is

$$\hat{Y}_i = 70.99 + 10.75(x_i - \bar{x}), \tag{10.3}$$

or equivalently, the fitted simple linear regression model equation (10.2) is

$$\hat{Y}_i = 70.99 - 10.75\bar{x} + 10.75x_i = 33.79 + 10.75x_i. \tag{10.4}$$

In Figure 10.12, we plot the data Y (the waiting time to the next eruption) against x (the length of an eruption just past) along with the fitted regression model.

We also point out that the WinBUGS software has a design tool called DoodleBUGS that can be used to graphically represent Bayesian models, see the pull-down menu item *Doodle*. To learn how to use this design tool, consult the DoodleBUGS manual that is available from the *Manuals* pull-down menu. In Figure 10.13, we show the graphical representation of the centered regression model 10.1. ◇

10.6 Further Uses of WinBUGS

In this section, we have shown how to implement Bayesian models using Win-BUGS, which is very commonly used in practice for Bayesian estimation. Advantages of WinBUGS are its overall ease of use and its flexibility for developing and comparing models.

Replacing the mathematical development and implementation in R by the automatic generation of full conditional densities for sampling in WinBUGS greatly decreases the time necessary to develop useful models. However, users must keep in mind the warning that comes with WinBUGS, which we quote in the first section of this chapter.

The use of WinBUGS can be integrated into R by using special libraries. The libraries R2WinBUGS and BRugs make it possible to run WinBUGS within R. The resulting advantage is that the simulated chains are stored within R, so that the user can graph and analyze these chains more flexibly within R. In particular, it is then possible to perform convergence diagnostics beyond what is available in WinBUGS. Also, running WinBUGS within R removes the need to interact directly with the WinBUGS interface and makes it easier to run longer chains.

Many excellent books on Bayesian inference using WinBUGS are currently available. Two books by Congdon [Con01] and [Con03] provide examples using WinBUGS to analyze various types of data, extending far beyond examples provided in the WinBUGS *Volumes of Examples*. Gelman and Hill [GH07] use R2WinBUGS to analyze data involving regression and hierarchical models. Also, an excellent general reference for further study of Bayesian data analysis is the book by Gelman, Carlin, Stern, and Rubin [GCSR04].

10.7 Problems

Problems for Section 10.2 (Estimating a Population Proportion)

10.1   In Example 10.1, run the program for 10 000 iterations with a burn-in period of 2000. What modifications are required in how the program of the example is run? Are the posterior estimate and the posterior 95% probability interval of π (denoted pp in the program) closer to the exact values computed in Chapter 8 than are the estimates in this chapter?

10.2   Run the program in Example 10.1 for 10 000 iterations with a burn-in period of 2000. Use a uniform prior on $(0, 1)$. Are the posterior point estimate and the posterior 95% probability interval of π (denoted pp in the program) consistent with the Agresti-Coull estimate and frequentist 95% confidence interval of Chapter 1?

10.3   In Example 10.1, suppose that only 100 voters were polled, with the same proportion favoring of Proposition A as in the example. That is, suppose there were $x = 62$ Yes responses among of the $n = 100$ voters polled.

a) Modify the program given in Example 10.1 to reflect this change in the data. Run the modified program, and report your results. Report the number of iterations you used and the resulting 95% Bayesian probability interval for the proportion of voters in favor of Proposition A.

b) What are the exact answers obtained for this smaller poll using the method presented in Chapter 8?

c) In view of the smaller sample, do you find an increase in the influence of the prior on the posterior?

Problems for Section 10.3 (Estimating Differences in Means)

10.4   Find a frequentist 95% confidence interval for the average time to recall a nonsense word. With data vectors `Choice` and `Simple` defined as in the `list` in the WinBUGS program of Example 10.2, use the R code `t.test(Choice, Simple, var.eq=T)`. Compare these results with the Bayesian analysis based on a flat prior. How much does the frequentist confidence interval change if you drop the assumption that the two populations have the same variance (by omitting the parameter `var.eq=T`)?

10.5   *Central Limit Theorem for medians.* The exponential distribution is often used to model reaction times. Suppose reaction times in Example 10.2 are exponential. Each subject in that experiment performed 90 trials. Perform the following simulation to see whether the distribution of the median of 90 exponential observations is nearly normal as claimed in the example. Briefly explain the program and interpret the results.

```
m = 5000;  n = 90;  x = rexp(m*n, 1/170)
med = apply(matrix(x, nrow=m),1, median)
hist(med, prob=T)
  xx = seq(min(med), max(med), length=100)
  lines(xx, dnorm(xx, mean(med), sd(med)))
shapiro.test(med)
```

Notes: Regardless of the value of the exponential mean, the demonstration is essentially the same. Recall that for moderate to large sample sizes, t procedures are reasonably robust to nonnormality. Moreover, for sufficiently large sample sizes, any goodness-of-fit test for normality tends to reject unless the population is precisely normal.

10.6   Below are IQ scores for 31 girls and 47 boys, all seventh-graders in the same Midwestern school district (data provided by Darlene Gordon, Purdue University, quoted in [Moo06]). Assume these students were chosen at random from some population. Modify the program of Example 10.2 to find a 95% Bayesian probability interval for the population difference between girls' and boys' IQ scores.

```
girls = c(114, 100, 104,  89, 102,  91, 114, 114, 103, 105,
          105, 130, 120, 132, 111, 128, 118, 119,  86,  72,
          111, 103,  74, 112, 107, 103,  98,  96, 112, 112,  93)
```

```
boys = c(111, 107, 100, 107, 115, 111,  97, 112, 104, 106,
         115, 109, 113, 128, 128, 118, 113, 124, 127, 136,
         106, 123, 124, 126, 116, 127, 119,  97, 102, 110,
         120, 103, 115,  93, 103,  79, 119, 110, 110, 107,
         105, 105, 110,  77,  90, 114, 106)
```

Note: For comparison, a frequentist 95% confidence interval for the difference in the population means, $\delta = \mu_{\text{girls}} - \mu_{\text{boys}}$, based on a pooled standard deviation, is $(-10.810, 1.145)$. There is no reason to believe the population variances are unequal, but if you want to try finding a Bayesian interval based on separate variances, the separate variances t procedure gives the longer interval $(-11.056, 1.391)$.

10.7   In 1999, as part of her master's thesis in education at CSU East Bay, Anne Nathan reported achievement test scores for 12 fourth-grade special education students and 14 regular fourth-grade students at a public school in the San Francisco Bay Area. Assume these students are randomly chosen from two populations. Scores for one type of test are shown below.

```
Reglr = c( 94, 112, 107, 118, 126, 103, 103,
          132,  98, 121,  88, 118, 100, 103)
Specl = c( 88,  82,  96,  87,  77,  80,
           98,  84,  88,  88,  90,  89)
```

a) Find a traditional 95% confidence interval for the difference in population means. A practical complication here is that the type of special education student studied here is relatively narrowly defined, so that the population variance for this population may be smaller than the population variance in the population of regular students.
b) Modify the program in Example 10.2 to find a 95% Bayesian probability interval for the difference in population mean scores. Assume equal population precisions. Use noninformative priors.
c) *Separate variances.* Modify the program of the example to allow for different population precisions. The two-sample problem with separate precisions is notoriously difficult to simulate using MCMC methods, especially when sample sizes are relatively small. Comment on the diagnostic plots. Do you believe the sampled values accurately reflect the posterior distributions of the parameters?
d) *Paired data.* Below are scores for an alternate kind of achievement test administered to the same 12 special education students (listed in the same order). Provide a Bayesian estimate of the population mean difference in scores between the two types of tests.

```
SpAlt = c(75,  86,  84,  80,  69,  81,
          89,  75,  84,  86,  85,  89)
```

10.8   *Changes in heights revisited.* Recall Example 9.2 (p224) in which the differences between morning and evening heights of 41 young men were measured. The data are shown and explored further in Problem 9.12 (p241), and are repeated below.

```
c( 8.50,  9.75,  9.75,  6.00,  4.00, 10.75,  9.25, 13.25,
  10.50, 12.00, 11.25, 14.50, 12.75,  9.25, 11.00, 11.00,
   8.75,  5.75,  9.25, 11.50, 11.75,  7.75,  7.25, 10.75,
   7.00,  8.00, 13.75,  5.50,  8.25,  8.75, 10.25, 12.50,
   4.50, 10.75,  6.75, 13.25, 14.75,  9.00,  6.25, 11.75,  6.25)
```

a) *Paired data.* Modify the WinBUGS program of Example 10.2 to find a 95% Bayesian probability interval for height difference in the population, using prior distributions equivalent to those shown in Example 9.2. Compare your numerical and graphical results with those obtained in Chapter 9. Also compare your modified program with the program of the example, mentioning similarities and differences.

b) *Informative prior.* Suppose you have previous experience with such height measurements. In particular, you believe the mean difference in heights is fairly sure to lie between 7 and 13 mm and that individual differences can be measured with a standard deviation between 1 and 5 mm. Use priors that reflect this experience, and report whether taking prior information into account markedly changes the results from those in part (a).

Problems for Section 10.4 (Variance Components)

10.9   In Problem 9.21, the model presented in Example 9.3 is used to analyze the dye data. Compare the results of the direct implementation of the model in the R code with the results of the WinBUGS implementation.

a) Run the code from Example 9.3 using the data given in Problem 9.21. Say which parameters have priors, and what priors are assumed for each.

b) Run the code from Example 10.3. What are the parameters with priors, and what are the assumed priors?

c) Are the posterior estimates of the common parameters similar, using R and using WinBUGS?

d) Does it appear that the different prior specifications in parts (a) and (b) influence the implemented Gibbs sampler?

Problems for Section 10.5 (Simple Linear Regression)

10.10   Compare the convergence properties of the two regression models presented in Section 10.5. Run the WinBUGS code given for the centered regression model (10.1) and the simple regression model (10.2).

a) Run the WinBUGS model from Example 10.4 and examine the autocorrelation in the history plots of simulated chains for α and β.

b) Modify the WinBUGS program from the example to implement the simple regression model given in equation (10.2). Run the new WinBUGS program and check the autocorrelation within the history plots of the α_0 chain and within the β chain.

c) Compare the correlation between the α_0 and β chains for the simple regression model with the same correlation for the α and β centered regression model. Use the plotting pull-down menu items *Inference* and *Correlations...* to examine the correlation between the simulated chains. Enter the *nodes* and click the *scatter* button to see a scatterplot, and click *print* to see the calculated correlation coefficient.

d) Examine the history plots from each model. Is there a difference in the convergence of Markov chains? Which model seems to converge more quickly?

e) Another method used to reduce the correlation in simulated Markov chains is *thinning*. Increase the number of updates to 40 000 and thin the chains from each model, using every 10th simulated value. Compare the results.

10.11 An alternative way of saving the data is to use the column format. Replace the data list in Example 10.4 with the following list and the data in column format, illustrated below.

```
list(xbar = 3.461, N = 107)

Y[] x[]
78   4.40000
74   3.90000
68   4.00000
76   4.00000
80   3.50000
84   4.10000
...
79   4.10000
END
```

To run this new WinBUGS program with the pull-down menus, follow the steps presented earlier in this chapter, except after highlighting the code list and clicking *load data* also highlight Y[] x[] and click *load data*. This will load the data in WinBUGS. Check that the results are similar to the results present in Section 10.5.

11

Appendix: Getting Started with R

11.1 Basics

This appendix focuses on some specific features and commands of R that
you will need in the first few chapters of this book. If you have never used
R before—or you need a review of the basics—we recommend starting here.
Throughout the book, we show the R code required for each new concept,
briefly explaining the new R commands involved. If you want more detailed
information, you can refer to the introductory and general reference manuals
available on the R website.

11.1.1 Installation

The R software package [RDCT09] can be downloaded and installed free
of charge from www.r-project.org. Versions are available to run under
Windows, Unix, or Macintosh operating systems. For example, here is how
to install R on a Windows machine from this web address (instructions that
as of this writing, have been stable for several years). In the left margin,
under Download, select CRAN. Then, in sequence, select the CRAN Mirror
site nearest you, Windows, base, and then finally the setup program, which
installs the latest release of R. After downloading is finished (which can take
seconds or minutes, depending on the speed of your connection), the R icon
should appear on your desktop. Select this icon to start R.

When R starts up, you will see the Console window—with a prompt > at
which you can type an instruction. This appendix shows introductory examples of R instructions. It is best read while seated at a computer with the
Console window open so you can interactively type R code shown here. Some
informal questions and formal problems are interspersed to help you reinforce
ideas in the examples. We recommend that you explore these to consolidate
what you have learned before going on to the next part. Learning to use R is
not a spectator sport.

E.A. Suess and B.E. Trumbo, *Introduction to Probability Simulation and Gibbs Sampling with R*,   275
Use R!, DOI 10.1007/978-0-387-68765-0_11, © Springer Science+Business Media, LLC 2010

11.1.2 Using R as a Calculator

To get the feel of how the Console works, try typing instructions for a few simple computations, pressing the Enter key after each. Use the asterisk (*) for multiplication and carat (^) for exponentiation. Here are some simple examples.

```
> 2 + 2
[1] 4
> (2 + 2)*5
[1] 20
> 2 + 2 * 5
[1] 12
> 2 + 3 * 5^2
[1] 77
> 3^2 + 4^2
[1] 25
> (4/2)^(1/2)
[1] 1.414214
> sqrt(3)
[1] 1.732051
```

After pressing Enter, if you discover you have made a mistake, it may be easier to use the up-arrow key (↑) to retrieve the previous instruction and edit it than to retype everything afresh. Below, after R returns 22 for the sum $1 + 2 + 3 + 4^2$, press ↑. This character does not print on-screen; instead the previous instruction appears at a fresh prompt. Then insert parentheses to obtain $(1 + 2 + 3 + 4)^2$, which equals 100.

```
> 1 + 2 + 3 + 4^2
[1] 22
> (1 + 2 + 3 + 4)^2
[1] 100
```

If you want to save the result of one computation before going on to the next, use the equal sign (=) to store the preliminary result as an *object*. Here the object is a number, which we choose to call **a**.

```
> a = 1 + 2 + 3 + 4
> a
[1] 10
> a^2
[1] 100
> sqrt(a)
[1] 3.162278
```

11.1.3 Defining a Vector

R is an object-oriented language. In this book, the most important objects are vectors. A vector is an ordered list of numbers called elements. The index of an element specifies its position in the list: index 1 for the first element, index 2 for the second, and so on.

Next we show a few ways to specify the elements of a vector. The symbol = is used to assign values. For example, y = 25 means that y is a vector with 25 as its only element. Below we show how this looks in the R Console window. But first we give some specific notes about making and printing vectors, which are also illustrated below.

- In other books containing R code, you may see <- used instead of = as the assignment operator. However, the simpler = now seems to predominate. Ordinarily, R is not fussy about where you put spaces, but if you use <-, then you must type the two symbols without a space between them.
- Names of objects in R are case-sensitive. Thus, y and Y are two different objects. In this book, to avoid confusion, we use mainly lowercase names.
- The name of a vector can be several characters long, but the first character must be a (capital or small) letter. Commonly, the additional characters are letters or numbers, but a period (.) may also be used.
- The symbol c is used to make a vector with several elements. You can think of c as standing for "combine." Because R ordinarily treats vectors as columns of elements (even though vectors are sometimes printed out in rows), you may prefer to think of c as standing for "column vector."
- To print a vector on the screen, just type its name. No special printing command is needed. Semicolons (;) can be used to separate two or more R commands on the same line. For clarity, these should be *closely related* commands. Below, we use a semicolon to define and print a vector in a single line of code.

```
> y = 25
> y
[1] 25
> b <- 1 + 2 + 3;  b
[1] 6
> b.test = b + 1;  b.test
[1] 7
> b.test = 0;  b.test
[1] 0
> Y
Error: Object "Y" not found
> d  <  -  1 + 2
Error: Object "d" not found
> v1 = c(7, 2, 3, 5);  v1
[1] 7 2 3 5
```

Here are a few convenient shorthand notations for making vectors. They are often used to make vectors with elements that follow a particular pattern, and they are especially convenient for long vectors that would be tedious to specify with the c-notation.

- The function `numeric` is often used to "initialize" a long vector containing all 0s that is to be used in a loop. Inside the loop, each element may be changed, one at a time, from 0 to some other value. (More about loops later.)
- The function `rep` (for *repeat*) is used to make a vector in which all elements are equal to a specified value (not necessarily 0).
- The function `seq` (for *sequence*) is used to make a vector that is a sequence of equally spaced values. Alternatively, if successive values differ by 1, such a vector can be defined using a colon (:).
- The c-symbol can be used to combine vectors as well as single numbers to form a longer vector.
- When long vectors are printed, the appearance of output on the screen depends on the width of the Console window, which you can adjust in the usual way (by selecting an edge or corner and dragging it to get the desired dimensions). When a vector is printed, the first line begins with [1]. The number in brackets at the start of each successive line is the index number of the first element printed on that line.

```
> v2 = numeric(4);   v2
[1] 0 0 0 0
> numeric(10)
 [1] 0 0 0 0 0 0 0 0 0 0
> v3 = rep(3, 4);   v3
[1] 3 3 3 3
> rep(4, 3)
[1] 4 4 4
> v4 = 1:4;   v4
[1] 1 2 3 4
> seq(1, 2.2, by=.1)
 [1] 1.0 1.1 1.2 1.3 1.4 1.5 1.6 1.7 1.8 1.9 2.0 2.1
[13] 2.2
> seq(1, 2.2, length=11)
 [1] 1.00 1.12 1.24 1.36 1.48 1.60 1.72 1.84 1.96 2.08
[11] 2.20
> v5 = c(v3, v4, 7);   v5
[1] 3 3 3 3 1 2 3 4 7
```

In each of its uses above, the function `seq` has three arguments. The first argument is the beginning number and the second argument indicates where to stop. These arguments are usually specified *by position*, appearing first and second in order within the parentheses. We prefer to specify the third argument *by name*. In one of the examples above, the additional argument is `by`. It is the increment between successive elements of the vector. In the other example of the function `seq` above, the additional argument is `length.out`, which may be shortened to `length` or even to `len`. It specifies the length of the resulting vector (that is, the number of elements it contains). The value of `by` or `length.out` is specified by using an equal sign (=).

Notes: In specifying arguments of an R function, you may *not* use `<-` as a substitute for `=`. Names of arguments can ordinarily be abbreviated as long as the abbreviation is not so severe as to cause ambiguity.

Problems

11.1   Predict the results of each of the following R statements, and then use R to verify the answers. The last three lines of code illustrate new statements. The R constant `pi` and functions `exp`, `log`, `log10`, `log2`, `sin`, and `tan` used here have their obvious meanings. (Because `pi` is a predefined, *reserved* constant in R, you should never assign another numerical value or vector to it.)

```
2 + 5^2 - 4^2;   2 / (5^2 - 4)^2
2 + (3^2 + 4^2)^(2/4);   2 * (3^2 - 4^2)^2 / 4
exp(1);   exp(1)^2;   exp(2);   log(exp(2))
log10(10^10);   log2(1024);   log10(10^(1:5))
pi;   sin(pi/2);   tan(pi/4)
```

11.2   In working with the following statements, experiment by adjusting the width of the Console window so that the output is easy to read. The last few lines of code illustrate some related ideas that we have not explicitly covered. A string included in quotes (") is a *character* object.

```
numeric(100);   numeric(10);   rep(0, 10);   rep(10, 10)
seq(0, 10);   seq(0, 10.5, by=1);   seq(0, 10, length=11)
0:9.5;   -.5:10;   0:10 - .5;   -1:9 + .5;   seq(-.5, 9.5)
-4:11;   4:-1;   4.5:10;   -4:-11.5
(10:22)/10;   10:22/10;   10/2:22;   (10/2):22
seq(1, 2.2, by=0.1);   seq(by=0.1, to=2.2, from=1)
seq(1, 2.2, length.out=13);   seq(1, 2.2, len=13)
r = 1:5;   s = -2:2;   s/r;   r/s;   s/s
r^0;   s^0;   s^.5;   1000^1000;   ?NaN
rep("I will not eat anchovy pizza in class.", 20)
rep(1:4, times=3);   rep(1:4, each=3)
```

11.3   At the Console > prompt, try typing each of `?Arithmetic`, `?log`, and `?seq` to see examples of the help screens that are immediately available while using R. Explore some of the variations you see there. (You can also use `help(log)` instead of `?log`, and similarly for other expressions.) Unfortunately for our purposes, sometimes you will find that a help screen gives you more technical detail than you want or need to know.

11.2 Using Vectors

In this section, we illustrate simple arithmetic operations with vectors, the use of indices to change and retrieve individual elements of vectors, and procedures for modifying and comparing vectors.

11.2.1 Simple Arithmetic with Vectors

Operations on vectors are performed element by element. In particular, when an operation involves a vector and a number, the number is used to modify each component of the vector as specified by the operation. Here are some examples.

```
> v1;  w1 = 3*v1;  w1
[1] 7 2 3 5
[1] 21  6  9 15
> w2 = v1/2;  w2
[1] 3.5 1.0 1.5 2.5
> w3 = 5 - v1;  w3
[1] -2  3  2  0
> w4 = w3^2;  w4
[1] 4 9 4 0
```

When arithmetic involves two vectors of the *same length*, then the operation applies to elements with the same index. In particular, it is important to remember that multiplication is element-wise. (In R, one can also compute vector products and do matrix multiplication, but we do not discuss these topics in this appendix.)

```
> w5 = w3 + w4;  w5
[1]  2 12  6  0
> (5:1)*(1:5)
[1] 5 8 9 8 5
> (5:0)^(0:5);  (5:1)/(1:5)
[1] 1 4 9 8 1 0
[1] 5.0 2.0 1.0 0.5 0.2
```

When the vectors involved are of *unequal lengths*, then the shorter vector is "recycled" as often as necessary to match the length of the longer vector. If the vectors are of different lengths because of a programming error, this can lead to unexpected results, but sometimes recycling of short vectors is the basis of clever programming. Furthermore, in a technical sense, all of the examples we have shown with operations involving a vector and a single number recycle the single number (which R regards as a one-element vector).

Here are several examples of operations involving vectors of unequal lengths. Notice that R gives a warning message if the recycling comes out "uneven"; that is, the length of the longer vector is not a multiple of the length of the shorter one. A warning message is warranted because this situation often arises because of a programming error.

```
> (1:10)/(1:2)
[1] 1 1 3 2 5 3 7 4 9 5
> (1:10)/(1:5)
[1] 1.000000 1.000000 1.000000 1.000000 1.000000 6.000000
[7] 3.500000 2.666667 2.250000 2.000000
```

```
> (1:10)/(1:3)
 [1]  1.0  1.0  1.0  4.0  2.5  2.0  7.0  4.0  3.0 10.0
Warning message: longer object length
        is not a multiple of shorter object length
        in: (1:10)/(1:3)
```

Problems

11.4   Using the definitions of vectors w3 and w4 given above, predict results and use R to verify your answers for the following: w3-w4, w4-w3, w4*w3, w4^w3, w4/w3, (1:4)^2, (1:4)^(0:3), and (1:2)*(0:3).

11.2.2 Indexes and Assignments

Sometimes we want to use only one element of a vector. To do this, we use the index notation [ ], which can be read as "sub." The simplest version of referencing by index is just to specify the index (position number within the vector) you want. However, the bracket notation can also be used along with the assignment operator = to change the value of an element of an existing vector. Similarly, it is possible to use or change the values of several specified elements of a vector. (We illustrate additional uses for the bracket notation later in this appendix.)

```
> w1;   w1[3]
[1] 21  6  9 15
[1] 9
> w5;   w5[9]
[1] 3 3 3 3 1 2 3 4 7
[1] 7
> v2;   v2[1] = 6;   v2
[1] 0 0 0 0
[1] 6 0 0 0
> v7 = numeric(10);   v7
 [1] 0 0 0 0 0 0 0 0 0 0
> v7[1:3] = 4:6;   v7
 [1] 4 5 6 0 0 0 0 0 0 0
> v7[2:4]
[1] 5 6 0
```

Problems

11.5   Which statements in the following lines of R code produce output? Predict and use R to verify the output.

```
x = 0:10;   f = x*(10-x)
f;   f[5:7]
f[6:11] = f[6];   f
x[11:1]
x1 = (1:10)/(1:5);   x1;   x1[8]
x1[8] = pi;   x1[6:8]
```

11.2.3 Vector Functions

We begin with vector functions that return a single number. The meanings of the functions in the following demonstration should be self-explanatory.

```
> w2;  max(w2)
[1] 3.5 1.0 1.5 2.5
[1] 3.5
> w3;  mean(w3)
[1] -2  3  2  0
[1] 0.75
> v1;  sum(v1)
[1] 7 2 3 5
[1] 17
> v4;  prod(v4)
[1] 1 2 3 4
[1] 24
> v5;  length(v5)
[1] 3 3 3 3 1 2 3 4 7
[1] 9
```

By using parentheses to indicate the sequence of operations, one can combine several expressions that involve functions. For example, here are two ways to find the variance of the elements of w3, followed by the computation of the standard deviation of this sample.

```
> (sum(w3^2) - (sum(w3)^2)/length(w3)) / (length(w3) - 1)
[1] 4.916667
> var(w3)
[1] 4.916667
> sqrt(var(w3));  sd(w3)
[1] 2.217356
[1] 2.217356
```

Some vector functions return vectors when they are applied to vectors. For example, sqrt takes the square root of each element of a vector. Other functions work similarly, for example exp, log, and sin. The function cumsum forms cumulative sums of the elements of a vector. Also, the vector function unique eliminates "redundant" elements in a vector, returning a vector of elements with no repeated values.

```
> sqrt(c(1, 4, 9, 16, 25))
[1] 1 2 3 4 5
> 1:5
[1] 1  2  3  4  5
> cumsum(1:5)
[1]  1  3  6 10 15
> cumsum(5:1)
[1]  5  9 12 14 15
```

```
> v5;   cumsum(v5)
[1] 3 3 3 3 1 2 3 4 7
[1]   3   6   9 12 13 15 18 22 29
> unique(v5)
[1] 3 1 2 4 7
> s = c(rep(3,5), rep(4,10));   s
 [1] 3 3 3 3 3 4 4 4 4 4 4 4 4 4 4
> length(s);   unique(s);   length(unique(s))
[1] 15
[1] 3 4
[1] 2
```

The function **round** with no second argument rounds values to integers.
With a second argument, it rounds to the requested number of digits.

```
> round(2.5)
[1] 2
> round(3.5)
[1] 4
> round(5/(1:3))
[1] 5 2 2
> round(5/(1:3), 3)
[1] 5.000 2.500 1.667
```

Problems

11.6 Predict and verify results of the following. The function **diff** of a
vector makes a new vector of successive differences that has one less element
than the original vector. The last line approximates e^2 by summing the first 16
terms of the Taylor (Maclaurin) series $e^x = \sum_{n=0}^{\infty} x^n/n!$, where $x = 2$.

```
length(0:5);   diff(0:5);   length(diff(0:5));   diff((0:5)^2)
x2 = c(1, 2, 7, 6, 5);   cumsum(x2);   diff(cumsum(x2))
unique(-5:5);   unique((-5:5)^2);   length(unique((-5:5)^2))
prod(1:5);   factorial(5);   factorial(1:5)
exp(1)^2;   a1 = exp(2);   a1
n = 0:15;   a2 = sum(2^n/factorial(n));   a2;   a1 - a2
```

11.7 The functions **floor** (to round down) and **ceiling** (to round up)
work similarly to **round**. You should explore these functions, using the vectors
shown above to illustrate **round**. See **?round** for other related functions.

11.2.4 Comparisons of Vectors

If two vectors are compared element by element, the result is a logical vector
that has elements TRUE and FALSE. Common comparison operators are ==
(equal), < (less than), <= (less than or equal to), != (not equal), and so on.
It is important to distinguish between the *assignment* operator = and the

comparison operator ==. Also, distinguish between > used as a prompt in the R Console window and as a comparison operator. Here are some examples. (Also, see the help screen at ?Comparison.)

```
> 1:5 < 5:1
[1]  TRUE  TRUE FALSE FALSE FALSE
> 1:5 <= 5:1
[1]  TRUE  TRUE  TRUE FALSE FALSE
> 1:5 == 5:1
[1] FALSE FALSE  TRUE FALSE FALSE
> 1:4 > 4:1
[1] FALSE FALSE  TRUE  TRUE
> 1:5 < 4
[1]  TRUE  TRUE  TRUE FALSE FALSE
> w4;  x3 = (w4 == 4);  x3
[1] 4 9 4 0
[1]  TRUE FALSE  TRUE FALSE
```

If R is coerced (that is, "forced") to do arithmetic on logical values, then it takes TRUE to be 1 and FALSE to be 0. The mean of a vector of 0s and 1s is the proportion of 1s in the vector. So the first result below shows that half of the elements of the numerical vector w4 are equal to 4; equivalently, half of the elements in the logical vector x3 are TRUE. The symbols T and F are reserved as abbreviations of TRUE and FALSE, respectively; they must never be used for other purposes.

```
> mean(x3)
[1] 0.5
> sum(c(T, T, F, F, T, T))
[1] 4
> mean(c(T, T, F, F, T, T))
[1] 0.6666667
```

Comparisons can be used inside brackets to specify particular elements of a vector. In such instances, it is convenient to read [ ] as "such that." From three of the statements below, we see different ways to show that exactly two elements of v5 are smaller than 3. The two statements on the last line are logically equivalent.

```
> v5;  v5[v5 < 3]
[1] 3 3 3 3 1 2 3 4 7
[1] 1 2
> length(v5[v5 < 3]);  sum(v5 < 3)
[1] 2
[1] 2
```

Up to this point, we have illustrated many of the fundamental rules about vector operations in R by focusing on individual statements. In the rest of this appendix, we begin to explore some elementary applications that involve putting statements together into brief "programs." If a program has more than a few lines, you may want to prepare it in a text editor (for example, Word-pad), "cut" multiple lines from there, and "paste" them at an R prompt (>). Treat each line as a paragraph; do not begin lines with >. (R has a built-in editor, but its unsaved contents will be lost if you make a mistake and crash R.)

Problems

11.8   Predict and verify the results of the following statements. Which ones produce output and which do not? The last two lines illustrate a *grid search* to approximate the maximum value of $f(t) = 6(t - t^2)$, for t in the closed interval $[-1, 1]$. Show that if t is chosen to have length 200, instead of 201, then the result is neither exact nor unique.

```
x4 = seq(-1, 1, by = .1);   x5 = round(x4);   x4;   x5
unique(x5);   x5==0;   x5[x5==0];   sum(x4==x5)
sum(x5==0);   length(x5);   mean(x5==0);   x5 = 0;   x5
t = seq(0, 1, len=201);   f = 6*(t - t^2)
mx.f = max(f);   mx.f;   t[f==mx.f]
```

11.3 Exploring Infinite Sequences

Of course, all vectors in R are of finite length. However, the behavior of an infinite sequence can often be illustrated or explored by looking at a sufficiently long finite vector.

Example 11.1. The Sum of the First n Positive Integers. One can show by mathematical induction that the sum s_n of the first n positive integers is $s_n = n(n+1)/2$. For example, $s_5 = \sum_{i=1}^{5} i = 1+2+3+4+5 = 15 = 5(6)/2$. For values of n up to 50, we can use R to illustrate this theorem with the following code.

```
> n = 1:50
> s = n*(n+1)/2
> cumsum(n)
 [1]    1    3    6   10   15   21   28   36   45   55
[11]   66   78   91  105  120  136  153  171  190  210
[21]  231  253  276  300  325  351  378  406  435  465
[31]  496  528  561  595  630  666  703  741  780  820
[41]  861  903  946  990 1035 1081 1128 1176 1225 1275
> s
 [1]    1    3    6   10   15   21   28   36   45   55
[11]   66   78   91  105  120  136  153  171  190  210
[21]  231  253  276  300  325  351  378  406  435  465
[31]  496  528  561  595  630  666  703  741  780  820
[41]  861  903  946  990 1035 1081 1128 1176 1225 1275
> mean(s == cumsum(n))
[1] 1
```

The last result indicates 100% agreement between two vectors: s, illustrating the formula for the sums, and cumsum(n), representing the sums themselves. In the last line, the comparison produces a logical vector of length 50. Taking the mean coerces its values to be numerical (0s or 1s, here all 1s). The mean of a sequence of 0s and 1s is the proportion of 1s. The demonstration below, for values of n up to 50 000, is similar.

```
> n = 1:50000
> s = n*(n + 1)/2
> mean(s == cumsum(n))
[1] 1
```

But in this case it is impractical to compare all the results by visual inspection, so the programmed comparison becomes crucial. ◇.

Of course, a demonstration in R is not a substitute for a formal proof. But if you had not seen the formula of Example 11.1 before and wondered whether it is true, this demonstration in R might encourage you that the result is credible and that it is worthwhile trying to construct a formal proof.

Example 11.2. A Sequence with Limit e. Using methods of calculus one can show that $e = 2.71828...$ is the limit of the infinite sequence $a_n = (1 + 1/n)^n$, where $n = 1, 2, 3, \ldots$. In R we can show how close to e the sequence becomes as n increases to 10 000. In the code below, the function cbind binds two column vectors together to make a matrix with 10 000 rows and two columns. For compact output, the expression in brackets prints out only a few designated rows of the matrix. That nothing follows the comma within the square brackets indicates that all (both) rows of the matrix are to be printed.

```
> n = 1:10000;   a = (1 + 1/n)^n
> cbind(n, a)[c(1:5, 10^(1:4)), ]
          n        a
[1,]      1 2.000000
[2,]      2 2.250000
[3,]      3 2.370370
[4,]      4 2.441406
[5,]      5 2.488320
[6,]     10 2.593742
[7,]    100 2.704814
[8,]   1000 2.716924
[9,] 10000 2.718146
```

The sequence a_n is monotone increasing (that is, each term in the sequence is larger than the one before). This can be illustrated by taking successive differences of the vector a. The result is a vector of length 9999, which can be used to show that all differences are positive. We also see that $a_{10\,000}$ provides an approximation to e that is accurate to three decimal places.

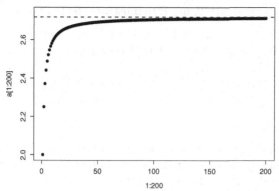

Figure 11.1. A plot of the first 200 terms of the sequence $a_n = (1 + 1/n)^n$, which has $\lim_{n \to \infty} a_n = e$ (indicated by the horizontal dashed line). See Example 11.2.

```
> da = diff(a)
> da[1:10]
 [1] 0.25000000 0.12037037 0.07103588 0.04691375
 [5] 0.03330637 0.02487333 0.01928482 0.01539028
 [9] 0.01256767 0.01045655
> mean(da > 0)
[1] 1
> exp(1) - a[10000]
[1] 0.0001359016
```

Throughout this book, we see that the extensive graphical capabilities of R are especially useful in probability and statistics. In particular, the `plot` function plots one vector against another vector of equal length. Here we use it to plot the first 200 values of a_n against the numbers $n = 1, 2, \ldots, 200$.

```
> plot(1:200, a[1:200], pch=19, main="A Sequence That Approaches e")
> abline(h = exp(1), col="darkgreen", lwd=2, lty="dashed")
```

The default plotting character in R is an open circle. Here we choose to substitute a smaller solid circle by specifying its code number 19. Also by default, R supplies a main "header" for the plot and labels for the axes, all based on names of the vectors plotted. Here we choose to specify our own more descriptive main label. Finally, we use the function `abline` to overlay a horizontal reference line at e, which we choose to draw twice the normal thickness (parameter `lwd`), dashed (`lty`), and in green (`col`). Results are shown in Figure 11.1. (Colors are not shown in print.) Within the resolution of the graph, it certainly looks as if the sequence a_n is approaching e. ◊

Problems

11.9   With $s_n = n(n+1)/2$, use R to illustrate the limit of s_n/n^2 as $n \to \infty$.

11.10   Modify the R code of Example 11.2, including the numerical and graphical displays, to illustrate that the sequence $b_n = (1 + 2/n)^n$ converges to e^2.

11.4 Loops

Because R handles vectors as objects, it is often possible to avoid writing loops in R. Even so, for some computations in R it is convenient or necessary to write loops, so we give a few examples.

Example 11.3. A Loop to Find the Mean of a Vector. We have already seen that the mean function can be used to find that the mean of the numbers $-2, 3, 2, 0$ is $3/4$ or 0.75. Recall from above that the vector w3 contains the desired elements and the statement mean(w3) returns 0.75.

Now, as a "toy" example, we show how we could find the mean of w3 using a loop. We initialize the object avg to have the value 0. Then, on each passage through the loop, we add (cumulatively) one more element of w3. Finally, we divide the accumulated total by the number of elements of w3 to find the mean of w3. We write this "program" without prompt characters (>) to make a clear distinction between the program and its output.

```
w3 = c(-2, 3, 2, 0)  # omit if w3 defined earlier in your R session
avg = 0              # when program finishes, this will be the mean
n = length(w3)
for (i in 1:n)
{
    avg = avg + w3[i]
}
avg = avg/n;  avg

> avg
[1] 0.75
```

Notice that R ignores comments following the symbol #. These are intended to be read by people. Because the mean function, which is widely used in statistics, is already defined in R (in terms of an internal loop), this simple program is only for purposes of illustration. ◇

Problems

11.11   How many times does the program in Example 11.3 execute the statement within the loop (between curly brackets { })? As you follow through the entire program, list all of the values assumed by avg.

Example 11.4. A Loop to Print Fibonacci Numbers. As far as we know, R does not come preprogrammed to print Fibonacci numbers—probably because they are not frequently used in probability and statistics. The following loop illustrates how one can find the first 30 elements of the Fibonacci sequence, $1, 1, 2, 3, 5, 8, 13, \ldots$. In this sequence, each element is the sum of the two previous elements. (In *Fi'·bo·na''·cci*, the *cci* sounds like *chee* in *cheese*.)

We begin by saying how many elements we want to display and initializing a vector fibo of appropriate length, which contains all 0s. Then we supply the two starting numbers of the sequence (two 1s) and use a loop to generate the rest. On each passage through the loop, the ith element of fibo is changed from its initial value of 0 to the appropriate Fibonacci number.

```
m = 30;  fibo = numeric(m);  fibo[1:2] = 1
for (i in 3:m)
{
  fibo[i] = fibo[i-2] + fibo[i-1]
}
fibo

> fibo
 [1]     1     1     2     3     5     8    13    21
 [9]    34    55    89   144   233   377   610   987
[17]  1597  2584  4181  6765 10946 17711 28657 46368
[25] 75025 121393 196418 317811 514229 832040
```

The sequence of ratios, obtained by dividing each Fibonacci number by the previous one, rapidly approaches the so-called Golden Ratio 1.618 used in ancient Greek architecture. The following four additional lines of R code find these ratios, print the $(m-1)$st value, and make a figure that uses the first 15 ratios to illustrate the rapid approach to a limit.

```
golden = fibo[2:m]/fibo[1:(m-1)]
golden[m-1]
plot(1:15, golden[1:15],
   main="Fibonacci Ratios Approach Golden Ratio")
abline(h=golden[m-1], col="darkgreen", lwd=2)

> golden[m-1]
[1] 1.618034
```

See Figure 11.2 for the plot. We could have found the ratios inside the loop, but in this book we make it a practice to put as little as possible inside loops and use the vector arithmetic built into R to do the rest. Also notice that it is alright to break a long R statement from one line to another—as long as the statement is syntactically complete only on the last line. Below, the plot statement is not complete until its right parenthesis) on the second line is encountered, matching the left parenthesis (at the beginning.

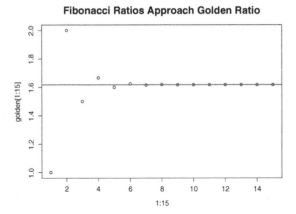

Figure 11.2. Rapid approach to a limit of the Fibonacci ratios. The Fibonacci numbers are generated using a loop in Example 11.4.

The Fibonacci sequence is not used elsewhere in this book, but if you are curious to know more about it, you can google "Fibonacci" to retrieve lots of comments on Fibonacci numbers. On the Internet, any fool fascinated by a topic can pose as an expert. Comments about mathematics posted on sites of universities, professional societies, and governmental organizations are generally most likely to be factual—but still be alert for errors. ◇

11.5 Graphical Functions

An important feature of this book is its many illustrative figures made using R. We have already seen a few of the extensive graphical capabilities of R. In the rest of this appendix, we introduce a few additional graphical methods.

Throughout the book, we show the most elemental code for making almost all of the figures. However, in order to make figures that are effective and attractive in print, we frequently needed to embellish the elemental code, and we omit details we feel many readers would find irrelevant or distracting. In contrast, in this appendix we show (almost) the full code for all the figures, and on the website that supports this book, we show more detailed code for most figures.

- Except in this appendix, we do not usually show how to make main headers or to label axes. In the main text, we mainly rely on captions to label and explain figures. We show code for main headers in this appendix because you may find it convenient to use headers in homework or projects.
- We seldom show parameters to govern the thickness (width) or style (dashed, dotted, etc.) of lines or to produce colors. These omissions mean

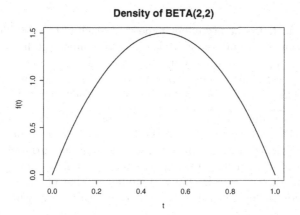

Figure 11.3. Graph of a continuous density curve, rendered by plotting many short line segments. See Example 11.5

that much of our elemental R graphics code also works with little or no modification in S-PLUS.

- To see an extensive list of colors that can be used for histogram bars, lines, and other features of graphs, type `colors()` at the prompt in the R Console. (If you want to make your own colors, see `?par`, about 80% of the way down under "Color Specification.")

Example 11.5. Plotting a "Continuous" Function. The `plot` function can be used to give the appearance of a smooth continuous function. For example, the following code graphs the parabola $f(t) = 6t(1 - t)$, where $0 \le t \le 1$. The argument `type = "l"` of the `plot` function connects 200 dots with line segments to make what is, for practical purposes, a smooth curve. (The symbol inside quotes is the letter *ell*, for *lines*—not the number *one*.) Use `?plot` to see other choices that can be specified for `type`. Also, we include additional arguments here to make the line double width and blue. Ordinarily, it is understood that the first argument is for the x-axis and the second for the y-axis. We think it is best to designate additional arguments by *name* (as with `type`, `lwd`, and so on), rather than by *position* (as for the first and second arguments).

```
t = seq(0, 1, length=200)
f = 6*(t - t^2)
plot(t, f, type="l", lwd=2, col="blue", ylab="f(t)",
    main="Density of BETA(2,2)")
```

The resulting plot is shown in Figure 11.3. ◇

Problems

11.12   (a) Without the argument `type="l"`, each of the 200 plotted points would be shown as a small open circle, with no connecting lines. Modify the program of Example 11.5 and try it. (b) If the first argument is also omitted, the remaining vector argument is plotted on the y-axis against the index on the x-axis. To illustrate, with `f` defined as above, show the result of `plot(f)`.

11.13   (a) The function $f(t) = 6t(1-t)$, for $0 < t < 1$, and 0 otherwise, is never negative and (together with the t-axis) it encloses unit area. Thus it qualifies as a density function of a continuous random variable—specifically a random variable with distribution BETA(2, 2). Make a plot of the density function of the distribution BETA(3, 2) similar to that of Example 11.5. The density function of the general beta distribution BETA(α, β) has the form

$$f(t) = \frac{\Gamma(\alpha + \beta)}{\Gamma(\alpha)\Gamma(\beta)} t^{\alpha-1}(1-t)^{\beta-1},$$

for t in the unit interval $(0, 1)$. Here we use the gamma function, which has $\Gamma(n) = (n-1)!$ for positive integer n and can be evaluated more generally in R with the function `gamma`. (b) Execute `gamma(seq(1, 4, by=.2))` and `factorial(0:3)`, and explain the results. (c) Use a grid search to find the mode of BETA(3, 2) (that is, the value of t at which the density function has its maximum value), and then verify the result using calculus.

Example 11.6. Making a Histogram of Data. Another important graphical function is the histogram function `hist`. As an illustration, suppose the IQ scores of 20 elementary school students are listed in the vector `iq` below. There are cleverer ways to import data into R, but simply making a list with the c-operator suffices throughout this book. The second statement below makes a suitably labeled histogram of these data.

```
iq <- c( 84, 108,  98, 110,  86, 123, 101, 114, 121, 131,
         90, 108, 105,  93,  95, 102, 119,  98,  94,  73)
hist(iq, xlab="IQ", col="wheat", label=T,
   main=paste("Histogram of IQ Scores of", length(iq), "Students"))
```

This histogram is shown in Figure 11.4. We have used the data to generate one part of the main header with the character function `paste` and used the argument `labels=T` to print counts atop the bars. The argument `col="wheat"` fills bars with a designated color, one of many predefined in R. ◇

Problems

11.14   In Example 11.6, look at the entries in the data vector `iq` to verify the heights of each of the histogram bars. That is, determine the number of observations in each of the intervals $[70, 80], (80, 90], \ldots (130, 140]$. Looking at the result of `sort(iq)` may help.

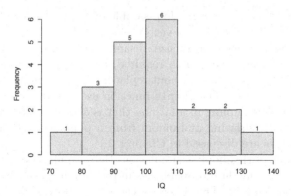

Figure 11.4. Histogram of IQ scores of students. This histogram illustrates several embellishments that are not defaults in R. See Example 11.6.

11.15 (a) In Example 11.6, what is the effect of omitting the arguments for xlab and col in the histogram statement? (b) What happens if you remove the argument labels=T? (c) What happens if you specify prob=T?

11.16 An algorithm in R determines, by default, how many intervals are used and what cutpoints (or **breaks**) separate the intervals. For many kinds of data, as in Example 11.6, this algorithm gives reasonable results. However, the **breaks** argument allows you to control what intervals are used. Experiment with hist(iq, breaks=10) to get a histogram with *roughly* ten bars, and hist(iq, breaks=seq(64.5, 134.5, by=10)) to specify exact cutpoints.

11.17 (a) A histogram should appear to "balance" at the sample mean (considering the bars to have "weights" in proportion to their heights). Estimate the mean of the 20 IQ scores from the histogram in Figure 11.4, and then use mean(iq) to find the exact sample mean $\bar{X}$ of these data. Also, use sd(iq) to find the standard deviation S, and perhaps use the formula for the sample standard deviation to get the same result with a hand calculator. (b) The so-called Empirical Rule suggests that, for many samples, all or almost all of the observations lie within an interval $\bar{X} \pm 3S$, extending three sample standard deviations on either side of the sample mean. Is that true here? (c) Also, find and explain the results of summary(iq).

11.18 In Example 11.6, we used the character function paste to put the number length(iq) into the header of the histogram. Similarly, use paste with round(golden[m-1], 3) in the program of Example 11.4 to put the value of the Golden Ratio into the main header of the plot.

11.6 Sampling from a Finite Population

The `sample` function selects a random sample of a specified size from a given population. Its first argument is a vector whose elements are the population. Its second argument is the sample size. If sampling is to be done with replacement, a third argument `repl=T` is required, otherwise sampling is without replacement. (When sampling without replacement, the sample size can't exceed the population size.) The `sample` function uses a pseudorandom number generator to do the sampling in a fashion that is almost impossible to distinguish from actually drawing at random from a population. (Pseudorandom number generators are discussed in Chapter 2.)

A major topic of this book is simulation. Here we illustrate the use of the `sample` function to simulate probabilities in two familiar random processes—dealing cards and rolling dice. Because the correct probabilities are known from simple combinatorics in these elementary examples, you can verify that the simulated results are accurate enough for practical purposes.

Example 11.7. Poker Hands. To deal a random poker hand, we take the population to be the 52 cards of a standard deck, from which we want to select 5 cards at random (without replacement, of course). The code in the first line below does the sampling. Because this is a random process, your answer will (almost certainly) not be the same as the one shown below.

In order to associate the output with actual playing cards, we imagine assigning numbers $1, 2, \ldots, 52$ to the 52 cards. If we want to count the Aces in a 5-card poker hand, it is convenient to associate the numbers $1, 2, 3, 4$ with the four Aces: $A\heartsuit, A\diamondsuit, A\clubsuit, A\spadesuit$, respectively. According to this convention, there is exactly one Ace in the particular result shown below. In general, the number of Aces can be determined in R by using the expression `sum(h < 5)`.

```
> h = sample(1:52, 5);   h
[1] 36 10  2 39 26
> h < 5;   sum(h < 5)
[1] FALSE FALSE  TRUE FALSE FALSE
[1] 1
```

If you repeat this random experiment several times in R, you will find that you often get no Aces in your randomly dealt hand and more rarely get one or several. What are the possible values that might be returned by `sum(h < 5)`? What would happen if you tried to sample 60 cards instead of five from a 52-card deck? Try it in R and see.

The program below uses a loop to "deal" 100 000 five-card hands, counts the Aces in each hand, and makes a histogram of the resulting 100 000 counts. In the `hist` function, we use two arguments to modify the default version.

- The argument `prob=T` puts "densities" (that is, relative frequencies or proportions out of 100 000) on the vertical axis of the histogram. These relative frequencies simulate the probabilities of getting various numbers of Aces.

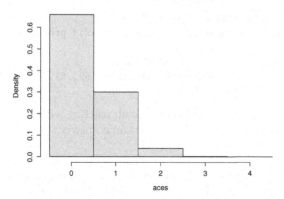

Figure 11.5. Simulated distribution of the number of Aces in a randomly dealt poker hand. See Example 11.7 for the simulation.

- An argument to specify the parameter **breaks** improves the break points along the horizontal axis to make a nice histogram. (Otherwise, R uses its own algorithm to make the breaks—an algorithm ill-suited to our integer data with many repeated values.)

Results are shown in Figure 11.5. At the end of the program, we tally how many times each number of Aces was observed. The command **as.factor** interprets **aces** as a vector of categories to be *tallied* by the **summary** statement. (What do you get from just **summary(aces)**?) Again, comments following the symbol # are explanatory annotations intended for human readers and are ignored by R.

```
set.seed(1234)
m = 100000
aces = numeric(m)          # m-vector of 0s to be modified in loop

for (i in 1:m)
{
  h = sample(1:52, 5)
  aces[i] = sum(h < 5)     # ith element of 'aces' is changed
}

cut = (0:5) - .5
hist(aces, breaks=cut, prob=T, col="Wheat",
     main="Aces in Poker Hands")
summary(as.factor(aces))        # observed counts
summary(as.factor(aces))/m      # simulated probabilities
round(choose(4, 1)*choose(48, 4)/choose(52, 5), 4)  # P{1 Ace}
```

```
> summary(as.factor(aces))          # observed counts
    0     1     2     3     4
65958 29982  3906   153     1
> summary(as.factor(aces))/m        # simulated probabilities
      0       1       2       3       4
0.65958 0.29982 0.03906 0.00153 0.00001
> round(choose(4, 1)*choose(48, 4)/choose(52, 5), 4)   # P{1 Ace}
[1] 0.2995
```

Every time you run a simulation program starting with the same seed and using the same software, you will get the same answer. ◇

Problems

11.19 Run the program of Example 11.7 for yourself. In this book, we often show the seed we used to get the sample results we print. This makes it possible for you to compare your results with ours. (Seeds are explained in more detail in Chapter 2.) In contrast, if you omit the `set.seed` statement when you run the program, you will get your own simulation with slightly different results. Your exact counts will differ, but your histogram should not look much different from ours. Find results from two runs.

11.20   The probability of obtaining exactly one Ace in a five-card poker hand is 0.2995 (correct to four places). This probability is obtained from the formula $\binom{4}{1}\binom{48}{4}/\binom{52}{5}$, where $\binom{n}{r} = \frac{n!}{r!(n-r)!}$, which is known as the number of combinations of n objects taken r at a time. In R, numbers of combinations can be evaluated using the function **choose**, as shown in the last line of code above. Starting with $r = 0{:}4$, write an additional statement, with three uses of **choose**, that gives exact values of all five probabilities simulated in Example 11.7.

Note: The probability of getting four Aces in a five-card hand is very small indeed: 1.847×10^{-5}. In Problem 11.20, you used combinatorics to verify this result. This probability is too small to be evaluated satisfactorily by our simulation—except to see that it must be very small. The absolute error is small, but the relative error is large. (Perhaps getting four Aces is somewhat more likely than this in amateur poker games where the cards left over from a previous game may not be sufficiently shuffled to imitate randomness as well as does the pseudorandom number generator in R. It takes about eight "riffle shuffles" done with reasonable competence to put a deck of cards into something like random order.)

Example 11.8. Rolling Dice. The code below simulates rolling two dice. For each roll, the population is `1:6`. Because it is possible for both dice to show the same number, we are sampling with replacement.

```
> d = sample(1:6, 2, repl=T);  d
[1] 6 6
```

In our run of this code in R, we happened to get two 6s. If you repeat this experiment several times, you will find that you do not usually get the same number on both dice.

What is the probability of getting the same number on both dice? You could determine whether this happens on a particular simulated roll of two dice with the code `length(unique(d))`. What would this code return in the instance shown above? In general, what are the possible values that might be returned?

Here is a program that simulates the sums of the numbers in 100 000 two-dice experiments. It tallies the results and makes a histogram (not shown) of simulated probabilities.

```
set.seed(1212)
m = 100000;   x = numeric(m)

for (i in 1:m)
{
  x[i] = sum(sample(1:6, 2, repl=T))
}

cut = (1:12) + .5;   header="Sums on Two Fair Dice"
hist(x, breaks=cut, prob=T, col="wheat", main=header)
summary(as.factor(x))

> summary(as.factor(x))
    2     3     4     5     6     7     8     9    10    11    12
 2793  5440  8428 11048 13794 16797 13902 11070  8337  5595  2796
```

Without looking at the program below, can you give the exact probabilities corresponding to each of the possible sums $2, 3, \ldots, 12$? (Hint: Each result can be expressed as a fraction with denominator 36.)

In this simulation of rolls of pairs of dice, we can avoid writing a loop, as shown in the program below. Imagine that one of the dice is red and the other is green. The first elements of the vectors `red` and `green` contain the results of the first roll, and the first element of the vector x contains the total number of spots showing on the first roll of the two dice. Each of these three vectors is of length 100 000.

```
set.seed(2008)
m = 100000
red = sample(1:6, m, repl=T);   green = sample(1:6, m, repl=T)
x = red + green
summary(as.factor(x))
sim = round(summary(as.factor(x))/m, 3)
exa = round(c(1:6, 5:1)/36, 3);   rbind(sim, exa)
hist(x, breaks=(1:12)+.5, prob=T, col="wheat",
     main="Sums on Two Fair Dice")
points(2:12, exa, pch=19, col="blue")
```

Sums on Two Fair Dice

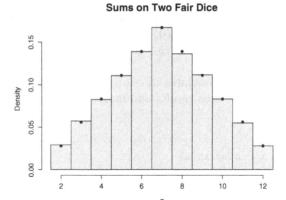

Figure 11.6. Histogram of the simulated distribution of the number of spots seen on a roll of two fair dice. The dots atop the histogram bars show exact probabilities. See Example 11.8.

```
> summary(as.factor(x))
    2     3     4     5     6     7     8     9    10    11    12
 2925  5722  8247 11074 13899 16716 13633 11166  8309  5495  2814
> sim = round(summary(as.factor(x))/m, 3)
> exa = round(c(1:6, 5:1)/36, 3);  rbind(sim, exa)
        2     3     4     5     6     7     8     9    10    11    12
sim 0.029 0.057 0.082 0.111 0.139 0.167 0.136 0.112 0.083 0.055 0.028
exa 0.028 0.056 0.083 0.111 0.139 0.167 0.139 0.111 0.083 0.056 0.028
```

Figure 11.6 shows the simulated probabilities (histogram) and the corresponding exact values (dots atop histogram bars). ◊

In the example above, we say that the second dice program is "vectorized" because it uses the capability of R to work efficiently with vectors in order to avoid writing an explicit loop. The second program runs faster because R executes its internally programmed loops for vector arithmetic faster than loops you write yourself.

Generally speaking, expert R programmers prefer to use vectorized code whenever possible. In this book, we occasionally write loops when we might have used vectorized code—especially if we think the code with loops will be substantially easier for you to understand or to modify for your own purposes.

Problems

11.21   (a) After you run the program in Example 11.8 with the seed shown, use the code cbind(red, green, x)[1:12,] to look at the first 12 elements of each vector. On the first dozen simulated rolls of two dice, how many times did you see a total of 7 spots? What is the "expected" result? (b) Also, consider these issues of syntax: What happens if you omit the comma within the square brackets? Explain the result of `mean(green==red)`. Why does `mean(green=red)` yield an error message?

11.22   (Continuation of Problem 11.20) As illustrated in the program of Example 11.8, a statement with `points` can be used to overlay individual dots onto a histogram (or other graph). Use `points`, together with the expression you wrote in Problem 11.20, to overlay dots at exact probabilities onto the histogram of Example 11.7. For this to work, why is it necessary to use `prob=T` when you make the histogram?

References

[AS64]   Abramowitz, Milton; Stegun, Irene A. (eds): Handbook of Mathemat-
            ical Functions with Formulas, Graphs, and Mathematical Tables, Na-
            tional Bureau of Standards, U.S. Government Printing Office, Wash-
            ington (1964) (10th printing, with corrections (1972)).

[AC98]   Agresti, Alan; Coull, Brent A.: Approximate is better than "exact"
            for interval estimation of binomial proportions, The American Statis-
            tician, **52**, 119–126 (1998).

[Bar88]   Barnsley, Michael: Fractals Everywhere, Academic Press, San Diego
            (1988)

[BKM55]   Blumen, Isadore; Kogan, Marvin; McCarthy, Phillip: Industrial Mobil-
            ity of Labor as a Probability Process, Cornell University Press, Ithaca,
            NY (1955).

[BT73]   Box, G.E.P.; Tiao, George: Bayesian Inference and Statistical Analy-
            sis, Addison-Wesley, Reading, MA (1973).

[Boy99]   Boyd, L.M.: The Grab Bag (syndicated newspaper column), The San
            Francisco Chronicle (July 17, 1999).

[BCD01]   Brown, Lawrence D.; Cai, T. Tony; DasGupta, Anirban: Interval esti-
            mation for a binomial proportion, Statistical Science, **16**, No. 2, 101–
            133 (2001).

[Bro65]   Brownlee, K.A.: Statistical Theory and Methodology in Science and
            Engineering, 2nd ed., Wiley, New York (1965).

[CG94]   Chib, S.; Greenberg, E.: Understanding the Metropolis-Hastings algo-
            rithm, The American Statistician, **49** 327–335 (1994).

[Con01]   Congdon, Peter: Bayesian Statistical Modelling, John Wiley & Sons
            Ltd., West Sussex, England (2001).

[Con03]   Congdon, Peter: Applied Bayesian Modelling, John Wiley & Sons Ltd.,
            West Sussex, England (2003).

[CJ89]   Courant, Richard; John, Fritz: Introduction to Calculus and Analysis,
            Vol. II. Springer-Verlag, New York (1989).

[CM65]   Cox, David R.; Miller, H.D.: The Theory of Stochastic Processes,
            Wiley, New York (1965). (The Tel Aviv weather data are summarized
            in Example 3.2.)

E.A. Suess and B.E. Trumbo, *Introduction to Probability Simulation and Gibbs Sampling with R*,   301
Use R!, DOI 10.1007/978-0-387-68765-0, © Springer Science+Business Media, LLC 2010

[Dav57]   Davies, O. L. (editor): Statistical Methods in Research and Production, 3rd ed., Oliver and Boyd, Edinburgh (1957).

[DVB85]   DeLong, Elizabeth R.; Vernon, W. Benson; Bollinger, R. Randall: Sensitivity and specificity of a monitoring test, Biometrics, **41**, 947–958 (December 1985).

[DEKM98]   Durbin, Richard; Eddy, Sean; Krogh, Anders; Michison, Graeme: Biological Sequence Analysis—Probabilistic Models of Proteins and Nucleic Acids, Cambridge University Press, Cambridge (1998).

[Fel57]   Feller, William: An Introduction to Probability Theory and Its Applications, Vol. 1, 2nd ed., Wiley, New York (1957). (Simulated coin tossing introduced in the 2nd ed.; see also 3rd ed. (1968).)

[Fry84]   Frydman, Halina: Maximum likelihood estimation for the mover-stayer model, Journal of the American Statistical Association, **79**, 632–638 (1984).

[Gas87]   Gastwirth, Joseph L.: The statistical precision of medical screening procedures: Applications to polygraph and AIDS antibody test data (including discussion), Statistical Science, **2**, 213–238 (1987).

[GS85]   Gelfand, Alan E.; Smith, A.F.M.: Sampling-based approaches to calculating marginal densities, Journal of the American Statistical Association, **85**, 398–409 (1990).

[GCSR04]   Gelman, Andrew; Carlin, John B.; Stern, Hal S.; Rubin, Donald B.: Bayesian Data Analysis, 2nd ed., Chapman & Hall/CRC, Boca Raton, FL (2004).

[GH07]   Gelman, Andrew; Hill, Jennifer: Data Analysis Using Regression and Multilevel/Hierarchical Models, Cambridge University Press, New York (2007).

[Gen98]   Gentle, James E.: Random Number Generation and Monte Carlo Methods, Springer, New York (1998).

[Goo61]   Goodman, Leo A.: Statistical methods for the mover–stayer model, Journal of the American Statistical Association, **56**, 841–868 (1961).

[Goo88]   Goodman, Roe: Introduction to Stochastic Models, Benjamin/ Cummings, Menlo Park, CA (1988).

[HRG83]   Hare, Tom; Russ, John; Gaulkner, Gary: A new pseudo-random generator, Call—A.P.P.L.E., 33–34 (January 1983).

[Has55]   Hastings, C., Jr.: Approximations for Digital Computers, Princeton University Press, Princeton, NJ (1955).

[HF94]   Herzog, B.; Felton, B.: Hemoglobin screening for normal newborns, Journal of Perinatology, 14, No. 4, 285-289 (1994).

[KG80]   Kennedy, William J., Jr.; Gentle, James E: Statistical Computing, Marcel Dekker, New York (1980).

[KvG92]   Kleijnen, Jack; van Groenendaal, Willem: Simulation: A Statistical Perspective. Wiley, Chichester (1992).

[LEc01]   L'Ecuyer, Pierre: Software for uniform random number generation: Distinguishing the good and the bad, Proceedings of the 2001 Winter Simulation Conference, IEEE Press, 95-105 (December 2001).

[Lee04]   Lee, Peter M.: Bayesian Statistics, An Introduction, 3rd ed., Hodder Arnold, London (2004).

[Liu01]   Liu, Jun S.: Monte Carlo Strategies in Scientific Computing. Springer, New York (2001).

[Man02]   Mann, Charles C.: Homeland insecurity, The Atlantic Monthly, 81–102
            (September 2002). (The quote in Problem 5.2 is from page 94.)

[MR58]   Majumdar, D.M.; Rao, C.R.: Bengal Anthropometric Survey, 1945: A
            statistical study, Sankhya, **19**, Parts 3 and 5 (1958).

[Moo06]   Moore, David: Basic Practice of Statistics, 4th ed. W. H. Freeman,
            New York (2006).

[Mor82]   Morton D.; Saah, A.; Silberg, S.; Owens, W.; Roberts, M.; Saah, M.:
            Lead absorption in children of employees in a lead-related industry,
            American Journal of Epidemiology, **115**, 549–555 (1982).

[Nun92]   Nunnikhoven, Thomas S.: A birthday problem solution for nonuniform
            frequencies, The American Statistician, **46**, 270–274 (1992).

[PG00]   Pagano, Marcello; Gauvreau, Kimberlee: Principles of Biostatistics,
            2nd ed., Duxbury Press, Belmont, CA (2000). (See Chapter 6 for a
            discussion of diagnostic tests.)

[Pit93]   Pitman, Jim: Probability, Springer, New York (1993).

[PC00]   Pitman, Jim; Camarri, Michael: Limit distributions and random trees
            derived from the birthday problem with unequal probabilities, Elec-
            tronic Journal of Probability, **5**, Paper 2, 1–18 (2000).

[Por94]   Portnoy, Stephen: A Lewis Carroll Pillow Problem: Probability of an
            obtuse triangle, Statistical Science, **9**, 279–284 (1994).

[PTVF92]   Press, William H.; Teukolsky, Saul A.; Vetterling, William T.; Flan-
            nery, Brian B: Numerical Recipes in C: The Art of Scientific Comput-
            ing, 2nd ed. Cambridge University Press, Cambridge (1992).
            (Also available at `lib-www.lanl.gov/numerical/bookcpdf.html`,
            website of Los Alamos National Laboratory.)

[RAN55]   The RAND Corporation: A Million Random Digits with 100 000 Nor-
            mal Deviates, Free Press, Glencoe, IL (1955). (Also available at
            `www.rand.org/publications/classics/randomdigits`.)

[Rao89]   Rao, C.R.: Statistics and Truth: Putting Chance to Work, Interna-
            tional Cooperative Publishing House, Fairland, MD (1989).

[RC04]   Robert, Christian P., Casella, George: Monte Carlo Statistical Meth-
            ods, 2nd ed., Springer, New York (2004).

[RDCT09]   R Development Core Team: R: A language and environment for sta-
            tistical computing, R Foundation for Statistical Computing, Vienna,
            Austria. ISBN 3-900051-07-0. (URL `http://www.R-project.org`.)

[Rob96]   Robinson, Kim Stanley: Blue Mars, Bantam Books, New York (1996,
            paper 1997). (See front matter. Sequel to Red Mars (1993) and Green
            Mars (1994).)

[Rosn93]   Rosenbaum, Paul R.: Hodges-Lehmann point estimates of treatment
            effect in observational studies, Journal of the American Statistical As-
            sociation, **88**, 1250–1253 (1993).

[Ros97]   Ross, Sheldon M.: Introduction to Probability Models, 6th ed., Acad-
            emic Press, San Diego (1997).

[SC80]   Snedecor, George W.; Cochran, William G.: Statistical Methods, 7th
            ed., Iowa State University Press, Ames (1980).

[Spa83]   Sparks, David: RND is fatally flawed, Call—A.P.P.L.E., 29–32 (Janu-
            ary 1983).

[Sta08]   Stapleton, James H.: Models for Probability and Statistical Inference:
            Theory and Applications, Wiley-Interscience, Hoboken, NJ (2008).

[STBL07]   Speigelhalter, David; Thomas, Andrew; Best, Nicky; Lunn, Dave: BUGS: Bayesian inference using Gibbs sampling. MRC Biostatistics Unit, Cambridge, England (2007).
http://www.mrc-bsu.cam.ac.uk/bugs

[Ste91]   Stern, Hal: On the probability of winning a football game, The American Statistician, **43**, 179–183 (1991).

[Ste92]   Stern, Hal: Pro Football Scores.
http://lib.stat.cmu.edu/datasets/profb (1992).

[TOLS06]   Thomas, Andrew; O Hara, Bob; Ligges, Uwe; Sturtz, Sibylle: Making BUGS Open, R News, **6**, 12–17 (2006).

[TSF01]   Trumbo, Bruce E.; Suess, Eric A.; Fraser, Christopher M.: Using computer simulations to investigate relationships between the sample mean and standard deviation, STATS—The Magazine for Students of Statistics, **32**, 10–15 (Fall 2001).

[Tru89]   Trumbo, Bruce E.: Statistics and Probability Demonstrations and Tutorials. COMPress, A Division of Wadsworth, Inc., Belmont, CA (1989).

[Tru02]   Trumbo, Bruce E.: Learning Statistics with Real Data, Duxbury, Pacific Grove, CA (2002).

[VR97]   Venables, W. N.; Ripley, B. D.: Modern Applied Statistics with S-PLUS, 2nd ed., Springer, New York (1997).

[Wei85]   Weisberg, Sanford: Applied Linear Regression, Wiley, New York (1985). (The Old Faithful data appear on pages 231 and 235, where they are used to illustrate regression. Partly because the data are strikingly bimodal, this is a very widely cited dataset. Modeled as a Markov chain in [Tru02], Exercise 12-18.)

Index

Headers in *italic type* refer to probability distributions. Names of R functions are set in `typewriter type`. Italicized page numbers refer to problems (often with explanations, definitions, or proof outlines). A plus sign (+) means there are consecutive pages on the same topic, sometimes beyond the obvious.